CEMENT AND MORTAR ADDITIVES

CEMENT AND MORTAR ADDITIVES

SECOND EDITION

Arnold J. Franklin

NOYES DATA CORPORATION
Park Ridge, New Jersey, U.S.A.
1976

Library of Congress Catalog Card Number: 76-24143
ISBN: 0-8155-0637-6
Printed in the United States

Published in the United States of America by
Noyes Data Corporation
Noyes Building, Park Ridge, New Jersey 07656

FOREWORD

The detailed, descriptive information in this book is based on U.S. patents issued since the middle sixties that deal with cement and mortar additives.

This book serves a double purpose in that it supplies detailed technical information and can be used as a guide to the U.S. patent literature in this field. By indicating all the information that is significant, and eliminating legal jargon and juristic phraseology, this book presents an advanced, technically oriented review of cement and mortar additives as depicted in U.S. patents. To round out the complete technological picture, we have included two reissues and one patent application published under the trial voluntary protest program initiated by the Commissioner of Patents and trademarks in January 1975.

The U.S. patent literature is the largest and most comprehensive collection of technical information in the world. There is more practical, commercial, timely process information assembled here than is available from any other source. The technical information obtained from a patent is extremely reliable and comprehensive; sufficient information must be included to avoid rejection for "insufficient disclosure." These patents include practically all of those issued on the subject in the United States during the period under review; there has been no bias in the selection of patents for inclusion.

The patent literature covers a substantial amount of information not available in the journal literature. The patent literature is a prime source of basic commercially useful information. This information is overlooked by those who rely primarily on the periodical journal literature. It is realized that there is a lag between a patent application on a new process development and the granting of a patent, but it is felt that this may roughly parallel or even anticipate the lag in putting that development into commercial practice.

Many of these patents are being utilized commercially. Whether used or not, they offer opportunities for technological transfer. Also, a major purpose of this book is to describe the number of technical possibilities available, which may open up profitable areas of research and development. The information contained in this book will allow you to establish a sound background before launching into research in this field.

Advanced composition and production methods developed by Noyes Data are employed to bring our new durably bound books to you in a minimum of time. Special techniques are used to close the gap between "manuscript" and "completed book." Industrial technology is progressing so rapidly that time-honored, conventional typesetting, binding and shipping methods are no longer suitable. We have bypassed the delays in the conventional book publishing cycle and provide the user with an effective and convenient means of reviewing up-to-date information in depth.

The Table of Contents is organized in such a way as to serve as a subject index. Other indexes by company, inventor and patent number help in providing easy access to the information contained in this book.

15 Reasons Why the U.S. Patent Office Literature Is Important to You –

1. The U.S. patent literature is the largest and most comprehensive collection of technical information in the world. There is more practical commercial process information assembled here than is available from any other source.

2. The technical information obtained from the patent literature is extremely comprehensive; sufficient information must be included to avoid rejection for "insufficient disclosure."

3. The patent literature is a prime source of basic commercially utilizable information. This information is overlooked by those who rely primarily on the periodical journal literature.

4. An important feature of the patent literature is that it can serve to avoid duplication of research and development.

5. Patents, unlike periodical literature, are bound by definition to contain new information, data and ideas.

6. It can serve as a source of new ideas in a different but related field, and may be outside the patent protection offered the original invention.

7. Since claims are narrowly defined, much valuable information is included that may be outside the legal protection afforded by the claims.

8. Patents discuss the difficulties associated with previous research, development or production techniques, and offer a specific method of overcoming problems. This gives clues to current process information that has not been published in periodicals or books.

9. Can aid in process design by providing a selection of alternate techniques. A powerful research and engineering tool.

10. Obtain licenses — many U.S. chemical patents have not been developed commercially.

11. Patents provide an excellent starting point for the next investigator.

12. Frequently, innovations derived from research are first disclosed in the patent literature, prior to coverage in the periodical literature.

13. Patents offer a most valuable method of keeping abreast of latest technologies, serving an individual's own "current awareness" program.

14. Copies of U.S. patents are easily obtained from the U.S. Patent Office at 50¢ a copy.

15. It is a creative source of ideas for those with imagination.

CONTENTS AND SUBJECT INDEX

INTRODUCTION

Cement is one of the most widely used materials in construction. Cement itself is a powdery complex calcined mixture of various inorganic calcium, aluminum and silicon compounds which can be made into a paste with water and which will cure or set into a solid mass when allowed to stand.

Concrete is composed of cement, aggregate (sand, gravel or broken stone, etc.) and various optional additives to modify or impart desirable characteristics or properties to the concrete. In general, concrete comprises about one volume of cement to two volumes of sand and four volumes of stone. These ratios can be varied, depending on the strength and other structural parameters desired. In general terms, concrete is cement-plus-aggregate.

Hydraulic cements are most commonly used to prepare concrete. The use of hydraulic cements of one form or another has been known since the Mesopotamian era and the age of early Egypt. Various limes were also employed by the early Greeks and later by the Romans as the principal component in cements and concrete. Portland cement, one type of hydraulic cement, was so named because of its resemblance to a natural limestone quarried on the Isle of Portland in England. It was first described in an early patent by J. Aspdin (1824).

Aqueous hydraulic cement slurries are frequently used for building construction, surfacing (roads, parking lots, etc.), in underground cementing operations (in geologic formations penetrated by a wellbore—usually employed therein for purposes of securing a casing in place or for water or brine shutoff), and in tunnel, dam or reservoir constructions. In many instances the ambient temperature in the cementing environment is sufficiently high to accelerate the normal rate of setting of the cement. In some instances the cement sets up so fast that there is not sufficient time for the preparation of the slurry and its proper emplacement in the location where it is to set up.

Conventional cement slurry compositions comprise, for example, (1) a hydraulic cement (Portland, aluminous, pozzolanic, expansive cements, or mixtures thereof which sometimes also contain gypsum or the like), (2) water, and (3) as desired, additional functional control agents to alter or control certain properties of the cement such as the rate of setting (usually a function of thickening rate), fluid loss to a porous material (e.g., a loosely consolidated geologic formation) in contact therewith during the setting period, and friction-reducing agents effective during movement through tubes, pipes, casings, and the like.

Many formulations, for example, have been developed which are particularly useful in the cementation or recementation of oil well pipes and cased bore holes. Cementation of cased bore holes has been a common practice in drill string, finishing and repairing operations for oil wells. The procedure comprises conveying a cement slurry to the bottom of a well through the pipe to be cemented and forcing it into the annular space between the outside of the pipe and the wall of the well, where it sets and hardens.

Once cement has been mixed with water, the process of setting begins and becomes faster as the ambient temperature and atmospheric pressure rise. As the cement slurry proceeds to the bottom of the well, conditions tend to accelerate the rate of setting. However, for successful operation, it is necessary to keep the cement slurry in a fluid and workable condition so that it can be pumped. This requires specific additives and formulation changes.

The growth of the cement industry in the United States was stimulated early in the nineteenth century by the discovery of natural argillaceous limestone deposits near the Erie Canal in New York and the Lehigh Canal in Pennsylvania. These limestone deposits were found to be highly suitable for making hydraulic cement. The introduction of the rotary kiln in 1899 paved the way for mass production of high-temperature Portland cements in the twentieth century.

The use of cement in the United States has grown from over 11 million barrels in 1900 to over 500 million barrels in the 1970s. World production figures indicate a market in excess of 3.5 billion barrels of cement. The USSR, United States, Japan, West Germany and Italy account for well over 50% of this consumption.

The approximate breakdown for cement use by markets in the United States is as follows:

	Percent
Residential	25
Streets and highways	18
Other public works	28
Industrial/commercial	20
Miscellaneous/farm	9

In the concrete-additives market, setting accelerators far outpace all others on a weight basis. Plasticizers, used to reduce the amount of water needed to impart plasticity to the fresh concrete, represent a sizeable and growing market. Other additives include set retarders, air-entraining agents, pigments, waterproofing additives and antifreeze agents. In recent years, considerable research effort has been devoted to the development of foamed and reinforced concrete composites.

This book describes over 220 processes providing data on several hundred compositions which have been tested in cement or mortar formulations, with major emphasis being placed on the more critical areas of set modifiers, air-entrainment and workability agents and general strength improvement. The description of these additives has, for clarity and continuity, been divided up into the main categories of cement, mortar and gypsum and well-cementing compositions. Other processes describe the use of additives to provide water-repellent, foamed, and aggregate cement mixtures.

SET MODIFIERS FOR CEMENT

RETARDERS

Boron-Containing Hydroxycarboxylic Acid

R.C. Martin; U.S. Patent 3,856,541; December 24, 1974; assigned to The Dow Chemical Company describes an improved aqueous hydraulic cement slurry and method of cementing in environments wherein the normal setting rate of the cement slurry should be retarded. The slurry contains as a required additive, an improved retardant to the rate of setting. The retardant does not have an adverse effect on the ultimate compressive strength of the slurry.

The retardant comprises (1) at least one water-soluble hydroxycarboxylic acid or metal salt thereof and (2) at least one of boric acid or a water-soluble salt thereof, e.g., a borate, for example, borax. This combination when added to a cement slurry in an effective amount results in a truly synergistic effect in retarding the setting rate thereof. In place of the above mixture a water-soluble boron-containing hydroxycarboxylic acid or alkali or alkaline earth metal salt thereof can be employed.

The retarded cement can be employed at temperatures as high as between about 200° and 400°F. Most commonly temperatures between 230° and about 340°F will be encountered; however, the retardant is effective at higher temperatures if the time existing between mixing and emplacement is not unduly short.

An effective amount of retardant is employed to retard the setting rate of a cement slurry sufficiently to allow its proper emplacement. The amount of the retardant, based on 100 parts by weight of dry hydraulic cement present can range from between about 0.5 and 5.0 parts by weight of component (1) and between about 0.5 and about 5.0 parts by weight of component (2) to make a total of both (1) and (2) of between about 1.0 and 10.0 parts based on the 100 parts dry weight of cement present. Preferably between about 1 and 4.0 parts of each of (1) and (2) are employed. From 0.1 to 10 parts, preferably 0.1 to 5.0 parts by weight per 100 parts by weight of dry cement of the water-soluble boron hydroxycarboxylic acid or alkali or alkaline earth metal salt thereof is employed.

Component (2) comprises at least one of boric acid or a water-soluble salt of boric acid. Of the latter natural minerals such as borax, kernite and ulexite are suitable. Other suitable water-soluble salts include, for example, alkali and alkaline earth metal salts, e.g., $LiBO_2 \cdot 8H_2O$; $NaBO_2 \cdot 4H_2O$; $Li_2B_4O_7 \cdot H_2O$; $Na_2B_4O_7 \cdot 10H_2O$; $LiB_5O_8 \cdot 5H_2O$; $NaB_5O_8 \cdot 5H_2O$; $KB_5O_8 \cdot 4H_2O$ and the like.

Component (1), the water-soluble hydroxycarboxylic acid or salt thereof, is any saturated or unsaturated hydroxycarboxylic acid or salt thereof (e.g., alkali or alkaline earth metal salt) containing one or more hydroxy groups and one or more carboxylic groups and which is sufficiently soluble in water to be effective as a retarder in the cement slurry.

Included by way of example are low molecular weight aliphatic materials such as, for example, tartaric acid, gluconic acid, citric acid or salt thereof and higher molecular weight materials such as carboxymethylcellulose, glucoheptonic acid or salt thereof and the like. The use of glucoheptonic acid or its salt as an additive for retarding the setting of cements is described in U.S. Patent 3,234,154.

The boron-containing hydroxycarboxylic acid or salt is the product formed by reacting a water-soluble hydroxycarboxylic acid or salt with, for example, boric acid. It is theorized that this reaction takes place in situ in the cement slurry when components (1) and (2) are added. Tests of preformed boron-containing hydroxycarboxylic acid or salt thereof as a retarder tend to support this theory since boron salts also are found to synergistically retard the set of cement slurries, especially under elevated temperatures. Suitable boron-containing compounds include, for example, those formed by the reaction of a hydroxycarboxylic acid or salt thereof and boric acid. Included by way of example are sodium borogluconate, calcium borogluconate and sodium borotartrate.

Sodium Gluconate

B.S.J. Ericsson and U. Palm; U.S. Patent 3,528,832; September 15, 1970; assigned to Mo och Domsjo AB, Sweden describe a composition for retarding the setting of inorganic binders with water, which comprises a water-soluble nonionic cellulose ether and a water-soluble salt of gluconic acid. An inorganic binder composition is also provided having an inorganic filler and the above set-retardant composition, together with a process for the slow controlled setting of an inorganic binder with water employing such a set-retardant composition.

Example: Four types of concrete were manufactured with coarse sand (gravel) having a grain size of 0 to 32 mm as aggregate material. The cement used was a mixed sample comprising equal portions of standard cement from three different manufacturers. The weight ratio between cement and gravel was 1:6.34. The set retardant was composed of 25% by weight ethyl hydroxyethyl cellulose, having a viscosity of 2,500 cp measured in a 2% aqueous solution, and 75% by weight sodium gluconate. It was added to the concrete mix at 0, 1.0, 2.0 and 2.5 grams per kg cement.

The fresh concrete was tested with respect to consistency, air content, volumetric weight and hardening sequence. A Proctor needle was used to determine the hardening sequence, and the test was carried out according to ASTM-C-403. The hardening sequence was followed only until a penetration resistance of approximately 25 kg/cm^2 had been reached.

The water and cement content were calculated from the bulk density and the weighed material amounts. The consistency was determined according to Mohs method (Kungl. Byggnadstyrelsens 1960:4, pages 9 and 10) according to which an open cylinder connected at the bottom with a semicylinder was filled with the concrete mix being tested, and repeatedly struck with a constant force against a solid foundation. An exposition of the test results is found in the table below.

Constituents cement: filler	Cement content, kg/m^3	Set-retardant, g./kg. cement	Water: cement, ratio	Consistency, moh units	Air content, percent	Weight by volume, kg/m^3	Setting time ASTM C-403, hours	Compressive strength 28 calendar days kg./cm.²	Compressive strength 28 calendar days percent
1:6.34	300	0	0.572	15	1.3	2.41	5	415	100
1:6.34	299	1.0	0.565	15	2.5	2.39	10	436	105
1:6.34	299	2.0	0.551	17	2.8	2.38	19	476	115
1:6.34	300	2.5	0.542	16	2.9	2.40	31	518	124

As can be seen from the data, the time taken for the concrete mix to set was increased by the set-retardant according to the process, 2.5 grams additive per kg cement gave six-fold increase of the setting time, in comparison with a concrete mass where no additive has been added. The compressive strength was at the same time increased by as much as 24%. The air content remained constant between 2.5 and 2.9% in the concrete mix which shows no detrimental effect on the air content.

Sodium Lignosulfate and Sodium Sulfate

V.H. Dodson and E. Farkas; U.S. Patent 3,351,478; November 7, 1967; assigned to W.R. Grace and Company describe concrete compositions containing a set retarding admixture composed of an organic set retarder, e.g., sodium lignosulfonate, and a water-soluble sulfate. Particularly preferred compositions are aqueous solutions containing:

		Parts by Weight
(A)	Sodium salt of lignosulfonic acid	4
	Sodium sulfate	1
(B)	Sodium salt of gluconic acid	6
	Sodium sulfate	5
(C)	Sodium salt of adipic acid	2
	Sodium sulfate	1
(D)	Sodium salt of tetrahydroxy adipic acid	2
	Sodium sulfate	1

Aluminum Phosphate

R.L. Angstadt, F.R. Hurley and C.F. Miller; U.S. Patent 3,409,452; November 5, 1968; assigned to W.R. Grace & Company describe a process for retarding the hardening rate of Portland cement by adding from 0.1 to 5 weight percent aluminum phosphate to the cement clinker in the process and manufacture. The following examples illustrate the process.

Example 1: Purified aluminum phosphate (chemical grade) was interground with a Type I Portland cement at the 1.0 and 2.0% level, based on the weight of the cement. The time of set was determined. The ASTM Standard Method of Test for Time of Setting of Hydraulic Cement by the Vicat Needle (ASTM C109-58) was followed. A 0.35 water to cement ratio was used. At least three tests were made in each instance. The results are shown below.

Additive	Time of Set
Blank	4 hr 45 min
1.0% $AlPO_4$	9 hr 9 min
2.0% $AlPO_4$	9 hr 15 min

Example 2: A variscite ore (crude $AlPO_4 \cdot 2H_2O$ containing 3.85% Fe_2O_3 and 34.8% Al_2O_3) was interground with four different Portland cements. Variscite in quantities of 1.0% by weight of Portland cement was added in each case. The time of set was determined following the ASTM Standard Method using the Vicat Needle. The times of set for both the blanks and for the aluminum phosphate interground materials are listed below. The water to cement ratio was 0.35. Two determinations were made in each case.

Cement and Additive	Time of Set
Cement A, blank	4 hr 40 min
Cement A, 1.0% variscite	7 hr 25 min
Cement B, blank	4 hr 50 min
Cement B, 1.0% variscite	9 hr 35 min
Cement C, blank	4 hr 50 min
Cement C, 1.0% variscite	8 hr 45 min
Cement D, blank	5 hr 0 min
Cement D, 1.0% variscite	8 hr 40 min

Ferrous Sulfate

M.R. Edelson and R.L. Angstadt; U.S. Patent 3,425,892; February 4, 1969; assigned to W.R. Grace & Company describe a process for retarding the setting time of a gypsum-containing cement by adding 0.1 to 5% ferrous sulfate to the cement. The following examples illustrate the process.

Example 1: In this example ferrous sulfate obtained as a by-product from the Glidden process for the manufacture of titanium dioxide was interground with four different Type I Portland cements. A retarder concentration of 1.0% by weight of the Portland cement was employed. The time of set was determined using the Vicat Cement Consistency Tester by the procedure of ASTM C187. Teh water to cement ratio was 0.35, and at least two experimental determinations were made to obtain each set time. The results obtained are shown below.

Cement	Retarder	Time of set, min.
A	Blank	300
A	$FeSO_4$	448
B	Blank	296
B	$FeSO_4$	481
C	Blank	291
C	$FeSO_4$	491
D	Blank	276
D	$FeSO_4$	511

Example 2: In this example the procedure of Example 1 was followed using reagent grade $FeSO_4$. The times of set obtained are shown below.

Cement	Retarder	Time of Set, minutes
A	Blank	300
A	$FeSO_4$	480
B	Blank	296
B	$FeSO_4$	526
C	Blank	291
C	$FeSO_4$	508
D	Blank	276
D	$FeSO_4$	512

Example 3: This example shows that the compressive strength of the Portland cement product is not impaired by the use of ferrous sulfate as a retarder. The seven day compressive strength of a mortar made using 740 grams of cement A, 2,035 grams of sand, and 481 ml of an aqueous one weight percent $FeSO_4$ solu- was measured, and was found to be 1,225 ± 100 psi. A similar mortar made using deionized water rather than the $FeSO_4$ solution was found to have a seven day compressive strength of 1,289 ± 100 psi, differing from the retarded mortar by a strength within experimental error.

Zinc Chloride

J.C. Keenum, Jr. and R.L. Angstadt; U.S. Patent 3,429,724; February 25, 1969; assigned to W.R. Grace & Company describe a process for retarding the setting time of Portland cement which involves adding from 0.1 to 5 weight percent zinc chloride or zinc nitrate. The following examples illustrate the process.

Example 1: In this example varying concentrations of zinc sulfate were mixed with two different Type I Portland cements. The zinc sulfate was added to the cement, dissolved in the mix water. A 0.40 water to cement ratio was used. The time of setting of the cements was determined by the ASTM method of test for time of setting of hydraulic cements by the Vicat Needle (ASTM C191-50). Three determinations were made for each additive, and the average value obtained is shown below.

Cement	$ZnSO_4$ Concentration, weight percent	Time of Set, minutes
A	0	341
A	0.10	355
A	0.25	618
A	0.50	700
B	0	331
B	0.10	324
B	0.25	652
B	0.50	762

As shown, the retardation obtained increases as the zinc sulfate concentration in the cement increases.

Example 2: In this example several different zinc salts were mixed with a Type I Portland cement. The zinc salts were added to the cement, and the cement composition was formed and tested by the procedure described in Example 1. The results obtained show that zinc acetate and zinc nitrate are also retarders for Portland cement hardening, and as the concentrations of these additives are increased, an increased retardation effect is obtained. Zinc chloride exhibits a similar effect.

Copper Acetate

J.C. Keenum, Jr. and R.L. Angstadt; U.S. Patent 3,429,725; February 25, 1969; assigned to W.R. Grace & Company describe a process for producing cement mixtures having reduced hardening rates which involves adding 0.1 to 5 weight percent of a copper acetate or copper nitrate to the Portland cement and mixing the ingredients. The following example illustrates the process.

Example: In this example varying concentrations of cupric acetate, cupric sulfate, cupric chloride, and cupric nitrate were mixed with several different Type I Portland cements. The copper salts were added to the cement, dissolved in the mix water. A 0.40 water to cement ratio was used.

The time of setting of the cements was determined by the ASTM method of test for time of setting of hydraulic cements by the Vicat Needle (ASTM C191-58). Three determinations were made for each additive, and the average value obtained is shown below.

	- - - - - - - - - - Retarder	- - - - - - - - - -	
Cement	Composition	Concentration, weight percent	Time of Set, minutes
A	Blank	0	292
A	Cupric acetate	0.30	342
A	Cupric acetate	0.40	433
A	Cupric sulfate	0.30	408
A	Cupric sulfate	0.40	447
A	Cupric nitrate	0.50	466
A	Cupric chloride	0.50	531
B	Blank	0	334
B	Cupric acetate	0.25	349
B	Cupric acetate	0.50	519
C	Blank	0	355
C	Cupric sulfate	0.25	392
C	Cupric sulfate	0.50	495

As shown, copper salts are active retarders for alite cements. Furthermore, the retardation obtained increases as the copper salt concentration in the alite cement increases.

Sodium Cellulose Acetate Sulfate

J.D. Curry, F.L. Diehl and J.P. Nirschl; U.S. Patent 3,951,674; April 20, 1976;

assigned to The Procter & Gamble Company describe cement compositions comprising a hydraulic cement and a water-soluble cellulose acetate sulfate additive at a cement:additive weight ratio in the range of 100:0.30 to 100:2.0, preferably 100:0.4 to 100:0.6. The sodium cellulose acetate sulfate preferred for use is Sulfacel. The following examples illustrate the process.

Example 1: A standardized compressive strength test using a well-defined mortar was carried out using ASTM test guidelines.

A mortar mixture consisting of one part Type I Portland cement and 2.75 parts sand (Ottawa Silica Sand Crystals) and having a water:cement weight ratio of 0.50 was formulated and tested for compressive strength. A compressive strength increase of about 23% over a control sample was noted when 0.5% (weight of cement) sodium cellulose acetate sulfate was added to the mortar mixture.

Example 2: A typical concrete containing the sodium cellulose acetate sulfate is as follows. Control and test samples of Portland cement were proportioned according to the Recommended Practice for Selecting Proportions for Concrete (ACI) and tested according to ASTM procedures.

The cement used was Portland cement, Miami, which conformed to ASTM Type I Portland cement. The coarse aggregate used was a high quality limestone with low absorption, approved for use in Portland cement concrete by the Kentucky Bureau of Highways in accordance with grading requirements for number 57 stone with a nominal size of one inch to number 4 sieve. The aggregate had a bulk specific gravity of 2.73 with the saturated surface dry, and an absorption of 0.64%.

The fine aggregate used consisted of a natural sand approved for use by the Kentucky Bureau of Highways. The fine aggregate had a bulk specific gravity of 2.64 with the saturated surface dry, an absorption of 1.96% and a fineness modulus of 2.83.

The air-entraining admixture used was neutralized Vinsol resin, a natural wood resin neutralized with sodium hydroxide. Vinsol resin was added to the Portland cement concrete mixture to give the final composition a 3 to 5% volume air content. Mixtures of the following compositions were tested and evaluated for compressive strength increase.

A typical mixture was as follows (parts by weight): Portland cement, 517; fine aggregate, 3,147; coarse aggregate, 1,748; water, 270. Control and test samples were prepared. Sodium cellulose acetate sulfate was added to the test sample. Immediately after mixing, the slump and air content were measured. Standard size cylinders (6 inches diameter x 12 inches height) were made from each mix. Four of the cylinders were subsequently capped and tested for compressive strength. The results were as follows:

Cement Mixture	Water:Cement Ratio	Percent Admixture	Slump	Compressive Strength, psi 7 Day	28 Day
Control	0.522	-	2¾"	3,335	4,609
Control + sodium cellulose acetate sulfate	0.452	0.5 (wt of cement)	3⅛"	4,547	5,556

As can be seen from the foregoing, the sodium cellulose acetate sulfate substantially increased the compressive strength of the test concrete sample over the control.

In the foregoing procedure, the sodium cellulose acetate sulfate is replaced by an equivalent amount of the following cellulose acetate sulfates, respectively, and equivalent results are secured: K^+ salt; NH_4^+; triethanol ammonium salt. Cement and concrete samples prepared in the foregoing manner also exhibit satisfactory tensile strength, setting time, and freeze/thaw stability.

Formaldehyde-Aminolignosulfonate Reaction Product

J.-G. Landry; U.S. Patent 3,689,296; September 5, 1972 describes a setting retarder for concrete which comprises the reaction product of formaldehyde and an aminolignosulfonate prepared by adding an amine to spent sulfite liquor. The retarder is added to the Portland cement clinker which is then ground.

Dialdehyde Starch

R.W. Previte; U.S. Patent 3,503,768; March 31, 1970; assigned to W.R. Grace & Company has found that the setting time of cement compositions is retarded by the addition of the product of periodic acid oxidation of starch.

The composition thus comprises cement and the product of the periodic acid oxidation of starch. This material is also referred to as an anionic polymeric dialdehyde derived by periodate oxidation of starch. The periodate oxidation of starch is very specific, cleaving the bond between carbon atoms 2 and 3 of the anhydroglucose unit to form two aldehyde groups. The following represents the aforementioned bond cleavage:

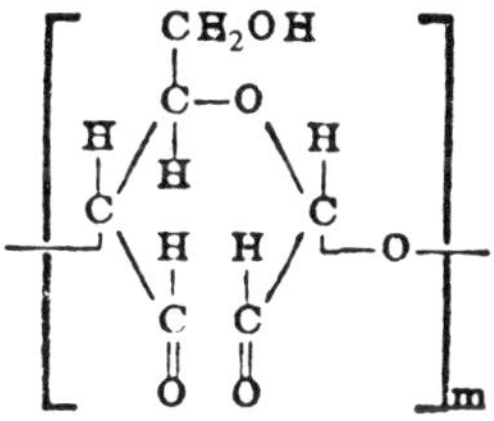

Although other oxidants for starch are known in the art, they do not produce the dialdehyde product achieved by the above described periodic acid oxidation of starch. Dialdehyde starch is known to the art and is commercially available.

The cements represented in the examples below contain 5½ sacks (517 lb) of cement per cubic yard. The compositions were tested for time of setting in accordance with ASTM C403-65T. Compressive strength measurements were carried out in accordance with ASTM C192-62T. The amount of the various mixtures will be expressed as a percentage of the dry cement.

The mixture of the process was added in solution with the water to the cement. For comparative purposes, conventional set retarders were also employed. The dialdehyde starch employed had the following typical properties: percent aldehyde content (number of original anhydroglucose units now in the dialdehyde form per 100 units in the chain), 80%; moisture, less than 10%; sulfated ash, 4%; and bulk density, 25 to 27 lb/ft^3.

TABLE 1: TYPE I PORTLAND CEMENT

Additive	Amount Added (%)	Initial Set (hr:min)	Final Set (hr:min)
Blank	-	5:00	6:40
Dialdehyde starch	0.05	7:00	8:40
Sodium hypochlorite oxidized starch	0.05	5:30	7:00
Lignosulfonate set retarder	0.237	6:10	8:00

TABLE 2: TYPE III PORTLAND CEMENT

Additive	Amount Added (%)	Initial Set (hr:min)	Final Set (hr:min)
Blank	-	6:00	8:00
Dialdehyde starch	0.075	7:30	9:35
Sodium hypochlorite oxidized starch	0.075	6:20	8:15
Lignosulfonate set retarder	0.237	5:05	7:10
Sodium gluconate	0.075	6:20	8:15

From the above it will be noted that the set retarder of this process provides greater setting efficiency than commercial set retarders as well as other oxidized starches, and at a lower addition level.

TABLE 3: TYPE I PORTLAND CEMENT

Percent Dialdehyde Starch	Initial Set (hr:min)	Final Set (hr:min)
0 (blank)	5:00	6:40
0.025	5:40	7:25
0.050	8:30	10:15
0.100	11:00	14:25
0.150	20:30	-

From the above it will be noted that the additive is effective over a relatively wide range.

N,N-Dimethyloldihydroxyethyleneurea

R.W. Previte; U.S. Patent 3,560,230; February 2, 1971; assigned to W.R. Grace & Company has found that the setting time of hydraulic cement compositions is retarded by the addition of N,N-dimethyloldihydroxyethyleneurea to the composition. The following example illustrates the process.

Example: Several mortar mixes were prepared according to ASTM C-403-68 procedures, using various types of Portland cement manufactured by different suppliers and containing, in varying amounts, N,N-dimethyloldihydroxyethyleneurea. The mixes were tested for set retardation by comparison of their initial and final setting times with the setting times of identical mixes which contained no additive. For comparison, mixes containing a commercial set retarder were tested with the following results.

Portland Cement Mortar Mix[1]

Additive	Amount added (percent)	Type I Portland cement: Supplier II: Initial[2] set, hrs.:min.	Supplier II: Final[3] set, hrs.:min.	Supplier II: Initial set, hrs.:min.	Supplier II: Final set, hrs.:min.	Supplier III: Initial set, hrs.:min.	Supplier III: Final set, hrs.:min.
Blank		4:45	6:50	5:30	6:50	4:45	6:10
Commercial retarder	0.24	5:40	7:35				
N,N,dimethyloldihydroxyethyleneurea	0.01					5:05	6:50
	0.06	6:15	8:00	6:35	8:30	5:30	7:20
	0.08					6:40	8:30
	0.09						

Additive	Amount added (percent)	Type III Portland cement: Initial set, hrs.:min.	Type III Portland cement: Final set, hrs.:min.
Blank		5:05	6:55
Commercial retarder	0.24	5:40	7:25
N,N,dimethyloldihydroxyethyleneurea	0.01		
	0.06		
	0.08		
	0.09	6:55	8:40

[1] Prepared according to ASTM C-403-68.
[2] Indicates the length of time expired before initial set.
[3] Indicates the length of time expired before final set.

An examination of the data shows the effectiveness of the N,N-dimethyloldihydroxyethyleneurea as a set retarder in various types of cement. Moreover, the compound was superior in retardation effect to the commercial set retarder even though the amount employed was as little as one quarter the amount of the commercial set retarder used.

Silicofluorides and Phosphoric Acid

O. Matsuda, T. Toki and N. Kudo; U.S. Patent 3,188,221; June 8, 1965; assigned to Onoda Cement Co., Ltd., Japan have found that a retarding agent consisting of at least one water-soluble silicofluoride selected from the group consisting of magnesium silicofluoride ($MgSiF_6$), zinc silicofluoride ($ZnSiF_6$), lead silicofluoride ($PbSiF_6$), aluminum silicofluoride [$Al_2(SiF_6)_3$], hydrofluorosilicic acid (H_2SiF_6), ammonium silicofluoride [$(NH_4)_2SiF_6$] and mixtures and at least one acid selected from the group consisting of inorganic acids such as normal phosphoric acid and boric acid and organic acids such as acetic acid and succinic acid has a remarkably distinguished effect and a high working stability for retarding the setting time of cement. A mixture of such inorganic acids and such organic acids may be used.

In related work *O. Matsuda and N. Kudo; U.S. Patent 3,317,327; May 2, 1967; assigned to Onoda Cement Company, Limited, Japan* describe a method of retarding the setting time of cement selected from Portland cement and a cement mixture comprising Portland cement by adding a retarding agent consisting of at least one silicofluoride compound as the first component and at least one organic surface active agent as the second component together with or without an additional member selected from inorganic acids and organic acids as the third component.

Tall Oil and Sucrose

F.G. Serafin, R.W. Previte and R.F. Stierli; U.S. Patents 3,885,985; May 27, 1975; and 3,865,601; February 11, 1975 both assigned to W.R. Grace & Co. describe an additive for imparting water repellancy to, and retarding the set of hydraulic cement compositions (e.g., Portland cement compositions such as masonry cements). The additive is an aqueous oil-in-water emulsion containing water, a water-insoluble, water-repelling acid component (e.g., tall oil), an emulsifier (e.g., a salt of such acid), and a setting time-retarding agent (e.g., sucrose).

The additive is ideally dispersible in water and preferably contains an air-entraining agent as an additional optional component. Intergrinding the emulsified additive with the cement component is the preferred manner of incorporating the additive in the cement composition. The following examples illustrate the process.

Example 1: An emulsion is formed containing the following ingredients in the proportions shown:

Component	Weight Percent
Tall oil	26.6
Retarder*	13.9
Base**	1.5
Air entraining agent***	1.1
Water	56.9

*Sucrose **Sodium hydroxide ***Octylphenoxy polyethoxy ethanol

The above composition resulted in a stable, water-dispersible emulsion which does not break up after storage at 40°F or after being frozen.

Examples 2 through 5: The following additional additive emulsions were prepared according to the process:

Component	Examples, Weight %			
	2	3	4	5
Tall oil	60.2	56.8	53.5	46.7
Retarder[1]	7.0	10.5	14.0	21.0
Base[2]	0.3	0.3	0.3	0.3
Air-entraining agent[3]	2.5	2.4	2.2	2.0
Water	30.0	30.0	30.0	30.0

[1]Sucrose
[2]Sodium hydroxide
[3]Octylphenoxy polyethoxy ethanol

Example 6: The emulsions of Examples 1 and 2 and a commercial air-entraining agent were interground at an addition rate of 0.1% with a masonry cement containing approximately 34% of Portland cement, 46% of limestone and 20% slag to an approximate surface area of 6,819 cm^2/g in the case of additive 1, 6,929 cm^2/g in the case of additive 2 and 6,743 cm^2/g in the case of the commercial air-entraining agent. The resulting compositions were tested for setting time and degree of air-entrainment according to ASTM procedures. The results were as follows.

ASTM C91 Air Entrainment

Additive	Water:Cement Ratio	Flow	% Air
Commercial additive	0.50	108	19.60
Example 1	0.47	113	19.95
Example 2	0.47	111	20.89

ASTM C266 Gillmore Setting Time

Additive	% Normal Consistency	Setting Time - Initial	Setting Time - Final
Commercial additive	26.0	2:18	3:25
Example 1	24.5	3:08	4:22
Example 2	24.5	2:52	3:58

Example 7: Several water-repellancy field tests according to ASTM SS-C-181e, Federal Specification for Masonry Cement were conducted on masonry cements in which the additive of Example 2 was employed in an amount of about 0.1% solids on the cement. In all of the field tests, such cements exhibited an acceptable water-repellancy of 5 grams or less water pickup.

ACCELERATORS AND HARDENERS

Melamine-Aldehyde Adduct

According to a process described by *R.B. Peppler and P.A. Rosskopf; U.S. Patents 3,864,290; February 4, 1975 and 3,767,436; October 23, 1973 both assigned to Martin Marietta Corporation* the rate of hardening of a Portland cement mix is accelerated by the addition of a low molecular weight, water-soluble hydroxylated adduct formed by reacting melamine and an aldehyde. The adduct is employed in an amount within the range of from about 0.01 to 1.00%, preferably from about 0.2 to 0.8% by weight of cement. The water content of the cement mix for a given consistency tends to be lowered by the incorporation of the adduct, providing an additional advantage from the practice of the process. The adduct may be employed as a concentrated solution in water and may be used in combination with known set accelerating agents and with set retarding agents where reduction in water content is desired without appreciable change in the rate of set of the cement mix.

The melamine-formaldehyde condensates are prepared, as known in the art, by mixing formaldehyde and melamine under alkaline conditions. The resulting melamine-formaldehyde reduction products, sufficiently unpolymerized to remain water-soluble, are available commercially in an 80% aqueous solution, i.e., about 80% by weight solute content based on the weight of solution.

In the process, melamine-formaldehyde adducts were used in hydraulic cement mixes comprising Portland cement, aggregate and sufficient water to effect hydraulic setting of the cement. The same type and brand of cement was used in each mix, and the kind and proportion of coarse and fine aggregate employed were likewise substantially the same.

The amount of water added in each instance to effect hydraulic setting was such as to produce concrete mixes of essentially the same consistency. The temperature of the concrete in which the adducts were tested was about 50°F.

For comparative purposes, the use of the melamine-formaldehyde adduct was compared with the equivalent plain cement mix and with such a mix containing the same percentage addition of a water-soluble, monomeric adduct of urea and formaldehyde having set accelerating properties.

The results set forth in the table below represent the average values of duplicate tests. In these tests, melamine-formaldehyde adducts sufficiently unpolymerized to remain soluble were employed in the form of a commercially available water solution having a solute content of 80% by weight of solution.

Mix No	Additive	Amt of Additive (% by wt of cement)	Water Content (gal/yd^3 of concrete)	Air Volume % Conc	Rate of Hardening Index, hr*
1	None	-	35.9	1.2	9⅜
2	Urea-formaldehyde adduct	0.20	34.9	1.8	7¼
3	Melamine-formaldehyde adduct	0.20	31.3	5.5	6¾

*The time referred to is the time that elapsed from the mixing of the concrete before a 30 pound pull is necessary to remove a steel pin of arbitrary dimension from the body of concrete, the development of the steel-concrete bond being proportional to the degree of hardening or setting of the concrete.

The results set forth in the table show that the melamine-formaldehyde adduct has a very significant effect in accelerating the rate of hardening of the concrete, providing for an appreciable reduction in the time required for set as compared with the time required for set when the same dosage of a urea adduct having desirable set accelerating properties was employed. Both additives were nontoxic and noncorrosive in nature. In addition, the use of the melamine-formaldehyde adduct resulted in a very considerable reduction in water content as compared with the plain mix containing no set accelerator, i.e., 31.3 gallons per cubic yard as compared with 35.9 gallons per cubic yard. The urea adduct, by comparison, produced a favorable reduction in water content compared with the plain mix, reducing the water content to 34.0 gallons per cubic yard, but considerably less reduction than was achieved with the adduct of the process.

R.B. Peppler and P.A. Rosskopf; U.S. Patent 3,785,839; January 15, 1974; assigned to Martin Marietta Corporation have also found that the rate of hardening of a Portland cement mix is accelerated by the addition of a monomeric, water-soluble hydroxylated adduct formed by reacting a urea and an aldehyde, the adduct being employed in an amount within the range of from about 0.01 to about 1.00%, advantageously about 0.2 to 0.8%, by weight of cement.

The water content of the cement mix for a given consistency tends to be lowered by the incorporation of the adduct, conveniently monomethylolurea and dimethylolurea, while avoiding toxic and corrosive effects normally associated with aldehyde-type set accelerators.

Lithium Salt and Hydroxylated Organic Acid

C. Hovasse and P. Allemand; U.S. Patent 3,826,665; July 30, 1974; assigned to Rhone-Progil, France have found that the inclusion of a composition comprised of (1) lithia or a water-soluble lithium salt and (2) a hydroxylated organic acid, or a salt or ester thereof, accelerates the setting and hardening times of aluminous cement and provides mortars, concretes, grouts, etc., derived from such modified high alumina type cements wherein the superior mechanical and handling properties generally characteristic of aluminous cements are not appreciably affected.

Example 1: In this example, a mortar composition was prepared according to the French standard NF.P 15.403 utilizing the following components: 1,350 grams of normal sand and 450 grams of aluminous cement.

The mortar was prepared according to the method described in the foregoing French standard with the exception that the ratio by weight of water to cement was decreased from 0.5 set forth in the standard to 0.45 inasmuch as it was found that the quantity of water must be decreased by about 10% in order to obtain a mortar which was not excessively fluid.

Examples 2 through 20: The mortar prepared in Example 1 was utilized as a reference blank for purposes of comparison and for preparing mortars containing the binary composition of this process. The quantities of lithia or lithium salt and hydroxylated organic acid, salt or ester thereof and the results of the comparative tests are set forth in tabular form on the following page.

In the table, spreading characterizes the fluidity of the mortar and is expressed as the ratio of the spreading coefficients of the mortar containing the composition of this process and that of the blank mortar (reference blank of Example 1). The spreading coefficient was measured using a volume of mortar which was subjected to compression by means of 15 successive shakings (flow test method), and it is expressed as the difference in centimeters between the diameter of the spread mortar and the diameter of the base of the frustum of a cone which is 8 cm. A reference mortar without any additives generally has a spreading of between 5.5 and 6.5 cm.

Mortar Realized with a Constant Ratio of Water to Cement Equal to 0.45

Example Number	Li Compounds, wt % Based on Cement	Organic Compounds, wt % Based on Cement	Spreading	Initial Setting	Final Setting
1	Reference blank	Reference blank	6	3 hr 50 min	4 hr 10 min
2	Lithia, 0.01	None	2.4	1 hr 25 min	1 hr 30 min
3	Lithia, 0.03	None	No spreading	30 min	35 min
4	Li chloride, 0.005	Na and K tartrate, 0.005	6	3 hr 10 min	3 hr 20 min
5	Li formate, 0.005	Citric acid, 0.001	6.2	3 hr	3 hr 20 min
6	Li chloride, 0.02	Ethyl lactate, 0.02	6	1 hr 0.5 min	1 hr 25 min
7	Li chloride, 0.1	Ethyl lactate, 0.1	6.5	30 min	45 min
8	Li chloride, 0.3	Ethyl lactate, 0.3	7.2	40 min	55 min
9	Li chloride, 1	Ethyl lactate, 1	8.2	55 min	1 hr 0.5 min
10	Lithia, 0.005	Na gluconate, 0.001	6.2	3 hr 30 min	3 hr 50 min
11	Lithia, 0.02	Na gluconate, 0.02	6.4	40 min	1 hr 30 min
12	Lithia, 0.1	Na gluconate, 0.1	6.7	1 hr 50 min	2 hr
13	Lithia, 0.5	Na gluconate, 0.5	6.1	2 hr	2 hr 10 min
14	Lithia, 1	Na gluconate, 1	7.3	5 min	45 min
15	Lithia, 0.2	Gluconic acid, 0.02	5.7	1 hr 20 min	1 hr 30 min
16	Lithia, 0.02	Na and K tartrate, 0.02	6.2	20 min	30 min
17	Lithia, 0.02	Tartaric acid, 0.02	6.9	40 min	55 min
18	Lithia, 0.02	Lactic acid, 0.02	6.5	15 min	25 min
19	Lithia, 0.02	Citric acid, 0.02	4.7	50 min	1 hr
20	Lithia, 0.02	Salicylic acid, 0.02	5.4	15 min	20 min

Mortar setting time was measured by means of a Vicat needle according to the French standard NF.P 15.431. It is apparent from a comparison of the results shown in the table that, compared to the reference blank (Example 1), the addition of lithium compounds alone significantly reduces the setting time but also causes the mortar to become rigid, and when the lithium content reaches 0.03% with respect to the cement, the mortar no longer spreads (Examples 2 and 3).

The simultaneous utilization of lithia or a lithium salt and an organic hydroxy acid or a derivative thereof in accordance with the process significantly accelerates the setting time without seriously altering the fluidity or handling characteristics of the mortar.

Further evidence of the advantages of the composition according to the process is provided by comparing the mechanical performance of the reference mortar with that of mortars containing the composition. Mechanical performance was determined by measuring the compression strength of the mortars on prismatic test tubes (4 x 4 x 16 cm) according to the French standard NF.P 15.401. The compression strength of the reference mortar changes from 25 bars after 5 hours to 833 bars after 24 hours, 938 bars after 7 days and 965 bars after 28 days.

For a mortar containing 0.02% lithia and 0.02% sodium gluconate, the compression strength values are 84 bars after 3 hours, 288 bars after 5 hours, 625 bars after 24 hours, 669 bars after 7 days and 730 bars after 28 days. With respect to a mortar containing 0.005% lithium chloride and 0.005% mixed sodium and potassium tartrate, the compression strength changes from 92 bars after 5 hours to 646 bars after 24 hours, 824 bars after 7 days and 950 bars after 28 days.

From the foregoing, it is evident that the addition of the composition of the process to mortar containing aluminous cement increases the mechanical performance of the mortar during the initial period following application, and also that although long-term setting performance is slightly decreased in comparison to the reference mortar, the long-term mechanical performance of mortars containing the composition of the process compare very favorably with the performance of mortars without additives.

3,6-Endomethylenehexahydro-o-Phthalic Acid-4-Sulfonic Acid

S. Ogura; U.S. Patent 3,870,533; March 11, 1975; assigned to Nippon Zeon Co., Ltd., Japan describes a hydraulic composition comprising 100 parts by weight of a hydraulic substance and 0.01 to 5 parts by weight of a 3,6-endomethylene-hexahydro-o-phthalic acid-4-sulfonic acid derivative of the formula

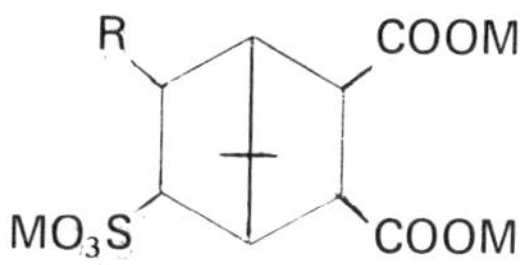

where R is a member selected from the group consisting of hydrogen, an alkyl radical of 1 to 4 carbon atoms and the —COOM group, and M is a member selected from the group consisting of hydrogen, an alkali metal atom, a half atom of an alkaline earth metal and the ammonium group, which M may be the same or different.

The hydraulic composition obtained by incorporating these additives in the hydraulic substance not only excels in its workability when it is to be mixed with water but also excels in its workability when the resulting mortar or concrete is to be used. Moreover, it can impart great strength to the hardened mortar or concrete at an early period.

The mechanism by which these effects are brought about in the hydraulic composition is not clear. However, it is presumed that in all probability these effects result from an increase taking place in the solubility of the hydrated alumina siliceous substances present at the particle surface of the hydraulic substance due to the fact that the 3,6-endomethylene-hexahydro-o-phthalic acid-4-sulfonic acid derivatives are highly soluble in water, with the consequence that the hydration reaction is promoted.

Example 1: Eight parts by weight of sodium hydroxide and 16.4 parts by weight of 3,6-endomethylene-1,2,3,6-tetrahydrophthalic anhydride were added to 100 parts by weight of water to obtain a homogeneous solution of disodium-3,6-endomethylene-1,2,3,6-tetrahydro-o-phthalate. On addition of 10.4 parts by weight of sodium hydrogen sulfite to the solution and stirring the mixture, the reaction proceeded exothermally. After the evolution of heat has stopped, the stirring was continued for a further two hours at 60°C. The water was then removed with the use of a rotating evaporator to obtain 33 parts by weight of a white powder of trisodium-3,6-endomethylene-hexahydro-o-phthalic acid-4-sulfonate (I).

A concrete test was conducted for clarifying the performance of this compound (I) when used as a concrete additive. The results obtained are shown in Table 1. The composition of the concrete used in the test was as follows: Portland cement, 300 kg/m^3; coarse aggregates (river gravel of maximum particle diameter of 25 mm), 1,200 kg/m^3; fine aggregates (river gravel of maximum particle diameter of 2.5 mm), 760 kg/m^3 (fine aggregates, 39%).

However, in the case of the control experiment not using the additive, the amount of fine aggregates was changed to 800 kg/m^3 (fine aggregates, 40%).

TABLE 1

Additive Name	Additive Amount, % Based on Cement	Water:Cement Ratio (%)	Slump (cm)	Amount of Air (%)	Compressive Strength (kg/cm^2) After 7 Days	Compressive Strength (kg/cm^2) After 28 Days
Compound (I)	0.125	45	6.5	2.2	230	359
	0.25	45	7.8	2.5	232	368
	0.50	45	9.0	3.0	256	405
Control	-	53	6.5	1.6	181	291
Lignin sulfonic acid type	0.25	45	6.5	4.1	224	335

Note: The slump, amount of air and compressive strength were determined in accordance with the JISA Methods 1101, 1116 and 1108, respectively.

It is apparent from the table that the performances of the concrete incorporated with the compound (I) of the process are much superior to the performances of the concrete incorporated with the commercial ligninsulfonic acid type additive not to mention the concrete not incorporated with an additive.

Example 2: By operating as in Example 1 but using 17.8 parts by weight of 5-methyl-3,6-endomethylene-1,2,3,6-tetrahydrophthalic anhydride instead of 16.4 parts by weight of 3,6-endomethylene-1,2,3,6-tetrahydrophthalic anhydride, 34.4 parts by weight of trisodium-5-methyl-3,6-endomethylene-hexahydro-o-phthalic acid-4-sulfonate (II) was obtained.

Operating in like manner but using 20.8 parts by weight of 5-carboxy-3,6-endomethylene-1,2,3,6-tetrahydrophthalic anhydride instead of 16.4 parts by weight of 3,6-endomethylene-1,2,3,6-tetrahydrophthalic anhydride and by using the sodium hydroxide in an amount of 12 parts by weight, 37.4 parts by weight of tetrasodium-5-carboxy-3,6-endomethylene-hexahydro-o-phthalic acid-4-sulfonate (III) was obtained.

For clarifying the performances of the compounds (II) and (III), when used as a concrete additive, a concrete test was conducted, with the results shown in Table 2. The composition of the concretes used in the test was the same as that of Example 1.

TABLE 2

Additive Name	Additive Amount, % Based on Cement	Water:Cement Ratio (%)	Slump (cm)	Amount of Air (%)	Compressive Strength (kg/cm^2) After 7 Days	Compressive Strength (kg/cm^2) After 28 Days
Compound (II)	0.25	45	7.1	3.0	224	356
Compound (III)	0.25	45	8.1	2.2	244	377
Control	-	53	6.5	1.5	220	337

It can be appreciated from Table 2 that, despite the fact that the amount used of water is less in the case of the concrete incorporated with the process compounds (II) and (III) than in the case of the concrete not incorporated with an additive, satisfactory fluidity is demonstrated. In addition, it is seen that the strength enhancement effect of the concrete incorporated with the process compounds is greater after the concrete has set.

Ethanolamine and Acetic Acid

A.T. Hersey and J.R. Tonry; U.S. Patent 3,689,295; September 5, 1972; assigned to Alpha Portland Cement Company describe a Portland cement composition which is capable of setting in a short period of time. The cement composition contains an additive composed of an ethanolamine and an acetate, acetic acid or acetic anhydride in amounts by weight between about 0.1 and 0.4%.

The cement composition includes sulfur combined as SO_3 in controlled amounts between about 1 and 2.5% and has a Blaine fineness of at least about 4,000 square centimeters per gram. Concrete formed from this cement composition will develop a compressive strength of at least about 125 psi within about two hours in mortar tests.

Ketocarboxylic Acid as Chelating Agent

H. Kokuta; U.S. Patent 3,664,854; May 23, 1972; assigned to Ajinomoto Co., Inc., Japan describe a quick setting and quick hardening cement. The cement contains aluminous cement, lime or lime-containing material, Portland cement, and a ketocarboxylic acid or its salt as a chelating agent. The chelating agent forms a complex selectively with Ca ion in the alkaline state and after a certain period, the complex will decompose or decrease. Therefore, the time to begin setting can be controlled to 1 to 90 minutes and sufficient hardening takes place in a short time.

As chelating agents, ketocarboxylic acids such as 2-ketogluconic acid, α-ketoglutaric acid, pyruvic acid, oxalacetic acid, and their salts, are preferably used. The setting time of a cement which is produced by mixing 100 parts by weight of aluminous cement, 30 parts by weight of dolomite plaster and 30 parts by weight of Portland cement, is somewhat longer than that of the flash setting cement which is produced by admixing aluminous cement, dolomite plaster and Portland cement clinker. The setting time of the former is from 30 seconds to 1 minute, however, the compressive strength after 30 minutes is no more than 60 kg/cm^2.

However, if the abovementioned chelating agent, for example, calcium 2-ketogluconate in the amount of only 1%, is added to the mixture of aluminous cement, lime and Portland cement to mask the Ca ion, the time to start setting becomes 12 minutes and compressive strength after 30 minutes becomes 110 kg/cm^2. If 1% of α-ketoglutaric acid is added, the time to start setting becomes 60 minutes and the compressive strength after two hours becomes 190 kg/cm^2.

Spodumene

R.L. Angstadt and F.R. Hurley; U.S. Hurley; U.S. Patent 3,331,695; July 18, 1967; assigned to W.R. Grace & Company describe a method for producing a cement mixture having an accelerated hardening rate comprising adding to an alite cement from about 0.1 to 10% spodumene based on the dry weight of the cement binder, and intimately mixing the ingredients to provide a uniform distribution of the spodumene throughout the cement.

Spodumene, $LiAlSi_2O_6$ has been found to be a superior alite cement hardening accelerator. Previously known cement hardening accelerators were water-soluble, and water solubility was thought to be a requisite characteristic of an accelerator. Unexpectedly, spodumene is an excellent accelerator even though it has a low solubility in water. Not only does the spodumene provide a rapid alite cement hardening rate, but the spodumene is a noncorrosive accelerator. When the alite cement is used in the presence of iron or other metal reinforcing materials, spodumene can be used without creating corrosion problems. Acceleration of alite cements can be obtained with from about 0.1 to 20% spodumene in the cement. Preferably, from about 1.0 to 15% spodumene is employed. The optimum range is from about 0.5 to 5% spodumene. These concentrations are expressed as percent of the dry weight of the cement binder.

Example 1: In this example, alpha-spodumene, a naturally occurring lithium ore having the chemical formula $LiAlSi_2O_6$ was interground with a Type III Portland cement. A 1,000 gram sample of the cement was blended for six hours in a laboratory ball mill with 20 grams of alpha-spodumene. A settable cement was prepared from the mixture using a water to cement ratio of 0.65 to 1. Two inch mortar cubes were prepared according to ASTM test C109-58, and their compressive strengths were measured after 12 hours. Nine cubes were tested. When no spodumene was used the compressive strength was 181 ± 116 psi. When 2.0 weight percent spodumene was used the compressive strength was 638 ± 91 pounds per square inch.

Example 2: In this example, beta-spodumene was employed as the accelerator. Alpha-spodumene undergoes irreversible phase transition to beta-spodumene at temperatures in excess of 900°C. Beta-spodumene was prepared for this test by heating alpha-spodumene at 980°C for four hours. A 1,000 gram sample of Type III Portland cement employed in Example 1 was interground with 20 grams of beta-spodumene. Cubes were prepared from the mixture and were tested by the method described in Example 1. When no spodumene was used the compressive strength was 181 ± 116. When 2.0 weight percent spodumene was used the compressive strength was 741 ± 119 psi.

Calcium Nitrite

R.L. Angstadt and F.R. Hurley; U.S. Patent 3,427,175; February 11, 1969; assigned to W.R. Grace & Company describe a method of accelerating the hardening rate of Portland cement which involves adding 0.1 to 10% calcium nitrite based on the dry weight of the cement. This accelerator is advantageous in that it also inhibits corrosion of the metals placed in the cement as reinforcing bars.

Amine Formates

A process described by *T.G. Kossivas; U.S. Patent 3,619,221; November 9, 1971; assigned to W.R. Grace & Company* involves a set-accelerating additive which comprises a water-soluble salt of formic acid and an amine. The water-soluble formate salts are obtained by adding the formic acid to the amine, or the amine to the acid at, for example, room temperature until a substantially neutral reaction product, with a pH of about 7, is obtained. It should be understood that while the formation of the water-soluble salt of formic acid and the amine is the primary and predominant product of the reaction, some formation of an amide of the amine and formic acid is probable, either initially during the reaction or upon standing of the reaction product over a period of time, and therefore, the formate additive of the process may contain a minor amount of amide in some instances.

Example: 80.0 parts by weight of primarily aliphatic and heterocyclic N-containing mono- and diamines obtained as a coproduct from a commercial process for the production of heterocyclic amines was neutralized (pH 7.0 to 7.2) with 19.2 parts by weight of formic acid. The major constituents of the amine coproduct mixture upon analysis were shown to be 4-(2-aminoethoxy(ethylmorpholine, 2-(4-morpholinylethoxy)ethanol, and bis-2-(4-morpholinyl)ethyl ester.

The amine formate product obtained above was tested for its effect on the setting time and the compressive strength of Portland cement concrete containing water and aggregate. The concrete was prepared and tested in this test and all of the tests which follow according to the procedures specified in ASTM C-494. A Type II Portland cement was used to make the concrete except where otherwise noted. For comparison, tests were conducted on a blank sample containing no additive as well as samples containing calcium formate as an accelerating additive. The results are reported below.

Additive	Addition Rate*	Initial Setting Time Acceleration Over Blank (hr:min)	1 Day Strength	7 Day Strength	28 Day Strength
-	-	-	1,650	4,008	5,315
Product of Example 1	0.3	0:10	2,041	4,862	5,816
	1.0	0:16	-	-	-
	2.0	0:29	-	-	-
Calcium formate	1.0	0:10	-	-	-
	2.0	1:05	1,900	3,800	5,006

*Weight percent, based on the weight of the dry cement.

From the data given above, it is seen that in addition to being an effective set-accelerator, the additive, unlike the calcium formate additive, increased the 7 and 28 day strength of the concrete composition.

The additives tested above were further tested to determine each additive's capability of offsetting the set-retardation effect of a water-reducing, set-retarding agent, corn syrup, in a Portland cement concrete composition. The data shown on the following page is the average setting time of six test runs for each additive on identical compositions and for the blank. The concretes were prepared using Types I, II, and III Portland cement. The data demonstrates the exceptionally high effectiveness of the additive of the process in offsetting the retardation effect of the corn syrup.

Additive*			
Corn Syrup	Product of Process	Calcium Formate	Initial Setting Time (hr:min)
0.18	-	-	6:15
0.18	-	0.18	5:30
0.18	-	1.50	0:30
0.18	0.18	-	2:00
0.18	0.42	-	1:20
0.18	1.15	-	0:15

*Weight percent, based on the weight of the dry cement.

8-Hydroxyquinoline

J. Chi-sun Yang; U.S. Patent 3,563,775; February 16, 1971; assigned to Johns-Manville Corporation has found that 8-hydroxyquinoline is an improved hydration accelerating agent for hydraulic setting cement containing calcium aluminate.

In the process accelerating early set or hydration of common hydraulic cement containing calcium aluminate, without incurring degrading influences, is achieved through the inclusion within the cementitious material, of only about 0.02% of 8-hydroxyquinoline based upon the weight of the total solids content of the cement material.

To facilitate effective distribution of such minimum proportions of additive to the total mass and therefore maximum uniformity as well as promptness of its influence therein, it is highly desirable that the 8-hydroxyquinoline be dispersed in the form of a dilute aqueous solution, for example, in concentrations of about 0.01 to 1.0%, or typically approximately 0.1% for addition to the cementitious material. Alternatively the 8-hydroxyquinoline can be premixed with the mixing water used for the preparation of the wet cement mixture which provides an equally effective and possibly more convenient procedure.

Sodium Aluminate and Tartaric Acid

B. Bonnel and C. Hovasse; U.S. Patent 3,656,985; April 18, 1972; assigned to Progil, France have found that the composition constituted by the association of an alkali metal aluminate, a rigidifying agent and setting accelerator, with an hydroxylated organic acid, its salts or its esters acting as fluidifying agent and setting retardant, in weight proportions with regard to cement of 0.5 to 5% of alkali aluminate and 0.05 to 2% of hydroxylated organic acid, its salts or its esters, represents a convenient setting accelerator for Portland cement and any other cement having a high content in tricalcium silicate. The alkali metal aluminates are preferably the ones in which the ratio of Me_2O/Al_2O_3 ranges between 1 and 2 (Me representing metal).

However it is possible to use, without any disadvantage, aluminates in which the precited proportion is comprised between 1 and 4. The hydroxylated organic acids are preferably tartaric, citric, gluconic, malic, lactic, salicylic acids, etc. or the salts of these acids, preferably alkali and alkaline earth metal salts and the esters of those acids in which acid or alcohol functions are esterified partly or wholly.

Practically the association of alkali metal aluminate and hydroxylated organic acid or its derivatives is mixed with a product, such as precipitated silica, having no influence, at use amount, on the cement behavior; this product is introduced for preventing any moisture retention by the mixture. The additive formulated in this way is preferably dispersed in tempering water; it may also be added directly to cement or in the mortars and concretes during their manufacture.

Alkali Metal Acid Carbonates, Alkali Metal Silicates and Organic Acid Salts

P. Allemand and C. Hovasse; U.S. Patent 3,782,984; January 1, 1974; assigned to Rhone-Progil, France have found that Portland type cements containing a high content of tricalcium silicate and mortars, concretes and the like derived from such cement are significantly improved by incorporating into such cements a composition comprised by weight with respect to cement, of about 0.5 to 5% of alkali metal acid carbonate, about 0.05 to 2% of hydroxylated organic acids in the form of the alkali metal salts thereof, and about 0.5 to 5% of alkali metal silicates having a molar ratio of SiO_2/M_2O between about 0.5 and 4, preferably 0.8 and 2, wherein M is an alkali metal.

The introduction of the composition of the process into Portland type cements accelerates the setting thereof and increases the mechanical performance of such cements by between about 50 to 120% measured after about 24 hours and further accomplishes the foregoing without materially decreasing the long-term characteristics of the products. Moreover, the handling qualities of cements containing the composition compare quite favorably with cements without additives and, therefore, are extremely well suited for use in all contemplated applications, and are especially advantageous in mass concrete casting.

In the examples below, the composition of the process was introduced into mortars containing Portland cement, each mortar comprised of the following. In the following, used standards are French standards.

Sand NF.P 15.403	1,350 g
Artificial Portland cement or similar	450 g
Water	225 g

The mortars were prepared according to the standard NF.P 15.403 and the composition was mixed with mixing water prior to addition to the mortar.

In the tables, mortar handling is expressed as the ratio of the spreading coefficients of the mortar containing the composition and that of a blank mortar. The spreading coefficient was measured using a volume of mortar which was submitted to compression by means of 15 successive shakings (flow test method), and is expressed as the difference in centimeters between spread mortar diameter and that of the cone trunk base which was 8 centimeters.

Mortar setting time was measured by means of a Vicat needle according to the standard NF.P 15.431 at 20°C in a medium saturated with moisture. Compression strength was measured on prismatic test tubes (4 x 4 x 16 cm) according to the standard NF.P 15.401.

Blanks utilized for purposes of comparison and for preparing mortars containing the composition of this process were prepared as follows.

Blank A: Mortar NF.P 15.403 with CPAC 325 (artificial Portland cement + fly ashes from steam generating station). Blank B: Mortar NF.P 15.403 with CPAL 325 (artificial Portland cement + blast furnace slag). Blank C: Mortar NF.P 15.403 with CPA 325 (artificial Portland cement).

In the compositions of the following examples, all percentages are by weight with respect to cement.

Example 1: The composition consists of Blank A, 2% of sodium silicate crystallized with 5 water molecules having a molar ratio $SiO_2/Na_2O = 1$, 2% of sodium acid carbonate, and 0.1% of sodium and potassium double tartrate.

Example 2: The composition consists of Blank A, 2% of sodium silicate crystallized with 5 water molecules having a molar ratio $SiO_2/Na_2O = 1$, 2% of sodium acid carbonate, and 0.06% of sodium gluconate.

Example 3: The composition consists of Blank B, 2% of sodium silicate crystallized with 5 water molecules having a molar ratio $SiO_2/Na_2O = 1$, 2% of sodium acid carbonate, and 0.1% of potassium and sodium double tartrate.

Example 4: The composition consists of Blank B, 2% of sodium silicate crystallized with 5 water molecules having a molar ratio $SiO_2/Na_2O = 1$, 2% of sodium acid carbonate, and 0.06% of sodium gluconate.

Example 5: The composition consists of Blank C, 2% of sodium silicate crystallized with 5 water molecules having a molar ratio $SiO_2/Na_2O = 1$, 2% of sodium acid carbonate, and 1% of sodium citrate.

Example 6: The composition consists of Blank C, 2% of sodium silicate crystallized with 5 water molecules having a molar ratio $SiO_2/Na_2O = 1$, 2% of sodium acid carbonate, and 2% of sodium lactate.

Example 7: The composition consists of Blank C, 2% of sodium silicate crystallized with 5 water molecules having a molar ratio $SiO_2/Na_2O = 1$, 2% of sodium acid carbonate, and 0.5% of sodium salicylate.

The results obtained utilizing the mixture in the preceding examples and the blanks prepared are given in the following table.

	Blank A	Ex. 1	Ex. 2	Blank B	Ex. 3	Ex. 4
Handling	1	>1	>1	1	>1	>1
Onset of setting (hr:min)	4:30	1:30	2:40	4:50	1:45	3:0
Setting completed (hr:min)	4:45	2:45	4:0	6:20	3:0	4:30
Compressive strength (kg/cm^2):						
1 day	96	197	210	63	95	110
3 days	259	284	295	187	205	220
7 days	321	325	336	292	315	330
28 days	382	410	437	406	444	465
90 days	455	440	455	467	460	475

	Blank C	Ex. 5	Ex. 6	Ex. 7
Handling	1	1.05	0.90	0.94
Onset of setting (hr:min)	4:45	2:0	1:10	1:25

(continued)

	Blank C	Ex. 5	Ex. 6	Ex. 7
Setting completed (hr:min)	6:0	4:30	1:30	1:45
Compressive strength (kg/cm^2):				
1 day	89.5	150	135	140
3 days	264	280	263	272
7 days	309	306	295	301
28 days	463	455	399	415

Calcium and Aluminum Oxide Clinkers

H. Uchikawa and S. Uchida; U.S. Patents 3,864,138; February 4, 1975; 3,864,141; February 4, 1975; and 3,867,163; February 18, 1975; all assigned to Onoda Cement Company, Limited, Japan describe a process for suitably regulating setting time of hydraulic cement which has high strength at the initial and later stages. The initial and later stages means less than 6 hours and more than 3 days, respectively, in the process.

When the mixed raw materials comprising calcareous, silicious and aluminous material as well as a small amount of halide such as calcium fluoride, calcium chloride, etc. are sintered, the initial crystallization region of calcium aluminate ($3CaO{\cdot}Al_2O_3$) is extremely narrowed and the clinker obtained will not contain calcium aluminate, but contains $11CaO{\cdot}7Al_2O_3{\cdot}CaX_2$ (X represents a halogen atom) as a stable phase, and thus the clinker containing $11CaO{\cdot}7Al_2O_3{\cdot}CaX_2$ as a stable phase, as well as $3CaO{\cdot}SiO_2$, $2CaO{\cdot}SiO_2$, $4CaO{\cdot}Al_2O_3{\cdot}Fe_2O_3$, etc. is obtained.

$11CaO{\cdot}7Al_2O_3{\cdot}CaX_2$ component has a high hydration activity and a greater hardenability, so that hemihydrate (or hemihydrate gypsum) is effective when added to the clinker containing $11CaO{\cdot}7Al_2O_3{\cdot}CaX_2$ component for retarding the setting time thereof and also insoluble anhydrite (or insoluble anhydrite gypsum) is added to the clinker containing $11CaO{\cdot}7Al_2O_3{\cdot}CaX_2$ for developing strength in the initial and the later stages.

However, as described in U.S. Patent 3,782,992 it has been found that if at least one of sulfates, nitrates and chlorides of potassium, sodium, magnesium, calcium, aluminum and ammonium (excepting $CaSO_4{\cdot}\frac{1}{2}H_2O$), is added to the clinker as a substitute of hemihydrate, setting time of cement thus obtained is retarded and the hardened matter has excellent strength, and that if the abovementioned additive is added to the clinker with anhydrite and hemihydrate, setting time of cement thus obtained is retarded and the cement will have good workability and a more excellent strength development property at the early and the later stages than that of the former cement.

An object of this process is to provide a process for regulating setting time of hydraulic cement comprising $11CaO{\cdot}7Al_2O_3{\cdot}CaX_2$ by using an additive other than the abovementioned additives. Another object of this process is to provide a process for preparing cement which has good workability and high initial strength and a hardened material thereof has high strength for a long period of time.

It has been found that in addition to the abovementioned additive, sodium hydrogen carbonate, water-soluble phosphates, silicofluorides, sodium silicates, sugars, carboxylic acids, ligninsulfonates, sulfuric esters of higher alcohols or alkylsulfonates are also effective.

According to the process, the setting time of mortar which is prepared from the clinker comprising 5 to 60% by weight of $11CaO \cdot 7Al_2O_3 \cdot CaX_2$ and more than 5% by weight of $3CaO \cdot SiO_2$, $2CaO \cdot SiO_2$, $4CaO \cdot Al_2O_3 \cdot Fe_2O_3$, etc. is not only regulated within a range of from 7 to 40 minutes, but also the mortar or concrete provides good workability and the hardened material develops excellent strength at the initial and the later stages.

Example: White clay, white bauxite, quicklime, copper slag and a small amount of gypsum as well as calcium fluoride and calcium chloride were ground by a shaft ball mill, 85 cm in diameter and 100 cm in length, and were mixed by means of a large-sized mixer so as to obtain clinker having a composition as shown in Table 1 and the resultant mixture was shaped by a rotating roll, 60 cm in diameter. The shaped material is sintered by a small-sized rotary kiln so that free lime in the obtained clinker was lower than 0.5% by weight. These results are shown in Table 1.

The obtained clinker was mixed with gypsum listed in Table 2, calcium ligninsulfonate, calcium alkylsulfonate and sulfuric ester of higher alcohol, sodium hydrogen carbonate, sodium tripolyphosphate, magnesium silicofluoride, respectively, at the ratio listed by Table 3 and each thus prepared cement was tested for the setting time and compressive strength of mortar in accordance with the test method JIS R 5,201. The results are summarized in Table 3. In Table 3, the setting time and the compressive strength of mortar prepared from cement in which gypsum alone was mixed with the clinker were listed for comparison with the above results.

TABLE 1

Kind of	Chemical Composition (%)									Free	Mineral Composition (%)	
Clinker	SiO_2	Al_2O_3	Fe_2O_3	CaO	MgO	SO_3	Na_2O	K_2O	Total	Lime	$C_{11}A_7{\cdot}CaX_2$	C_3S
A	15.9	15.7	2.3	61.9	0.5	1.2	0.08	0.07	97.6	0.2	X = F 27	51
B	17.8	15.1	2.0	61.5	0.4	1.0	0.03	0.05	97.9	0.1	X = Cl 25	44

Note: Mineral composition was determined by means of x-ray diffraction analysis.

TABLE 2

Kind of Gypsum	Ig. Loss	SiO_2	$Al_2O_3+Fe_2O_3$	CaO	MgO	SO_3	Total
Hemihydrate	6.1	1.8	0.9	37.5	0.2	53.6	100.1
Anhydrite	-	1.9	1.0	40.0	0.3	57.2	100.4

TABLE 3

Amount of addition (% by weight)				Setting time of mortar (min.)		Compressive strength of mortar (kg/cm²)					
Clinker	Anhydrite	Hemihydrate	Additive	Initial	Final	3 hrs.	6 hrs.	1 day	3 days	7 days	28 days
A 85	13	2	—	16	24	105 (20.4)	124 (30.2)	217 (48.8)	249 (50.3)	333 (62.5)	450 (88.4)
A 85	15	—	Surface active agent comprising mainly calcium lignin-sulfonate 0.3*	17	22	118 (25.4)	127 (29.8)	204 (40.4)	308 (65.4)	354 (70.7)	495 (89.9)
A 85	15	—	Sodium hydrogen carbonate 0.8*	20	25	108 (24.4)	136 (30.1)	239 (49.5)	288 (53.3)	318 (68.2)	466 (89.7)
A 85	15		Sodium tripoly-phosphate 0.5*	19	26	110 (25.3)	126 (27.4)	250 (55.6)	289 (57.9)	345 (72.4)	453 (87.1)
A 84	12	2	Magnesium silico-fluoride 2.0 (based on F)	25	30	136 (31.6)	166 (36.5)	262 (55.6)	310 (60.0)	357 (73.1)	491 (90.6)
A 85	15	—	Surface active agent comprising mainly sulfuric ester of higher alcohol 0.3*	23	30	130 (30.5)	154 (33.6)	246 (52.3)	276 (57.4)	343 (71.5)	484 (90.4)
B 84	14	—	Magnesium silico-fluoride 2.0	26	34	125 (25.4)	145 (30.4)	224 (49.9)	275 (56.8)	324 (62.5)	478 (82.6)
A 85	15	—	Surface active agent comprising mainly calcium alkyl-sulfonate 0.3*	15	25	124 (27.6)	150 (34.2)	230 (50.9)	293 (59.6)	326 (69.5)	476 (88.1)

Note: (1) Values in parentheses show bending strength (kg/cm²)
(2) *Shows that the compound is solved in water kneaded together with cement

Calcined Aluminum Sulfate

T. Burge; U.S. Patent 3,782,991; January 1, 1974; assigned to Sika AG, Switzerland describes an additive for improving the properties of building materials, such as cement, concrete and mortar, especially for increasing the early strength, and the ability to carry out concreting at low temperatures and for volume control, without impairing the other qualitative properties of the building material. The additive, apart from extenders and/or other additive substances consists of calcinated, anhydrous or water-free aluminum sulfate. In the examples the quantitative amounts given (parts, percent) are always based upon weight.

Example 1: One thousand parts commercially available aluminum sulfate [$Al_2(SO_4)_3 \cdot 12H_2O$] are heated in an electric furnace equipped with a flue upon sheet metal shelves during the course of 2 hours to a temperature of 450° to 490°C. This temperature is maintained for 24 hours. Such is then cooled while excluding air and after adding 5 parts calcium stearate ground in a ball mill to cement fineness. This product can be maintained for unlimited time in closed vats or containers.

100 parts diatomaceous earth are homogeneously admixed with 5 parts triethanolamine in a ball mill. 35 parts sodium silicate are added to this premixture and such is again homogenized. There is then obtained 140 parts mixture. To this there is added 60 parts of the calcinated aluminum sulfate as obtained above.

Example 2: Calcinated aluminum sulfate is produced according to the procedure of Example 1. 100 parts of the water-free product are ground in a ball mill with 100 parts broken gypsum. A different additive is obtained if there is ground 200 parts gypsum.

Example 3: Initially calcinated aluminum sulfate is fabricated according to the mode of operation described above in connection with Example 1. 100 parts of the obtained water-free product are ground in a ball mill with 100 parts broken limestone until reaching cement fineness.

Example 4: 10 grams of water and 20 grams cement (or cement plus additive) are stirred for 3 minutes and permitted to stand during the desired hydration time. After expiration of the desired time the hydration is interrupted by the addition of ethanol (20 ml). The water-alcohol mixture is separated from the cement by centrifuging. There is removed 20 ml from the mixture; after the addition of 10 ml benzol the liquid is heated (60° to 70°C), until the initial turbidity has disappeared. During cooling the temperature is measured at which the turbidity again appears.

By means of a calibrating plot it is possible to determine the quantity of water corresponding to the relevant temperature, that is, there can be determined the difference between the original amount of water (10 ml) and the quantity of water remaining after completion of hydration.

The described trial was carried out with two samples. The first sample consisted of pure cement, the second of 20 grams cement plus 0.4 grams of an additive consisting of 35 parts sodium silicate, 10 parts triethanolamine and 55 parts calcinated aluminum sulfate.

Hydration was monitored at 0° to 5°C according to the above described techniques. The results have been tabulated in the table below from which there can be recognized the much more rapid hydration, that is, the development of the earlier strength in the cement prepared with the additive of this process.

	Grams of Bound Water per 20 Grams Cement	
Hydration After–	Without Additive	With Additive
Hours:		
0.5	0.15	0.34
1	0.17	0.42
2	0.21	0.55
3	0.23	0.63
4	0.26	0.76
5	0.29	0.825
6	0.325	0.89
7	0.36	0.93

Example 5: This example illustrates the development of the incipient or early strength properties by virtue of the process additive, and specifically while using different cements.

There is produced a heavy or dense concrete from 230 parts sand, with a grain size of 0 to 8 millimeters, 252 parts gravel, with a grain size of 8 to 30 millimeters, 75 parts cement and 37.5 parts water. The water-cement ratio (W/Z) therefore amounted to 0.5. The additive consisted of 35% sodium silicate, 5% triethanolamine and 60% calcinated aluminum sulfate. The results have been set forth together with those of the control trials in the table below.

Concrete Compressive Strengths in kg/cm²

After–	Normal Portland Cement (Z 375)[1,a]	High-Grade Portland Cement (Z 550)[2,a]	Normal Portland Cement Plus 2% $CaCl_2$	Normal Portland Cement Plus 2% of the Process Additive, Based on Cement and $Al_2(S_4O_3)$
Hours:				
6	- - -	8	19	10
9	14	31	38	53
12	29	80	69	86
15	60	132	81	117
18	85	180	102	144
21	112	193	114	163
24	146	212	142	173
48	218	290	222	225

[1]ASTM: Type I
[2]ASTM: Type III
[a]Norms according to *Cement Standards*, Cembureau, Paris (1968)

The rapid development of the early strength, that is to say, 9 to 15 hours after preparation can be clearly recognized from this table, likewise the fact that the additives do not cause any loss in strength.

Replacement of Sodium Oxide by Calcium Oxide

R.J. Murray and A.W. Brown; U.S. Patent 3,942,994; March 9, 1976; assigned to The Associated Portland Cement Manufacturers Limited, England describe an early strength hydraulic cement comprising: from 15 to 90% by weight based on the cement, of an orthorhombic or tetragonal phase $XC_{14}A_5$ which is $NC_{14}A_5$ as further defined or the corresponding phase in which the place of the sodium oxide is at least partly taken by at least one other alkali metal oxide; the balance being predominantly calcium silicates.

It is known that conventional Portland cements after mixing with water, although ultimately yielding high strengths, are initially slow to set in comparison with some other cementitious materials and that the early strength and rate of strength development during this setting period is poor. In some applications, such as where movement of shuttering or the turnover of molds in a short time is essential or where early exposure to mechanical stresses is required such as in the patching of roads or airport runways, the availability of a cement with many of the properties of a Portland cement but with a rapid and controlled set would be a distinct advantage.

Although the rate of setting and strength development of Portland cement can be increased by addition of a monocalcium aluminate (CA) cement and/or accelerators and can be marginally improved by increasing its C_3A content, it is difficult to obtain consistent and reproducible results by these methods, particularly where the process involves pumping a slurry of the cement or a concrete containing it or when used in conjunction with an aggregate contaminated with or containing substances capable of acting as accelerators or retarders of the set of ordinary Portland cements.

Cements containing $C_{12}A_7$ which provide rapid but reproducible and controllable setting times can be produced for such applications where reproducibility and controlled setting times are necessary. These compositions, however, require the use of a high aluminous material such as bauxite in their raw feed preparation, which increases their cost. Moreover, it is generally necessary to intergrind two separately prepared clinkers to obtain the desired phase composition and physical properties.

The objective of this process is the provision of cements which will have comparable, and in some instances improved, properties to those which have been described above but which do not necessitate the use of expensive raw materials, such as bauxite, and which in certain instances can be prepared from a single clinker.

It is known that small amounts of potassium or sodium oxide and other compounds which are present as trace impurities in Portland cement compositions enter into solid solutions in certain of the clinker phases during burning and may affect the hydraulic properties of these phases. It has also been reported during investigations of the CaO-Al_2O_3-Na_2O ternary system that if additions of Na_2O exceed the solid solution limit of Na_2O in the cubic C_3A phase then it leads to the formation of an orthorhombic or possibly tetragonal sodium calcium aluminate phase (or phases).

Although the exact composition of this compound is disputed in the literature, it is now generally accepted that the initially proposed ternary compound NC_8A_3 (where N represents Na_2O, C represents CaO and A represents Al_2O_3) is incorrect and that the quantity of Na_2O required to substitute in place of CaO in the cubic C_3A structure in order to form the orthorhombic form is of the order of 4.5%. This corresponds to a compound with the formula $(C_{1-x}N_x)_3A$ where x lies between 0.06 and 0.07 which approximates to $NC_{14}A_5$. This nomenclature $NC_{14}A_5$ will for convenience be used to refer to the orthorhombic and/or tetragonal sodium calcium aluminate phase (or phases) referred to above. The $NC_{14}A_5$ may contain Na_2O or traces of other oxides in solid solution.

For the purpose of identification of the phase concerned, it has been found and reported elsewhere that the cubic C_3A phase will take up Na_2O in solid solution to the extent of about 2% by weight. The presence of greater amounts of Na_2O results in the conversion of C_3A into $NC_{14}A_5$, this conversion being complete when the Na_2O content reaches about 4.5% by weight.

A further increase in Na_2O content results in a solid solution of Na_2O in $NC_{14}A_5$ until, when the Na_2O content reaches about 6% by weight, yet a further phase forms resulting in a decrease of the $NC_{14}A_5$ content. Phases containing more than 2% or less than 10% by weight of Na_2O contain sufficient $NC_{14}A_5$ to be useful in cements in accordance with the process. It should be noted that when the solid solution limit of Na_2O in $NC_{14}A_5$ is exceeded an additional alumina molecule is required for each additional Na_2O molecule, to maintain saturated solid solution, forming an alkali modified calcium aluminate increasingly richer in alumina or a solid solution of sodium aluminate. Some Na_2O also enters into solid solution in the C_2S phase and these two chemical substitutions together eventually preclude the simultaneous formation of C_3S.

Reported hydration studies of $NC_{14}A_5$ or its solid solutions suggested that it has less hydraulic activity than pure C_3A alone. However, it has been found that by utilizing $NC_{14}A_5$ or related compounds in which the sodium is mainly or partially replaced by other alkali metals, to provide a substantial proportion of the hydraulic constituents of an otherwise primarily siliceous cement, together if necessary with retarders and/or accelerators selected and proportioned to provide a desired delay in setting to enable working and placing, it is possible to obtain a rapidly setting and hardening cement giving early strength but at the same time having a setting time which is controllable even in the presence of contaminants which are known to have a retarding effect upon conventional Portland cement.

Thus, the process provides an early strength hydraulic cement comprising from 15 to 90% by weight based on the cement, of an orthorhombic or tetragonal phase $XC_{14}A_5$ which is $NC_{14}A_5$ or the corresponding phase in which the place of the sodium oxide is at least partly taken by at least one other alkali metal oxide; the balance being predominantly calcium silicates. The $XC_{14}A_5$ may contain Na_2O or other alkali metal oxide, or traces of other oxides, in solid solution. C_3A with or without alkali metal oxides or traces of other oxides in solid solution may be present.

Preferably calcium sulfate in the form of natural or synthetic gypsums or soluble anhydrites is incorporated in the cement in order to obtain optimum strength development. Such an addition slightly retards the set but further small additions of retarders and/or accelerators effective to provide a desired setting time sufficient for the purpose for which the cement is intended are possible. These additional retarders are preferably conventional organic retarders for Portland cements such as citric acid, lignosulfonates and boric acid or special retarders such as sodium bicarbonate and disodium hydrogen orthophosphate. In some instances the addition during grinding of a small quantity of water may also be used to provide a retarding action.

The preferred accelerators can be incorporated either as additions made to the raw feed, such as alkalies which are retained as alkali metal sulfate within the clinker during burning; or as additions either made at the grinding stage or blended into the final cement, such as potassium sulfate, conventional accelerators for Portland cement or grinding aids such as a mixture of triethanolamine and acetic acid.

The $NC_{14}A_5$ or related compound may be incorporated into the cement in several ways. In one method the essentially pure compound may be separately prepared and interground with a Portland cement clinker or thoroughly blended in a finely divided form with a Portland cement.

A second method involves the preparation of a clinker rich in $NC_{14}A_5$ (or related compound) and sparse in C_3S which is then interground with a Portland cement clinker or thoroughly blended in a finely divided form with Portland cement in order to provide a final product having a desired concentration of $NC_{14}A_5$ or related compound. This method is useful where the risk is present of excess alkali inhibiting the formation of C_3S in the clinker.

A third method involves the preparation of the final cement directly by burning selected raw materials in the presence of an appropriate amount of an alkali metal compound such as sodium carbonate, the initial Al_2O_3 content of the raw materials being sufficient to permit formation of the desired amount of a $NC_{14}A_5$ or related phase during burning. In the following examples all percentages and proportions are by weight.

Example 1: A clinker rich in alkali-modified C_3A was prepared as follows. In order to avoid a heterogeneous distribution of soluble sodium carbonate during drying of prefabricated pellets of raw material, a source of largely insoluble Na_2O was prepared by blending a china clay whose principal constituents were SiO_2, 48.2%; Al_2O_3, 36.0%; Fe_2O_3, 1.1% and CaO, 0.1% with a finely divided silica (89.1% SiO_2) and a sodium carbonate in the approximate proportions (dry basis) 55% clay, 22% silica, and 23% sodium carbonate. The mix was sintered in an oil-fired furnace at 1000°C for approximately five minutes to yield a product of the analysis:

	Percent
SiO_2	56.0
Al_2O_3	21.0
Fe_2O_3	0.6
CaO	0.2
Na_2O (water-soluble)	5.5
Na_2O (water-insoluble)	10.7

This alkali-frit product was ground in a ball mill to a residue of 10% on a BS 90 μm sieve and the Na_2O source thus produced was blended to form a raw feed with the above china clay, and a whiting with the analysis:

	Percent		Percent
SiO_2	1.2	SO_3	0.09
Al_2O_3	0.2	K_2O	0.04
Fe_2O_3	0.1	Na_2O	0.06
CaO	54.7	Mn_2O_3	0.04
MgO	0.3	P_2O_5	0.06
L.O.I	0.4	TiO_2	0.01
CO_2	42.6		

in the approximate proportions (dry basis) 75.7% whiting, 17.1% china clay, 7.2% Na_2O source, these proportions being such as to allow for approximately 15% loss of Na_2O during firing. The raw feed was sintered in an oil fired furnace at about 1420°C to produce a free lime content as determined by the hot ethylene glycol extraction method of 1.3%. The final clinker analysis was as follows:

	Percent
SiO_2	20.5
Al_2O_3	12.0
Fe_2O_3	0.5
CaO	63.7
MgO	0.5
SO_3	0.03
K_2O	0.2
Na_2O	1.5

The lime saturation factor as defined in BSS 12 (1971) of this clinker was 0.883, the silica ratio (S/A + F) 1.64 and the alumina ratio (A/F) 24.0. The potential phase analysis of this clinker as calculated from its oxide analysis, and assuming all the Na_2O reacted with the calcium aluminate phase to form an $NC_{14}A_5$-Na_2O solid solution is:

	Percent
C_3S	21.4
C_2S	38.1
$NC_{14}A_5$ (with Na_2O in solid solution)	31.2
C_4AF	1.5

together with other minor phases. The presence of the above principal phases was confirmed by x-ray diffraction techniques. A cement was prepared by grinding this clinker with gypsum and citric acid to a surface area of 468 m^2/kg measured by the air permeability method according to BSS 12 (1971). The quantity of gypsum added was such as to give a total SO_3 content attributable both to the added gypsum and to that present in the clinker of 2.5% as determined by analysis, while the amount of citric acid added was 1.5%.

Example 2: The cement of Example 1 was tested for setting time according to BSS 12 (1971) and gave a time to initial set (% consistency water 30.3%) of 20 minutes. The pumpability time of a paste having a water/cement ratio of 0.5 was ten minutes. A concrete for compressive strength tests was made up from 1 part cement, 2.5 parts Mount Sorrel granite, 3.5 parts of Curtis sand and 0.6 parts water. The test results were as follows for the compressive strengths of 100 mm concrete cubes:

	psi
After 1 hour	50
2 hours	765
4 hours	860
8 hours	910
24 hours	1090
3 days	1535
7 days	1745
After 28 days	2420
3 months	3305

Example 3: The cement of Example 1 was used as a binder for a coal shale, the overall composition of the mix being 1 part cement, 6 parts dry shale and 2 parts water. The setting time of the slurry produced was approximately ten minutes and the compressive strengths of 100 mm cubes of the mix were:

	psi
After 2 hours	110
4 hours	110
24 hours	110
3 days	155
7 days	245
28 days	265

Calcium Sulfate-Calcium Fluoroaluminate

T. Mizunuma and T. Yoshida; U.S. Patent 3,856,540; December 24, 1974; assigned to Denki Kagaku Kogyo KK, Japan describe a cement additive consisting mainly of a mixture of calcium sulfate and calcium fluoroaluminate having the chemical formula $3CaO \cdot 3Al_2O_3 \cdot CaF_2$. The calcium fluoroaluminate is produced by mixing lime, bauxite and fluorspar in such an amount that the mineral of the resulting product becomes $3CaO \cdot 3Al_2O_3 \cdot CaF_2$ and burning the resulting mixture at a temperature of 1200° to 1400°C.

If the temperature is lower or higher than the above range and the cooling rate of the resulting clinker is not proper, side reactions occur and therefore the desired results are not obtained. The above ternary compound $3CaO \cdot 3Al_2O_3 \cdot CaF_2$ is known in the art.

The mixing ratio of the above described calcium fluoroaluminate and calcium sulfate is 80 to 10% by weight of calcium fluoroaluminate and 20 to 90% by weight of calcium sulfate. When the resulting mixture is combined with conventional Portland cement in an amount of 1 to 30% by weight, the set time of the cement may be shortened in a controllable way and the cement will develop a high strength upon setting, and simultaneously the cement composition will develop an effective expansion.

Example 1: Commercially available calcium carbonate, commercially available alumina and fluorspar were mixed in the following proportions. 43.8% by weight of calcium carbonate (purity 99.0%), 44.3% by weight of alumina (purity 99.5%), and 11.9% by weight of fluorspar (purity 95.0%). The resulting mixture was charged into a platinum crucible and burned at 1350°C for 2 hours in an electric furnace. The composition of the resultant clinker was confirmed by x-ray analysis as $3CaO \cdot 3Al_2O_3 \cdot CaF_2$. The chemical analysis according to JIS R5202 is shown in Table 1.

TABLE 1

LOI	SiO_2	Fe_2O_3	Al_2O_3	CaO	TiO_2	MgO	CaF_2	Total	Free CaO
0.8%	7.4%	3.8%	45.8%	30.4%	0.5%	0.4%	10.9%	99.8%	0.4%

The obtained clinker contained 0.4% by weight of free CaO. The clinker was mixed with anhydrous calcium sulfate, which was obtained by burning calcium sulfate dihydrate at 1000°C for one hour, and the resulting mixture was ground to a specific surface area of 2,930 cm^2/g to obtain a cement additive. In the above mixing, 25% by weight of the clinker ($3CaO·3Al_2O_3·CaF_2$) and 75% by weight of anhydrous calcium sulfate were mixed.

The cement additive and conventional Portland cement were mixed in a weight ratio of 11:89% to prepare an enriched cement, and the expansion rate of the enriched cement was measured according to JIS A1125. The measuring method is as follows. The enriched cement and sand were mixed in a weight ratio of 1:2 to prepare a mortar, and the mortar was mixed at a water/cement ratio of 65%, formed in a metal mold having a dimension of 4 x 4 x 16 cm, cured for one day in humid air and taken out from the metal mold. The length of the mortar sample was used as a base length.

The mortar sample was further cured in water at 20°C for the predetermined period of time as shown in the following Table 2, and the length of the sample was compared with the original length, whereby the expansion rate of the sample was measured. The obtained result (in percent) is shown in Table 2.

For comparison, the above experiment was repeated, except that $12CaO·7Al_2O_3$ clinker and $CaO·Al_2O_3$ clinker were used instead of the $3CaO·3Al_2O_3·CaF_2$ clinker of the process. The obtained results are also shown in Table 2.

TABLE 2

	This Process	Conventional Additive	
	$3CaO·3Al_2O_3·CaF_2$:$CaSO_4$ 25:75*	$12CaO·7Al_2O_3$:$CaSO_4$ 25:75*	$CaO·Al_2O_3$:$CaSO_4$ 25:75*
1 Day	0.053	0.034	0.012
3 Days	0.129	0.025	0.007
7 Days	0.247	0.022	0.013
14 Days	0.295	0.029	0.019
28 Days	0.315	0.031	0.021

*Mixing ratio (percent by weight)

Example 2: A cement additive was produced from the $3CaO·3Al_2O_3·CaF_2$ clinker obtained in Example 1 and anhydrous calcium sulfate in the same manner as described in Example 1, except that 33% by weight of the clinker and 67% by weight of the calcium sulfate were mixed and the mixture was ground to a specific surface area of 5,200 cm^2/g. The cement additive and conventional Portland cement were mixed in a weight ratio of 7:93% to prepare a mortar. The result of the strength test of the mortar according to JIS R5201 is shown in Table 3 together with the result in the case of mortar of conventional Portland cement alone.

TABLE 3

	Bending Strength (kg/cm^2)			Compressive Strength (kg/cm^2)		
	3 Days	7 Days	28 Days	3 Days	7 Days	28 Days
Conventional Portland Cement	30.3	49.5	79.3	121	229	385
This Process	36.7	65.7	85.6	172	295	402

As seen from the examples, when the cement additive of the process, which is composed of calcium fluoroaluminate represented by the chemical formula $3CaO \cdot 3Al_2O_3 \cdot CaF_2$ and calcium sulfate, is mixed with cement, the expansion rate of the resulting cement composition is about ten times or more as compared with the case when conventional cement additive is used, and moreover the strength of the cement composition can be increased by about 30% or more.

CEMENT ADDITIVES AND COMPOSITIONS

WATER CONTROL

Pyrazolones

R.H. Cooper; U.S. Patent 3,867,160; February 18, 1975; assigned to The Dow Chemical Company describes a method of producing improved cement mortar, concrete, soil-cement and soil-lime systems which exhibit the properties of increased strength, decreased water absorption or enhanced resistance to scaling. This is accomplished by mixing into the system a small amount of a N-heterocyclic compound. Typical of the useful compounds are:

4-(2-methoxy-5-nitrophenylazo)-1,3-diphenyl-2-pyrazoline-5-one;
4-[x-(2,5-dichlorophenylazo)salicylidiene] -1,3-diphenyl-2-pyrazoline-5-one;
5-oxo-3-phenyl-2-pyrazoline-1-thiocarboxamide;
4-isopropylidene-1,2-diphenyl-3,5-pyrazolidinedione;
3,5-dioxo-1,2-diphenyl-4-pyrazolidinecarboxanilide;
1,2-diphenyl-3,5-pyrazolidinedione;
4-acetonyl-1,2-diphenyl-3,5-pyrazolidinedione.

Example 1: Dried silty clay soil, 497.6 grams (Code No. 13C) was mixed with 19.2 grams of Portland cement. This mixture was wetted for compaction by spraying 61.4 grams of water into the mixture while subjecting it to further mixing. After a uniform mix was obtained, 85 gram aliquots were placed in a cylindrical molding tube 3 cm in diameter and compressed from both ends in a hydraulic press at a pressure of 740 psi until a static condition was attained. The molded specimen was then ejected from the molding tube and placed in a 100% RH atmosphere at 73°F for 14 and 28 day cures. This resulted in a series of soil-cement specimens (3 cm x 6 cm) containing 4% Portland cement based on the dry weight of soil.

This procedure was repeated except that the appropriate quantity of pyrazolone was dissolved in acetone, added to the water and sprayed onto the soil-cement mixture. In this manner identical specimens as above were produced containing various percentages of the pyrazolone compounds. After curing the specimens they were allowed to stand for 1 day at room temperature and conditions; then they were soaked in water for 24 hours. Next, the core strengths of the immersed specimens were tested on a Soiltest U-160 unconfined compressive strength tester. The cores were loaded through their primary axis at a rate of 0.07 in/min. The maximum load supported was divided by the cross-sectional area to give the immersed unconfined compressive strengths shown in the table below.

Unconfined Compressed Immersed Strengths of Modified Soil-Cement Systems

Compound	Percentage	Unconfined Compressed Immersed Strength (psi)	Increased Strength over Control, Times Greater
Control	None	30	—
1	0.10	81	2.7
2	0.10	85	2.8
3	0.10	97	3.2
3	0.05	89	2.9
4	0.10	74	2.4
5	0.10	74	2.4
6	0.10	88	2.9
6	0.05	53	1.8
6	0.025	51	1.7
7	0.10	81	2.7
7	0.05	52	1.7
8	0.10	70	2.3
9	0.10	54	1.8
10	0.10	63	2.1
11	0.10	54	1.8

The increased strengths of the treated specimens are generally associated with a decrease in moisture content; hence the ability of these pyrazolone compounds to inhibit moisture absorption in soil-cement or soil-lime systems is established.

Example 2: To 1 part of cement was added 2.75 parts of graded, standard sand by weight (ASTM designation: C-10-9-64). The mixing was done mechanically in a mixmuller in accordance with the ASTM procedure given in Section 6 of Method C-305-64T. The cement mortar mix consisted of 500 grams Portland cement, 1,375 grams graded sand and 240 grams of water.

The mixing water was placed in the mixing bowl, followed by cement, powdered pyrazolone and sand while subjecting the mixture to mixing according to C-305-64T procedure. The cement mortar was placed into two 3-gang, 2" cube molds according to ASTM procedure C-109-64. The specimens were covered with a wet cloth and allowed to harden overnight at room temperature. They were removed and placed in a 100% RH atmosphere for 14 and 28 day cures followed by air drying at 50% RH, 73°F, for 24 hours.

The cubes were then subjected to slow freeze-thaw tests according to ASTM procedure C-292. The dried cubes were covered with a 3% NaCl solution to a depth of about one-half inch above the cubes. After 24 hours of soaking the cubes were drained, washed with fresh water, wiped dry and weighed. They were again covered with the 3% NaCl solution to a depth of about one-half inch over the cubes and placed in a freezer at 0°F overnight. Next, the cubes were thereafter allowed to thaw in the brine solution at room temperature whereupon the above procedure was repeated. Every five cycles data were taken and fresh brine introduced.

The following table shows the curability of the cement on concrete system cubes. At the rating intervals the cubes were visually noted according to the scale and system taken from Axon, et al, "Laboratory Freeze-Thaw Tests vs Outdoor Exposure Tests," *Highway Research Record,* No. 268, August 1969, p 35. The marked reduction in scaling in the modified cement on concrete systems proves the utility of these pyrazolone compounds as scale inhibitors without significant increase in brine absorption. The cement used in all the preceding examples was Portland Cement Type 1.

Durability of Pyrazolone Cement Mortar Cubes During Slow Freeze-Thaw Cycles

	Scale Rating* 14 Day Cure						
	Number of Cycles						
	1	3	5	10	15	20	30
Unmodified	0	0	½	1	2	3	4
(2) at 0.10%	0	0	0	0	0	½	½
(2) at 0.25%	0	0	0	0	0	0	½
(3) at 0.10%	0	0	0	0	0	½	½
(3) at 0.25%	0	0	0	0	0	1	1
Air entrained	0	0	0	0	½	1	2

	28 Day Cure, Number of Cycles																
	1	3	5	10	15	20	25	30	35	40	50	60	70	80	90	100	110	130
Unmodified	0	0	0	½	1	2	2	3	3	3	3	3	3	3	4	4	4	4
(2) at 0.10%	0	0	0	0	0	0	0	0	0	0	½	1	1	2	2	2	2	3
(2) at 0.25%	0	0	0	0	0	0	½	½	½	½	½	½	½	½	½	½	½	1
(3) at 0.10%	0	0	0	0	½	½	½	½	½	½	1	1	1	1	2	3	3	3
(3) at 0.25%	0	0	0	0	0	0	0	½	½	½	½	½	1	1	1	2	2	2
Air entrained	0	0	0	0	0	1	2	3	3	3	3	3	3	4	4	4	4	4

*Scale rating
0 = No scale
1 = Slight scale
2 = Slight to moderate scale
3 = Moderate scale
4 = Moderate to heavy scale
5 = Heavy scale

Inositol

R.B. Peppler and P.A. Rosskopf; U.S. Patent 3,725,097; April 3, 1973; assigned to Martin Marietta Corporation have found that the water content of hydraulic cement mixes for a given plasticity is reduced, and the strength of the hardened concrete is increased by the incorporation of inositol. Early strength is increased, and the retardation of the rate of set is less than that normally encountered in obtaining the desired advantages in the properties of the mixes.

The inositol is employed in an amount within the range of from about 0.005 to about 0.75% by weight, preferably from about 0.10 to about 0.25% by weight of cement. Set accelerators can also be incorporated in the cement mix, the properties of the accelerators and the inositol being additive so that an increase in the compressive strength of the hardened mix is achieved while, at the same time, increasing the rate of set of the mix including the inositol and accelerators.

Inositol, $C_6H_6(OH)_6 \cdot 2H_2O$, is also known as hexahydroxycyclohexane and as cyclohexanehexol and cyclohexitol. It is a commercially available product prepared from corn steep liquor by precipitation and hydrolysis of crude phytate. Inositol exists in the form of white crystals, odorless up to 224° to 227°C, anhydrous at 100°C, and soluble in water.

Glyoxal

M.J. Link; U.S. Patent 3,365,319; January 23, 1968; assigned to Union Carbide

Corp. describes a method for improving the physical properties of Portland cement compositions by incorporating from 0.01 to 2% by weight glyoxal based on the dry weight of the Portland cement in the composition. In the following examples, the mixing procedure set forth in detail below was utilized in all cases and evaluation of the cement compositions was made in accordance with accepted test procedures of the ASTM.

Mixing Procedure – Concrete samples were prepared in accordance with the following schedule of steps. In each instance where glyoxal was utilized, it was added to the water charged in step (3).

(1) The mixer was buttered with a ⅓ size batch.

(2) The butter batch was discarded.

(3) ¾ of the anticipated water requirement was charged to the mixer.

(4) All of the sand and stone was charged to the mixer (with the percent moisture contents on each being determined subsequently by oven drying of samples).

(5) The mixer was operated for 15 seconds.

(6) All of the cement was charged to the mixer.

(7) The mixer was operated for a 2 minute period during which additional water was added in an amount sufficient to give a 3 ± ½ inch slump.

(8) The mix was permitted to stand for a 2 minute period.

(9) The mixer was operated for 2 minutes.

(10) Mixing was stopped and tests run for slump, air content and unit weight.

(11) The tested material of step (10) was returned to the mixer and remixed.

(12) The entire batch was discharged into a metal tub and specimens for the proposed tests were cast. After casting, all specimens were aged at 74°F and 95+% RH.

Example 1: A series of concrete mixes containing various concentrations of glyoxal were prepared in accordance with the mixing procedure described above and the time of set and compressive strengths determined. Consideration of the results indicated that the addition of glyoxal brought about a substantial increase in the compressive strength of the concrete and a substantial reduction in the total water required. For example, the addition of 0.10% glyoxal, based on the dry weight of cement, resulted in an increase in the 28 day compressive strength of almost 40% and a reduction in the total water required of 17%.

Example 2: Two concrete mixes, one a conventional mix and the other containing a small amount of glyoxal, were prepared in accordance with the mixing procedure described above and both the compressive and tensile strengths were determined.

Consideration of the results indicated that the addition of glyoxal resulted in a substantial increase in the compressive strength of the concrete regardless of the aging period to which the specimen was subjected. Thus, the addition of 0.15%

glyoxal, based on the dry weight of cement, resulted in a 41% increase in 7 day strength, a 52% increase in 28 day strength, and a 38% increase in 180 day strength. The 28 day tensile strength was increased 12% by addition of 0.15 weight percent glyoxal. It was also noted that increasing the concentration of glyoxal up to 1%, based on the dry weight of cement, resulted in an increase in the time of set, i.e., the setting of the concrete is retarded.

Tricalcium Silicate

G.H. Sadran; U.S. Patent 3,645,750; February 29, 1972; assigned to Ciments Lafarge SA, France describes aluminous cements containing tricalcium silicate. The tricalcium silicate acts as a delayed-action water-binding agent, which removes water from cement mortars in excess of that required to hydrate aluminous cement to the stable crystalline form of tricalcium aluminate hexahydrate.

In related work *P. Stiglitz; U.S. Patent 3,649,318; March 14, 1972; Ciments Lafarge SA, France* describes a method for preparing stable concrete from aluminous cements. This is accomplished by controlling the water/cement ratio between 0.25 and 0.4 when the cement is mixed with water. This ratio insures use of the water for hydration of the binding paste into cubic aluminate.

Polyvinylpyrrolidone and Sodium Naphthalene Sulfonate

A process described by *C.F. Weisend; U.S. Patent 3,359,225; December 19, 1967* involves an additive for Portland type cements which reduces the friction encountered as the cement mixture is flowed or pumped into place, and which also permits the utilization of decreased quantities of water in the cement mixture, thus appreciably increasing the strength of the hardened and cured cement.

The cement additive comprises polyvinylpyrrolidone having an average MW of the order of magnitude of 40,000, combined with the sodium salt of naphthalene sulfonate condensed with formaldehyde and having an average MW of at least 1,500, or of the magnitude of 1,500.

The polyvinylpyrrolidone is present in quantities of ½ to 10% by weight of the additive, the balance of the cement additive being essentially the sodium salt of naphthalene sulfonate condensed with formaldehyde, or the composition including 90 to 99.5% by weight of such sodium salt. The materials are merely mixed together and then may be added to the dry Portland type cement for thorough mixing before incorporation into a cement mixture or slurry.

The sodium salt functions to reduce the friction encountered when the cement slurry is flowed or pumped into place, and as an example, permits the proper cementing of casing in a well bore utilizing greatly reduced pressures, flow rates and flow velocities while achieving the necessary turbulence to assure that the cement slurry moves fully and completely into the desired locations.

The polyvinylpyrrolidone prevents the separation of free water from the cement slurry, not just as a bulking material, but apparently through a protective colloid action, and accordingly, the water present in the slurry tends to be retained therein until full hydration of the Portland type cement has occurred.

The cement additive may be added to the dry cement in the quantities of 0.1 to

3% by weight of the dry cement, less than 0.1% being more or less ineffective, and more than 3% being merely economically wasteful, and larger increases ultimately affecting the cement disadvantageously.

The usual range of incorporation would be 0.3 to 1% by weight, and very effective results have been obtained by the utilization of 0.75% by weight of the cement additive based on the weight of the dry cement. After the additive has been incorporated thoroughly into the dry cement, the latter is handled in conventional fashion, being made up in the desired mix or slurry.

Nonylphenol Polyethylene Glycol Ether-Triethanolamine Mixtures

A process described by *H.C. Fischer; U.S. Patent 3,287,145; November 22, 1966; assigned to Union Carbide Corporation* involves incorporating a chemical additive which improves the bleeding characteristics.

The chemical mixture consists of the product resulting from the intimate mixing at atmospheric conditions of suitable nitrogen base and alkyl aryl polyoxyalkene ethers, alcohols and mixtures thereof where the weight percentage ratio of the latter to the former is between 0.5 and 4.0. The alkyl aryl polyoxyalkene portion may be considered to be of the general formula:

$$R(OCH_2CH_2)_xOH$$

or

$$R{-}O{-}(C_2H_4O)_xH$$

where R represents an alkyl aryl portion consisting of benzene ring with at least one alkyl group, having 4 to 18 carbons substituted thereon. A multiplicity of polyalkene units, that is, the degree of polymerization, is indicated by x, being of the order of 6 to 30.

Suitable nitrogen bases which are combined with the alkyl aryl polyoxyalkene portion include mono- and polyaliphatic, aralkyl, carbocyclic and heterocyclic amines whether of primary, secondary or tertiary character; also the alkylolamines, whether mono- or mono-substituted or poly- or poly-substituted.

Example 1: 60 parts by weight of nonylphenol polyethylene glycol ether (9.5 polyethylene units) were added at room conditions to 40 parts by weight of triethanolamine in the presence of a variable amount of water, the latter assisting in promoting intimate contact. The resulting product was suitable as a chemical mixture for the purpose of the process. This product is referred to a product A.

Example 2: The mixture was prepared as in Example 1, but the formulating proportions were 71 parts of nonylphenol polyethylene glycol ether (9.5 polyethylene units) and 29 parts of triethanolamine. This product is referred to as product B.

Example 3: The mixture was prepared as in Example 1, but the formulating proportions were 33 parts of nonylphenol polyethylene glycol ether (9.5 polyethylene units) and 67 parts of triethanolamine. This product is referred to as product C.

Example 4: The mixture was prepared as in Example 1, but the formulating proportions were 75 parts of nonylphenol polyethylene glycol ether (9.5 polyethylene units) and 25 parts of triethanolamine. This product is referred to as product D.

Example 5: The mixture was prepared as in Example 1, but the formulating proportions were 50 parts of nonylphenol polyethylene glycol ether (9.5 polyethylene units) and 50 parts of triethanolamine. This product is referred to as product E.

Water-reduction characteristics or maintenance of consistency, as measured by slump, at reduced water contents are important process variables. Although the effects of such water reductions are exhibited in the hardened products, it is in the plastic state that the chemical mixtures of this process affect consistency.

For the purpose of illustrating water-reduction characteristics and the effect of the chemical mixtures relating to this process on consistency, data regarding Examples 1 through 5 (products A through E) are set forth in the table below. Tests were performed on concrete proportioned 1:2.1:2.8 by weight natural sand (fineness modulus, 290; cement factor, nominal 6 sacks per cubic yard): ASTM Type I Portland cement: river gravel (maximum size ¾").

Mix No.	Mixture	Dosage, % by Wt Cement	Water Content Pounds	Water Content Percent	Slump, Inches	Air Content, Percent
1	None	-	333	100	3.5	1.5
2	A	0.142	284	86	3.5	4.7
3	A	0.142	333	100	6.5	4.2
4	B	0.093	277	74	3.5	4.3
5	B	0.093	333	100	7.25	4.0
6	C	0.065	300	91	3.5	4.0
7	C	0.065	333	100	5.5	4.4
8	D	0.182	280	85	3.5	5.0
9	D	0.182	333	100	7.0	4.2
10	E	0.210	290	88	3.5	4.0
11	E	0.210	333	100	5.75	4.2

It can be seen that a decided decrease in consistency or increase in fluidity, as measured by slump, has been brought about by use of the various mixtures at equal water content (mixes 3, 5, 7, 9 and 11). In order to maintain the slump equal to that of the untreated mix, a significant reduction in water content was necessitated for each treated mix (mixes 2, 4, 6, 8 and 10). Such a water reduction is a significant step in the right direction toward the enhancement of concrete properties. The effect of the chemical mixture of this process on the bleeding characteristics of typical concrete mixes was tested.

The bleeding of mixes treated with the composition was decidedly less than that of the other mixes of comparable mix proportions. It was also noted that lesser quantities of bleed water are involved at the time of cessation of bleeding.

Salts of β-Naphthalenesulfonic Acid-Formaldehyde Condensate

S. Nishi, A. Oshio and E. Kiyomitsu; U.S. Patent 3,677,780; July 18, 1972; assigned to Onoda Cement Company Limited, Japan describe a method of producing mortar or concrete products which have high strength by subjecting to autoclave curing shaped articles of mortar or concrete incorporating therein a salt of β-naphthalenesulfonic acid-formaldehyde condensate as represented by the formula

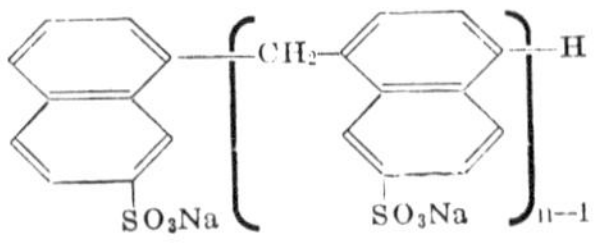

where n is a number greater than 2.

Representative results of the tests are indicated in the following tables and accompanying drawings. In the tests, normal Portland cement and sand were mixed in the weight ratio of 1:2 and the mixture was added with an aqueous solution of either one of the three water reducing agents, i.e., a sodium salt of β-naphthalenesulfonic acid (β-NS)-formaldehyde condensate containing at least 5 β-NS units in a molecule, Maginon 100 (the main component of which is a calcium alkylarylsulfonate) and Pozzolith No. 5L (the main component of which is calcium ligninsulfonate), in an amount as to provide mortar of flow value of 160 to 180 mm.

The mortar was mixed in the manner in accordance with JIS R 5201-1964 for test of the strength of cement, cast into a frame of 4 x 4 x 16 cm and, after 3 days, the mortar was taken out of the frame to obtain a test piece. The aggregate used was a sand from the Kinu-gawa River having maximum grain size of 5 mm (FM = 2.72). The test piece was put in an autoclave and was heated over 5 hours to 183°C (corresponding to a saturated water vapor pressure of 10 atm), maintained under these temperature and pressure conditions for 5 hours for effecting curing and then cooled over 5 hours to room temperature. The test piece was then taken out of the autoclave and subjected to a strength test. The results of the tests were summarized in Table 1 and Figure 2.1. In Figure 2.1, the curves **1**, **2** and **3** are respectively of a test piece made out of a mortar incorporated with a sodium salt of β-NS-formaldehyde condensate, Maginon 100 or Pozzolith No. 5L.

TABLE 1

Water reducing agent	Amount added (percent)	Flow (mm.)	Water/cement ratios (percent)	Bending strength (kg./cm.²)	Compressive strength (kg./cm.²)	Relative compressive strength	Remarks
Sodium salt of β-NS-formaldehyde condensate	0.00	169	37.3	170	900	100	
	0.25	168	37.7	163	911	101	
	0.375	167	36.3	169	1000	111	
	0.50	178	34.3	182	1062	118	
	0.625	171	33.0	209	1132	126	
	0.75	162	32.2	191	1390	154	
	0.875	178	31.3	192	1430	159	
	1.25	180	28.0	203	1450	161	
	1.5	171	27.0	199	1456	162	
	2.0	183	27.7	208	1479	165	
	2.2	173	27.9	200	1440	160	
	2.5	178	28.0	189	1220	136	
	3.0	186	28.3	175	1118	125	
	4.0	176	28.3	166	1143	127	
Maginon 100	0.15	157	37.6	172	954	106	
	0.30	153	34.5	183	1040	116	
	0.45	158	33.4	189	1080	120	
	0.75	170	32.0	195	1080	120	Viscosity of mortar increased in the step of mixing.
	1.50	188	29.8	153	810	90	
Pozzolith No 5L	0.20	162	35.3	204	1040	116	
	0.30	157	34.8	200	1020	113	
	0.50	159	35.7	200	1010	112	Mortar stiffened in the step of mixing.
	1.00	159	40.0	150	800	88.8	

FIGURE 2.1: HIGH STRENGTH MORTAR OR CONCRETE

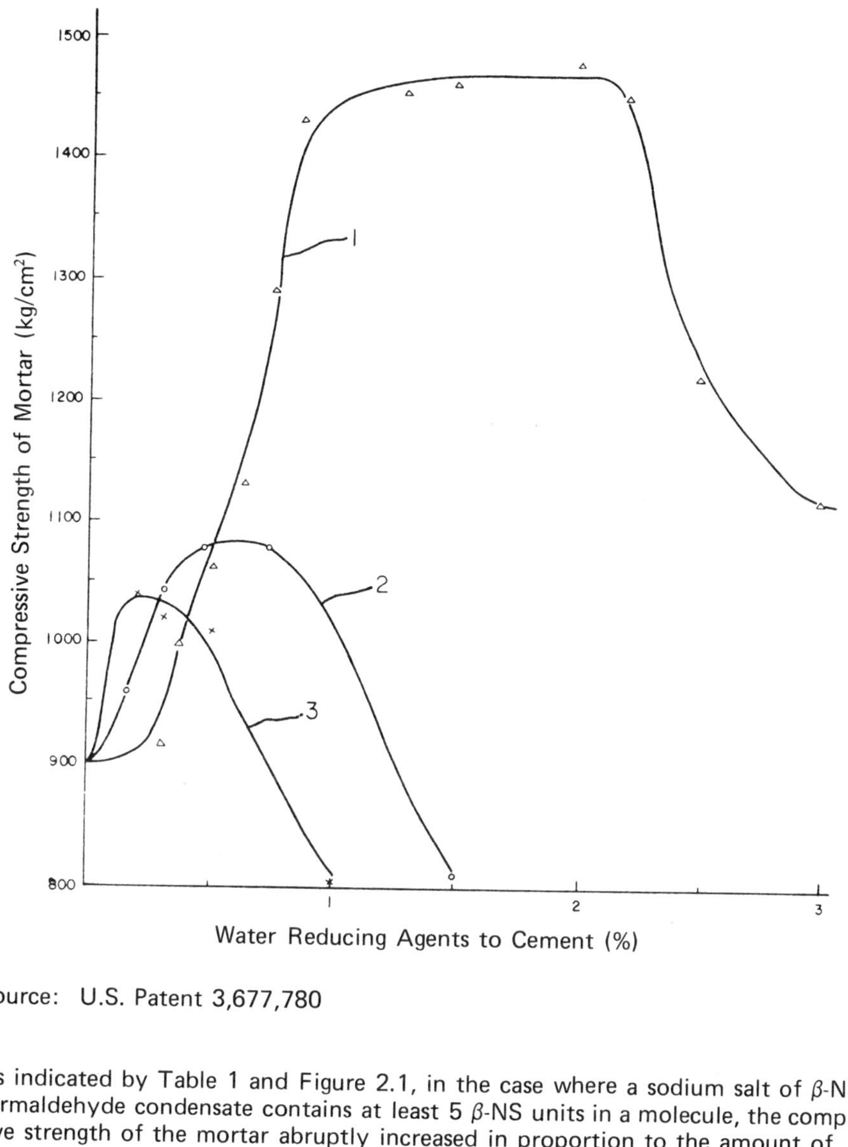

Source: U.S. Patent 3,677,780

As indicated by Table 1 and Figure 2.1, in the case where a sodium salt of β-NS-formaldehyde condensate contains at least 5 β-NS units in a molecule, the compressive strength of the mortar abruptly increased in proportion to the amount of the agent added until the amount of 0.75% and took the maximum value of 1,400 to 1,480 kg/cm² at the amount added within the range of from 0.75 to 2.3%, and then decreased at the amount of over 2.3%. The maximum compression strength is 55 to 65% higher than that of plain mortar.

On the contrary, in cases of Maginon 100 and Pozzolith No. 5L, the compression strength of mortar incorporated with one of them also increased in proportion to the amount added, but the maximum value was about 1,070 kg/cm², at the amounts added of 0.45 to 0.75% in case of Maginon 100 and about 1,040 kg/cm² at the amounts added of 0.2 to 0.5% in case of Pozzolith No. 5L. These values are by far smaller than the value of the maximum compressive strength obtained by incorporation of the sodium salt of β-NS-formaldehyde condensate, and the range of the amount added necessary for obtaining the maximum compressive strength is considerably narrow compared with that in the case of the sodium salt of β-NS-formaldehyde condensate.

Next, a gravel from the Abe-kawa River (maximum grain size 20 mm) as a coarse aggregate, a sand from the Kinu-gawa River as a fine aggregate and normal or moderate heat Portland cement were mixed in the proportions listed in Table 2 and the mixture was added to water and a sodium salt of β-NS-formaldehyde condensate containing at least 5 β-NS units in a molecule, in the amounts listed in Table 2, mixed and cast in a cylindrical frame having a diameter of 10 cm and a height of 20 cm. After 24 hours, the concrete formed was taken out of the frame, heated over 8 hours to 183°C (saturated steam pressure at 10 atm), maintained at this temperature for 5 hours and cooled over 8 hours to room temperature. The compressive strengths of the products thus obtained are shown in Table 2.

TABLE 2

Cement	Normal Portland Cement			Moderate Heat Portland Cement		
Amounts added of sodium salt of β-NS-formaldehyde condensate (%)	0	0.75	1.25	2.6	0	0.75
Amounts of cement per unit volume (kg/m³)	480	480	480	480	480	480
Amounts of water per unit volume (kg/m³)	173	141	130	131	175	135
Water-cement ratios (%)	36.3	29.4	27.0	27.4	36.5	28.2
Sand percentage (%)	43	41	41	41	43	41
Amounts of aggregates per unit volume (kg/m³):						
25-5 mm	1,035	1,100	1,117	1,116	1,034	1,111
Under 5 mm	779	760	772	771	778	773
Slump (cm)	6.5	6.0	7.0	8.0	6.0	7.0
Compressive strength (kg/cm²)	700	1,080	1,150	920	637	991
Relative compressive strength	100	155	164	132	100	156

As indicated by Table 2, the addition of the sodium salt of β-NS-formaldehyde condensate to concrete resulted in an increase in compressive strength of the concrete within a certain range of amount added as in the case of mortars.

Although it was ascertained by the above tests that the sodium salt of β-NS-formaldehyde condensate was one of the most effective water reducing agents, the following tests were made for the sake of ascertaining that the increase in compression strength was caused under high pressure steam curing or by decrease in water/cement ratio.

To the cement and sand mixture as used in the above tests were added water and the sodium salt of β-NS-formaldehyde condensate as used in the above tests, in various proportions as listed in Table 3, and the resulting mortars were molded under the same conditions as mentioned above into test pieces. The test pieces were, after curing in water at 20 ± 1°C for 7 days, subjected to measurement of compressive strength. The results obtained were summarized in Table 3.

TABLE 3

Water reducing agent	Sodium salt of β-NS-formaldehyde condensate				
Amounts added (percent)	0	0.5	1.0	2.0	3.0
Flow (mm.)	160	176	180	184	175
Water/cement ratio (percent)	37.2	34.3	30.3	28.3	29.5
Compressive strength of mortar (kg./cm.^{2})	618	668	780	767	673
Relative compressive strength	100	108	125	124	109

The above results indicate that, in case of mortars which have been cured in water under atmospheric pressure, the sodium salt of β-NS-formaldehyde condensate gave a maximum compressive strength of 760 to 780 kg/cm^2 at the amounts added of 1.0 to 2.0%, but the increase in maximum compressive strength compared with the maximum compressive strength of a plain mortar is only 25%. On the other hand, the cement and sand mixture of the same composition as mentioned above was added with water alone in variable water/cement ratios and the resulting mortars were molded into test pieces under the same conditions as mentioned above. The test pieces were then cured in an autoclave under the same conditions as mentioned above and subjected to measurement of compressive strength. The results obtained were as listed in Table 4.

TABLE 4

Water reducing agent	Water/cement ratio	Compressive strength (kg./cm.^{2})	Relative compressive strength
None	37.0	900	100
	35.0	910	101
	32.0	955	106
	29.0	981	109

As understood from the above results, as the water/cement ratio decreased the compressive strength of mortar became higher, but the compressive strength of mortar prepared by using no water reducing agent is by far lower than that of mortar of the same water/cement ratio as listed in Table 1 which was prepared using a sodium salt of β-NS-formaldehyde condensate. Therefore, it is believed that the increase in compressive strength of mortar as indicated by Table 1 and Figure 2.1 resulting from the addition of the sodium salt of β-NS-formaldehyde condensate is caused by a certain special interaction between the sodium salt of β-NS-formaldehyde condensate and cement during curing in an autoclave.

Example: Normal Portland cement and sand from the Kinu-gawa River (maximum grain size 5 mm, FM = 2.72) were mixed in the weight ratio of 1:2. The mixture was added with water in the amount to provide a water/cement ratio of 27% and with 0.9%, based on the weight of cement, of a sodium salt of a β-NS-formalin condensate (containing 5 to 10 β-NS units in a molecule) and mixed in a mixer for 2 minutes to obtain a mortar of a flow value of 170 mm.

The mortar was immediately cast in a frame of 4 x 4 x 16 cm and, after curing at 20°C for 24 hours in a moist chamber at RH of 95%, taken out of the frame. The molded mortar was put in an autoclave and the autoclave was heated to 183°C (10 atm) over 2 hours, maintained at the temperature for 4 hours and then cooled and released from pressure over 2 hours. The compressive strength of mortar product thus obtained was 1,390 kg/cm^2.

Formaldehyde-Naphthalenesulfonic Acid Condensate and Gluconic Acid Salts

K. Hattori, T. Yamakawa and A. Tuji; U.S. Patent 3,686,133; August 22, 1972; assigned to Kao Soap Co., Ltd., Japan describe a composition of matter, which is useful as a cement dispersing agent. The mixture consists of (a) a salt of a high molecular weight condensation product of naphthalenesulfonic acid and formaldehyde, containing not more than 8 wt % of unreacted naphthalenesulfonic acid and not less than 70 wt % of high molecular weight condensates having more than 5 naphthalene nuclei, and (b) a salt of gluconic acid. The cement dispersing agent is mixed with a hydraulic cement composition in order to form cement paste, mortar, concrete or the like.

The high molecular weight condensation product of naphthalenesulfonic acid-formaldehyde which may be used in this process can be prepared in such a manner as described below. That is to say, naphthalenesulfonic acid and formaldehyde (formalin) are first condensed in the presence of a sulfuric acid catalyst in a conventional manner (e.g., by the method described in PB Report, FIAT Final Report No. 1141) and, upon the solidification of the reaction mixture due to the progress of the condensation reaction, appropriate amounts of water, formalin and the catalyst are added to the mixture and the reaction is carried out further until a highly condensed product is obtained.

The condensation product has a residual content of the mononuclear compound (unreacted naphthalenesulfonic acid) of not more than 8% by weight, or preferably not more than 5% by weight, and the content of high molecular weight condensates having 5 or more naphthalene nuclei is not less than 70 wt %, based on the total amount of the condensation product.

For example, such a high molecular weight condensation product can be synthesized by using 1.8 mols of concentrated sulfuric acid and 1 mol of formaldehyde for each mol of naphthalene in the following manner.

128 grams of naphthalene is melted with heat and is kept at a temperature between 120° and 125°C. 128 grams of 98% sulfuric acid (1.84 SG) is added dropwise to the melt over a period of about 1 hour. After this, the mixture is caused to react at 160°C for 3 hours, and then cooled to 120°C and 96.6 grams of water is added. Next, 51 grams of 98% sulfuric acid is added, and the temperature is lowered to 80°C. While the mixture is maintained at a temperature between 80° and 85°C, 81.8 grams of 37% formalin is added dropwise over a period of 3 hours. Following the addition of the formalin, the temperature of the mixture is increased to 95° to 100°C over a period of about 1 hour, and then the reaction is continued for 25 hours at that temperature. The numbers of naphthalene nuclei in the condensate synthesized in this manner were as follows:

Number of naphthalene nuclei	Content in the product (percent by weight)	
1	1.5	7.4
2	0.8	
3	1.6	
4	3.5	
5	4.5	92.6
6	4.0	
7	84.1	

Before use, the condensation product should be converted either to a calcium salt by liming or to a sodium salt by soda addition. The salts of gluconic acid, which are to be used in the composition, include sodium, lithium, potassium and calcium salts.

The addition of the dispersing agent composition is extremely beneficial to the dispersibility of hydraulic cement, and permits a remarkable reduction in the amount of the mixing water required to form a hydraulic cement-based mixture for use as mortar and concrete. This process is illustrated by the following examples, in which the high (molecular weight) condensation product of naphthalene acid-formaldehyde used is the product prepared in the process as above described and consists of 1.5% of mononuclear compound, 5.9% of 2 to 4 nuclear compounds, and 92.6% of 5 and more polynuclear compounds. All percentages are by weight.

Example 1: Flow tests of ordinary Portland cement compositions containing mixtures of different proportions of (a) a high (molecular weight) condensation product of naphthalenesulfonic acid-formaldehyde and (b) a gluconate were conducted.

To 500 gram portions of ordinary Portland cement were added 145 cc portions of water containing 0.1, 0.2, 0.25 and 0.5% by weight, on the basis of the cement weight, respectively, of mixtures of different proportions of the sodium salt of the high condensation product of naphthalenesulfonic acid-formaldehyde and sodium gluconate. Each mixture was mixed in a mortar mixer for 3 minutes. Then, it was placed on a flow table and was subjected to a total of 15 vertical up and down oscillations at a rate of 1 oscillation per second to see how the cement paste spread, according to ASTM standard C-124-39. The results of the obtained flow values of cement are listed in the table below.

Composition (percent)		Amount added to cement (percent by weight)			
Sodium salt of high condensation product of naphthalenesulfonic acid-formaldehyde	Sodium gluconate	0.1	0.2	0.25	0.5
		Millimeters			
100	0	178	190	207	289
90	10	173	180	229	297
80	20	191	252	270	295
70	30	180	233	250	279
60	40	190	201	238	270
50	50	189	197	224	253
40	60	189	195	221	234
30	70	180	188	214	220
20	80	184	186	205	209
10	90	184	186	203	209
0	100	184	185	198	199
0	0	168			

As can be seen from the above table, the combined use of the sodium salt of the high condensation product of naphthalenesulfonic acid-formaldehyde and sodium gluconate leads to a great improvement in the dispersion effect as compared with the cases where either component is used alone.

Example 2: In the same manner as described in Example 1, cement flow tests were conducted with the following additive compositions to compare their dispersion effects. The results are shown in the following table.

Additive	Amount added, percent (on basis of cement weight) 0.1	0.2	0.3	0.4	0.5	0.6	0.7	0.8	0.9	1.0
A, mm	182	277	296	[1] >300	>300	>300	>300	>300	>300	>300
B, mm	180	262	284	>300	>300	>300	>300	>300	>300	>300
C, mm	178	203	236	268	276	282	288	>300	>300	>306
D, mm	177	182	189	192	190	190	189	188	187	188
E, mm	-	198	212	214	212	208	206	203	200	198

[1] The diameter of the flow table used in the test for determining flow values of cement was 300 mm and, accordingly, the flow values more than 300 could not be determined, as the spread cement paste will run over the table in such cases and such flow values were merely indicated as >300 in the table.

The additive compositions are as follows.

(A) A mixture of 80 parts (parts are by weight) of the sodium salt of high condensation product of naphthalene-sulfonic aicd-formaldehyde and 20 parts of sodium gluconate

(B) A mixture of 80 parts of the calcium salt of high condensation product of naphthalenesulfonic acid-formaldehyde and 20 parts of sodium gluconate

(C) Sodium salt of high condensation product of naphthalenesulfonic acid-formaldehyde

(D) Sodium gluconate

(E) Calcium ligninsulfonate

As will be apparent from the above table, the dispersion effects which the compositions A and B according to the process can achieve for cement are much greater than those which can be attained by the individual components of the compositions alone and the effect by a typical known cement dispersant, i.e., calcium ligninsulfonate.

Example 3: Concrete tests were performed with the following admixtures or additive compositions.

(F) A mixture of 70 parts of the sodium salt of a high condensation product of naphthalenesulfonic acid-formaldehyde and 30 parts of sodium gluconate

(C) Sodium salt of high condensation product of naphthalenesulfonic acid-formaldehyde

(G) Commercially available water-reducing agent of ligninsulfonic acid type

The materials used were ordinary Portland cement; fine aggregate sand (2.58 SG and 3.00 FM); coarse aggregate gravel (2.56 SG). The composition of concrete is as shown in the following table.

Additive	Amount added: Grams	Amount added: On basis of cement (percent)	Water/cement ratio (percent)	Sand/aggregate ratio (percent)	Water (kg.)	Cement (kg.)	Fine aggregate (S) (kg.)	Coarse aggregate (G) (kg.)
None	-	-	42.2	40	178	420	702	1,039
F	840	0.20	39.3	40	165	420	704	1,047
	1,050	0.25	38.1	40	160	420	712	1,062
	1,260	0.30	37.4	40	157	420	722	1,075
C	840	0.20	40.5	40	170	420	689	1,024
	1,050	0.25	40.2	40	169	420	689	1,027
	1,260	0.30	39.0	40	164	420	715	1,065
G	840	0.20	40.5	40	170	420	689	1,024
	1,050	0.25	40.2	40	169	420	689	1,027
	1,260	0.30	39.0	40	167	420	691	1,029

The test results were as shown in the following table.

Additive	Amount added on basis of cement (percent)	Slump (cm.)	Air content (percent)	Water reducing (percent)	Compression strength ($kg./cm.^2$) 3rd day	7th day	28th day
None		4.2	1.6		234	322	477
F	0.20	4.0	2.0	7.0	303	440	536
	0.25	4.0	2.4	10.0	315	446	559
	0.30	4.6	2.5	12.0	328	448	589
C	0.20	4.7	1.6	4.5	292	406	518
	0.25	4.1	1.8	5.0	300	413	522
	0.30	5.3	1.5	8.0	313	415	535
G	0.20	4.1	2.1	4.5	295	415	513
	0.25	4.0	2.0	5.0	291	389	409
	0.30	4.2	2.0	6.0	299	397	514

As clearly indicated in the above table, the compositions according to the process are remarkably advantageous in enhancing both the water-reducing and cement strengthening effects over the ligninsulfonate-type water-reducing agent (G) in widest use and the high condensation product of naphthalenesulfonic acid-formaldehyde (C) alone.

According to a process described by *T. Kitsuda, C. Yamakawa and K. Hattori; U.S. Patent 3,788,868; January 29, 1974; assigned to Kao Soap Co., Ltd. and Japanese National Railways, Japan* the fluidity of cement compositions can be maintained by adding to the cement composition a nonretarding, nonair-entraining cement dispersing agent selected from the group consisting of

(a) water-soluble salts of condensates having not less than 1,500 MW, and obtained by condensing with formaldehyde sulfonated products of monocyclic or fused polycyclic aromatic benzenoid hydrocarbon compounds and

(b) water-soluble salts of sulfonated products of fused polycyclic aromatic benzenoid hydrocarbon compounds having at least 3 benzene rings.

The dispersing agent is added at at least two chronologically spaced intervals or continuously so that the fluidity of the cement composition can be maintained over an extended period of time.

Phosphonic Acid

D.L. Schmidt, R.H. Cooper and G.H. Brandt; U.S. Patent 3,794,506; Feb. 26, 1974; assigned to The Dow Chemical Company describe an improved cementitious composition curable by hydration and having enhanced properties, such as less water absorbency, decreased spalling or greater strength retention. The process comprises a cementitious material and a small but effective amount of a phosphonic or phosphinic acid or soluble salt thereof containing various hydrophobic substituents.

Typical of the useful compounds are hexyl phosphonic acid, octyl phosphonic acid, n-decyl phosphonic acid, dodecyl phosphonic acid and octadecyl phosphonic acid. Representative of the phosphinic acids are dihexyl phosphinic acid, diheptyl phosphinic acid, didecyl phosphinic acid, dihexadecyl phosphinic acid and dioctadecyl phosphinic acid.

Alkenyl-Substituted Succinic Anhydride

K.B. Bozer and R.H. Cooper; U.S. Patent 3,817,767; June 18, 1974; assigned to The Dow Chemical Company describe a cementitious composition curable by hydration and having enhanced properties, such as less water absorbency and greater strength. The composition comprises a cementitious material and a small but effective amount of alkenyl-substituted succinic acid or anhydride.

The active agent is an alkenyl succinic anhydride of the formula

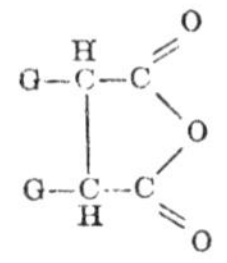

or an alkenyl succinic acid or salt of the formula

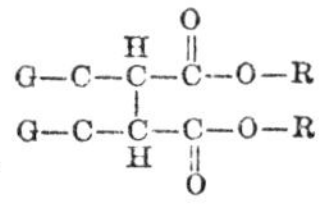

in either of which formulas at least one G represents alkenyl of from 6 to 16, both inclusive, carbons, the other G independently represents hydrogen or alkenyl of from 8 to 16, both inclusive, carbons and either R is independently hydrogen, ammonium or alkali metal.

Among the compounds that are articles of commerce and are usable in the process are the following: a mixed hexadecenylsuccinic anhydride represented by the manufacturer as being an isomeric mixture; a relatively pure 1-decenylsuccinic anhydride having a refractive index (n) at 20°C for the D line of sodium light of 1.4691; a mixed dodecenylsuccinic anhydride as a viscous liquid boiling at 180° to 182°C under 5 mm mercury pressure absolute; a pure 1-dodecenylsuccinic anhydride as a crystaline solid melting at 38° to 40°C; a 1-hexadecenylsuccinic anhydride melting at 59° to 60°C; a tetradecenylsuccinic anhydride melting at 53° to 56.5°C; and a 1,1,3,5-tetramethyl-2-octenylsuccinic anhydride, supplied as a viscous yellow liquid. Other substances equally adapted to be used include octenyl succinic anhydride and didodecenylsuccinic anhydride.

Particulate Adsorbents

According to a process described by *H.N. Babcock; U.S. Patent 3,890,157; June 17, 1975* shrinkage of aqueous hydraulic cement mixtures is eliminated by incorporating therein a porous particulate material, such as an adsorbent such as activated alumina, activated bauxite, activated silica gel, and activated carbon. The particulate adsorbent material has a volume of entrapped gas and is capable of releasing at least a major portion of such gas during setting and early hardening of the cement mixture while in contact with water. The adsorbent gas release is fairly rapid making it particularly advantageous for fast setting cementitious compositions.

In the following examples, the performance of the additive was judged by the expansion and contraction of the cementitious system as soon as it was mixed with water and cast in a cylindrical mold with approximately 10% of exposed surface. The expansion and contraction of the cast was determined by the vertical movement of the top surface. For the purpose of higher accuracy, a light test was used to measure the movement of the top surface.

The test consists of using a focused light beam to project a shadow of the top surface onto a screen equipped with a vertical graduation. The magnification is 72 times. The movement of the top surface on the screen is recorded every 10 to 20 minutes for each cast until final set, which usually takes about 3 to 4 hours with longer setting materials and less than 60 minutes with fast setting cementitious composition.

A thin layer of water was added to the mold for cast setting under no evaporation condition. To facilitate the detection of the movement of the top surface, a marble was placed on top of the surface and the expansion or contraction of the cast was determined by the movement of the apex of the shadow projected on the screen.

Example 1: In this example, the no evaporation condition was used for setting a Portland cement paste prepared by mixing Allentown-Type III cement with water in a ratio of 4 gallons per sack of cement. Silica gel added thereto is activated and has a particle size in the range of 6 to 12 mesh. The results of the described light test on various casts using different amounts of silica gel are tabulated below.

% Silica	Volume Change After 3 Hours	Volume Change After 24 Hours (dry)
0.1	-2.7	-2.2
0.3	+0.8	+0.75
0.5*	+0.7	-0.1
1.0	+2.0	+1.3
1.0**	+0.6	0.0

*Some drying took place during this run which may account for the discrepancy between the runs with 0.3 and 0.5% admixture.
**The gel was ground to 50-100 mesh.

Example 2: A Portland cement paste similar to Example 1 was used in this example. The additives used were as follows:

(A) Activated bauxite

(B) Silica gel

(C) Synthetic bone char

(D) Activated carbon

(E) Magnesia-silica gel

Only 1% by weight of cement of these additives was used. The results of the light tests are tabulated in the following table. In contrast to those results, an aqueous sand-cement mixture prepared in the same manner as those used in the examples exhibited a negative volume change, that is, shrinkage.

	Volume Change After 3 Hours	After 24 Hours (dry)
A	+0.6	+0.8
B	+0.6	+0.9
C	+0.2	+0.25
D	+0.1	+0.15
E	+0.3	+0.13

The foregoing discussion has particular relevance to particulate material capable of releasing comparatively large amounts of gas upon adsorption of water and includes the adsorbents described in use with cementitious systems having varied setting times. Additionally, it has been found that the adsorbents are particularly advantageously useful with fast setting cementitious systems. For example, a hydraulic cement substantially devoid of gypsum has the following setting times with various amounts of gypsum added thereto and is of the type for which the process is suitable.

Percent of Gypsum	Water to Cement Ratio	Initial Set
0.72%	6:1	10 min
0.8%	3.9:1	11 min
0.8%	5:1	22 min
0.8%	6:1	25 min
1.03%	3.9:1	25 min
1.03%	5:1	40 min
1.2%	4.8:1	45 min
1.8%	4.5:1	60 min

Pressure-calcined gypsum, also known as alpha gypsum, may be used in combination with hydraulic cement to make a fast setting cementitious composition. The pressure-calcined gypsum may be of the type marketed as Hydro-Stone. The amount of pressure-calcined gypsum that may be used for preparing a fast composition may range within a wide range, 5 to 100%, by weight of the cement being suitable with 25 to 75% by weight of the cement being particularly advantageous.

The adsorbent-like porous particulate materials described herein may be advantageously used in cementitious compositions that include calcined gypsum. While the presence of the calcined gypsum aids in elimination of long-term shrinkage and thus the elimination of such cracking causes, the adsorbent-like material results in shrinkage inhibition during fast setting.

Example 3: In this example, an Allentown-Type III cement was mixed with water in a proportion of 400 grams cement to 142 grams water and cast as described above. Identical mixes had respectively included in them various adsorbents as set forth below and were also cast as described above. The cast compositions were observed using the light test set forth above. The observations made at various times were recorded and are compiled in the following table. While the time observations made for specific materials vary within a matter of a few minutes such does not affect the trend shown. The adsorbent materials were added in the amount of 4 grams, approximately 1%, and are as follows:

(F) Activated silica gel (1.2 grams only added)

(G) Activated carbon (not dry)

(H) Activated carbon (dry)
(I) Activated magnesia-silica gel
(J) Magnesia (Sea Sorb 43)
(K) Magnesia (Sea Sorb 53)

Time After Casting (min)	Control	Volume Change (F)	(G)	(H)	(I)	(J)	(K)
0	0	0	0	0	0	0	0
10	-	-	-1.0	-0.1	0	-0.15	-0.2
15	-1.0	0	-	-	-	-	-
20	-	-	-1.0	-0.1	0	-0.25	-0.25
30	-1.5	+0.5	-	-	-	-0.3	-0.15
40	-	-	-1.1	-0.15	0	-	-
45	-1.5	+0.5	-	-	-	-0.3	-0.15
50	-1.5	+0.6	-	-	-	-	-
60	-	-	-1.1	-0.15	0	-0.3	-0.1

The results set forth in the above table demonstrate that shrinkage may be inhibited during fast setting, even within the first 10 to 30 minutes of setting, according to this process. In addition, it is seen that controlled shrinkage inhibition may be obtained through the addition of selected adsorbents. For example, the two types of magnesia show different values of shrinkage inhibition as do the differing adsorbents themselves. The activated carbon which was not dry did not work successfully as seen from the above table. Particle size may also be varied for control, and the various mesh sizes are given thusly, 60/100.

Ammonium Hydroxide and Carbohydrate

According to a process described by *C.W. Humphrey; U.S. Patent 3,857,715; December 31, 1974* a hydraulic cement-containing product is made by employing a hydraulic cement mixture prepared by homogeneously mixing into a dry mix of a hydraulic cement and aggregate, first, an aqueous solution of a water-soluble alkali compound, such as ammonium hydroxide, and then a water-soluble carbohydrate, such as molasses, and water sufficient to provide a mixture that is cohesive and plastic. Because of the low water content of the hydraulic cement mixture, the resulting products have good ultimate properties and can be made by an extrusion process in addition to the conventional molding process.

Digestion of Carbohydrates

A process described by *J.C. Steinberg and K.R. Gray; U.S. Patent 3,332,791; July 25, 1967; assigned to Rayonier Incorporated* relates to hydraulic cement compositions such as Portland type cement slurries containing organic dispersants.

It has been suggested in many publications relating to cement products that various types of carbohydrates such as reducing sugars, partially hydrolyzed starches, and the like, be used to disperse, render more workable and retard the setting time of slurries of hydraulic cement. Such compounds are known to be effective retardants. None, however has proven itself practical in general use.

The retarding effect of such carbohydrates on cement slurries is so great that any quantity of the carbohydrate that is effective as a dispersant and water-reducing agent also retards the setting time beyond permissible limits. As a result, rather than being considered useful in cement slurries they are usually considered to be harmful and are carefully removed where they occur in combination with other types of dispersants such as spent sulfite liquor solids.

After considerable experimental work a method was found for overcoming the undesirable characteristics of water-soluble carbohydrates such as mono-, di- and oligosaccharides when used as dispersants for cement slurries while retaining and enhancing their desirable qualities. By the method a portion of the carbohydrate is converted into complex products and the resultant blend of unchanged carbohydrate and complex products may be used as such, or concentrated if desired. This blend solution is an unexpectedly effective dispersant and water-reducing agent for slurries of hydraulic cement.

In the process the product is formed by digestion of a suitable water-soluble carbohydrate at an elevated temperature in a solution of a water-soluble alkali metal or ammonium sulfite or bisulfite until from 15 to 75% of the carbohydrate present is converted into water-soluble complexes. Suitable carbohydrates include glucose, sucrose, mannose, xylose and arabinose among the sugars.

In addition, water-soluble carbohydrates such as partially hydrolyzed starches and hemicelluloses such as those obtained by the steam hydrolysis of wood chips, etc., are also suitable. Digestion is accomplished by heating a 20 to 70% aqueous solution of the carbohydrate with sufficient sulfite or bisulfite to provide from 0.05 to 0.40 part SO_2 per part of carbohydrate at a temperature of from 120° to 170°C until a substantial proportion of the SO_2 is consumed.

For example, an operation may be carried out by digesting corn syrup having a dextrose equivalent of 41.5 to 44.5 with sodium bisulfite in an amount equivalent to 0.10 part SO_2 per part of corn syrup solids for 30 minutes at 140°C, the reactants comprising a solution having a concentration of 45% solids. This normally takes from 20 to 40 minutes after the desired digestion temperature has been attained. Digestion under these conditions converts from 15 to 75% of the soluble carbohydrate present into partially unidentified complex water-soluble organic compounds.

The extent of conversion into complex water-soluble organic compounds is measured by the disappearance of total carbohydrate, the total carbohydrate being measured by chromatographic determination of monosaccharide following acid hydrolysis [see method described by J.E. Jeffery, E.V. Partlow and W.J. Polglase, *Anal. Chem 32,* 1774 (1960)]. The product liquor containing its blend of carbohydrate and conversion products can be used in this form or it can be concentrated if desired.

It is the preferred practice not to take it to dryness, because of the hygroscopic nature of the conversion products. The dispersant is preferably added to the cement slurry in an amount varying from 0.01 to 0.30% by weight, on a dry basis, based on the weight of the cement. The following examples illustrate the preparation of the products of the process and their use and utility in hydraulic cement slurries as dispersants and water-reducing agents.

Example 1: 1,333 grams of glucose with an 8.8% moisture content and 378 grams of $Na_2S_2O_5$ with a 3% moisture content were dissolved in 1,794 grams of water. The resulting solution had a solids content of 43.6%, a pH of 4.55 and a viscosity of 9.2 cp at 25°C. It was placed in a suitable closed stainless steel digester and heated to 140°C in 8 minutes. After 20 minutes digestion at that temperature it was cooled to room temperature.

The product solution contained 41.3% total solids, had a pH of 2.9 and viscosity of 8.8 cp at 25°C. The residual SO_2 had been reduced to 1.8% on a dry basis. Before use as a dispersant in cement dispersions it was neutralized to a pH of 7.0 with NaOH. Analysis of the foregoing dispersant showed that 34.3% of the glucose had been converted to complex water-soluble conversion products. This dispersant was designated as Product A.

Example 2: 232 grams of $Na_2S_2O_5$ with a 4.0% moisture content and 1,500 grams of sucrose were dissolved in 1,756 grams of water. The resulting solution had a solids content of 49.0%, a pH of 4.3 and a viscosity of 14 cp at 25°C. It was placed in a suitable closed stainless steel digester and heated to 165°C in 13 minutes. After 10 minutes digestion at that temperature it was quickly cooled to room temperature.

The digested product solution had 47.7% total solids, a pH of 2.7, a viscosity of 12 cp at 25°C, and 0.1% residual SO_2 on a dry basis. After neutralization to a pH of 9.0 a small amount of fungicide was added. Analysis showed that 58.4% of the sucrose had been converted to complex water-soluble conversion products. This product solution was designated as Product B.

Example 3: 2,295 grams of light corn syrup containing 76.2% total solids of which 92% were free of polymeric glucose, 187.6 grams of Na_2SO_3 with a moisture content of 8.2% and 136.2 grams of $Na_2S_2O_5$ with a moisture content of 4.6% were dissolved in 1,496 grams of water. The solution contained 48.9% total solids and has a pH of 6.6 and a viscosity of 22 cp at 25°C.

It was placed in a closed digester and heated to 165°C in 8 minutes, digested for 30 minutes at that temperature and then cooled to room temperature. The resultant product solution had a total solids content of 41.6%, a pH of 4.6, a viscosity of 12 cp at 25°C, and a residual SO_2 content of 0.1%. Before use it was neutralized to a pH of 7 with NaOH solution. Analysis showed that 47.5% of the glucose had been converted to other materials during the digestion stage. This dispersant solution was designated Product C.

Example 4: Concrete test samples were prepared with and without the addition of Products A through C of Examples 1 through 3, respectively. In addition, comparative test samples were prepared containing equivalent amounts of the undigested starting materials from which Products A through C were prepared. The cement used was a standard construction grade Type I Portland cement.

The sand was air dried and classified into the following fractions: 100% passed a 4 mesh screen, 15% was retained on an 8 mesh screen, 15% on a 16 mesh screen, 30% on a 30 mesh screen, 24% on a 50 mesh screen and the balance of 16% on a 100 mesh screen. The gravel was classified into the following fractions: 100% passed a 0.75" screen, 73.8% was retained on a 0.375" screen and the balance was retained on a 4 mesh screen. The gravel was soaked in water and drained for 5 minutes prior to use.

Individual batches of concrete were prepared by mixing 20,070 grams of the foregoing air dried sand for 2 minutes with 6,400 grams of the cement. To this mixture was added 21,220 grams soaked, drained gravel (equivalent to 20,790 grams of gravel saturated with water but dry on the surface and 430 grams of water, 330 grams of which were necessary to saturate the sand and 100 grams of excess which was available as mixing water) and water in an amount such that the net mixing water was as shown in column 4 of the table that follows. (The term net mixing water refers to the difference between the total water content of the concrete and the water required to render the sand and gravel saturated but surface dry.)

The materials listed in column 2, and termed admixtures, were dissolved in and added with the mixing water. This slurry was mixed for 2 minutes, permitted to stand 3 minutes and remixed for 1 minute. The amount of water added in each case was that required to give the plastic concrete a nominal slump of 2.5" as determined by ASTM methods. Air content, time to initial and final set and the 7 and 28 day compressive strengths for each concrete test sample were also determined by standard ASTM methods and recorded in the following table.

	----- Admixture -----				- Time to Set -		- Compressive-Strength, psi	
Concrete Batch No.	Name	Dosage, percent of cement (dry basis)	Net Mixing Water, g	Air Content, percent by volume	Initial, hr	Final, hr	7 Day Cure	28 Day Cure
1	None	-	3,450	1.8	4.1	6.3	2,290	4,170
2	Glucose*	0.19	3,131	2.0	8.2	11.2	3,470	5,207
3	Product A	0.19	3,145	2.1	6.4	8.2	3,610	5,805
4	Sucrose*	0.19	3,071	2.2	14.6	23.0	3,370	4,588
5	Product B	0.19	3,144	2.2	6.7	8.5	3,590	6,005
6	Corn syrup*	0.19	3,074	2.0	7.5	9.5	3,680	5,140
7	Product C	0.19	3,094	2.3	5.6	7.5	4,070	6,135

*No digestion.

The results illustrate the syngeristic advantages obtained both in strength of product and reduced retarding effect that results from the digestion of the carbohydrates with soluble salts of sulfurous acid. These results demonstrate that digestion of the carbohydrate with soluble salts of sulfurous acid increases the strength of the products formed from a cement slurry even beyond that normally attained by the addition of the undigested carbohydrate while simultaneously preventing excessive retardation of the setting time.

Partial Hydrolysate of Wood Chips

In a process described by *J.C. Steinberg, K.R. Gray and J.K. Hamilton; U.S. Patent 3,520,707; July 14, 1970; assigned to Rayonier Incorporated* the setting of Portland cement concrete and other Portland cement slurries is retarded, the water content of the slurry is reduced and the strength of the concrete is increased by incorporating in the slurry composition a small quantity of a water-reducing and retarding mixture comprising a hydrolysate obtained by the partial hydrolysis of the hemicellulose constituents of wood chips.

The hydrolysate consists predominantly of short chain polymers of the noncellulose carbohydrate constituents resulting from the hydrolysis of wood chips, and advantageously may be subjected to an alkaline treatment to reduce the free (monomeric) sugar content. The following example illustrates the preparation of a partial hydrolysate from a softwood.

Example 1: Southern yellow pine logs of three species, longleaf pine *(Pinus palustris)*, slash pine *(Pinus elliottii)* and loblolly pine *(Pinus taeda)*, were debarked and passed through a chipper to provide chips having the approximate size ½" x ¾" x ⅛". The chips were loaded into a digester and heated to 175°C in 55 minutes by means of direct, saturated steam. During the next 45 minutes, while holding the temperature at 175°C, saturated steam was passed into the bottom of the digester and vapor was relieved from the top of the digester.

Steaming was then discontinued, the digester was relieved and liquid (partial hydrolysate) was drained from the bottom of the digester. This hydrolysate, having a solids concentration of 8.5% solids, was concentrated under vacuum (i.e., at a temperature of 50° to 60°C) in a forced circulation flash-type evaporator. The concentrated liquor, containing 63.1% solids, was designated Product A. The modification of a softwood partial hydrolysate by alkaline treatment under the preferred conditions is illustrated by the following example.

Example 2: 220 lb of Product A (from Example 1) and 28.1 lb of 49.5% NaOH solution were mixed. The resulting solution, having a temperature of 58°C, was transferred to a steam-jacketed, agitated reactor, heated to 90°C with indirect steam, retained at 90°C for 60 minutes, cooled to 25°C and removed from the reactor. The reaction product, containing 56.9% solids, was diluted with 8.4 lb of water to give Product B, having a solids concentration of 55.1%.

Analyses of effective wood hydrolysate products obtained by steam treatment of southern pine chips followed by concentration only (Product A of Example 1) or by concentration and mild alkaline treatment (Product B of Example 2) follow:

	- - Product - - -	
Constituent, percent (dry basis)	**A**	**B**
Glucose, mannose and galactose in polymeric form	61	36
Free hexoses	8	3
Free pentoses	8	1
Lignin	8	8
Lactonizable hydroxy acids	4	10
Sodium salts of volatile acids	1	13
Other constituents	10	29

In the production of concrete, the hemicellulose-containing wood hydrolysates are generally added to the mixing water in the form of concentrated solutions at dosages corresponding to 0.10 to 0.30% solids, based on cement. The use of such products in concrete is described in the following example.

Example 3: To demonstrate the utility of the products of the process in the production of concrete, an unmodified hydrolysate product (Product A of Example 1) or an alkaline-modified hydrolysate product (Product B of Example 2) are preferably added to the mixing water, whereupon the resulting water reducing and retarding mixture solution is mixed into a preblended mixture of sand, cement and gravel to form the concrete batch. In order to obtain a set retardation of about two hours (ASTM specification for water reducing and retarding mixtures, classification Type D, is 1 to 3 hours' retardation), mixtures A and B are added to the mixing water in amounts corresponding respectively to 0.20

and 0.25% (on a dry basis) of the cement in the batch. Typical results obtained through the use of these products at the noted dosages and at two slump (workability) levels are given below.

	Product Dosage, Percent of Cement	Slump, inch (±0.5")	Water Reduction Percent of Mixing Water	Retardation, hr Initial Set	Final Set	Comprehensive Strength, psi 7-Day Cure	28-Day Cure
No admixture	0.00	4.5	0.0	0.0	0.0	2,840	4,190
Product A	0.20	4.5	7.7	2.0	2.1	3,395	4,665
Product B	0.24	4.5	11.5	2.0	2.1	3,785	5,315
No admixture	0.00	6.5	0.0	0.0	0.0	2,275	3,215
Product A	0.20	6.5	7.9	2.2	2.3	3,410	4.985
Product B	0.25	6.5	9.7	2.3	2.3	3,380	4,885

Digested Wood Chips

J.C. Steinberg and K.R. Gray; U.S. Patent 3,530,112; September 22, 1970; assigned to ITT Rayonier Incorporated describe a dispersant for hydraulic cement slurries which is prepared by reacting

(a) a concentrated hydrolysate containing predominantly short-chain polymers of the noncellulose carbohydrate constituents resulting from the partial hydrolysis of coniferous wood, with

(b) a water-soluble salt of sulfurous acid.

The amount of such water-soluble salt is equivalent to 0.10 to 0.25 part of SO_2 per parts solids in the hydrolysate. The reaction is carried out at a temperature of 160° to 180°C until the resulting reaction product is substantially free of sugars.

Example 1: Wood chips of mixed southern yellow pines were treated with direct steam for 45 minutes at 170°C and the condensate containing partially hydrolyzed hemicelluloses was drained from the digester. This liquor was concentrated under vacuum to 65% solids. Analysis of this concentrate by paper chromatography showed 41.9% combined sugars and 17.8% free sugars, on a dry basis. Portions of the concentrate were digested with solutions of soluble salts of sulfurous acid to yield products of low sugar content and having solids concentrations of 40 to 50%.

	Digestion Chemical		Digestion Conditions			
Product	Name	Amount, % of Solids in Concentrated Hydrolysate	Max Temp, °C	Time at Max Temp, min	Sugar % (dry basis) Combined	Free
A	Na_2SO_3	10.0	165	30	14.7	0.0
B	$(NH_4)_2SO_3$	7.5	165	40	0.9	7.7
	$(NH_4)HSO_3$	7.5	-	-	-	-
C	$(NH_4)_2SO_3$	15.0	165	40	6.5	4.1
D	$(NH_4)_2SO_3$	10.0	165	50	0.1	2.0
	$(NH_4)HSO_3$	10.0	-	-	-	-
E	$(NH_4)_2SO_3$	20.0	165	50	6.4	1.2

Example 2: This example shows that the digestion conditions of Example 1 result in products which are classed as nonretarding chemical mixtures for concrete (i.e., mixtures retarding the setting time 1 hour or less), whereas the undigested material is a retarding mixture.

Concrete test samples were prepared with and without the addition of Products A through E of Example 1. In addition, a comparative test sample was prepared containing the same amount of undigested starting material (F) from which Products A through E were prepared. The method of preparing and testing the concrete samples was as described in Example 8 of U.S. Patent 3,332,791.

Product No.	Dosage, % of Cement Dry Basis	Retardation of Final Concrete Set, hr
None	0.00	0.0
A	0.20	0.6
B	0.20	0.5
C	0.20	0.7
D	0.20	0.4
E	0.20	0.4
F*	0.20	2.5

*Partially hydrolyzed hemicelluloses from which Products A through E were prepared.

The agents of the process are useful as Type A admixtures (i.e., the ASTM classification of water-reducing chemical mixtures for concrete). When added to concrete at dosages comparable to those at which other admixtures are frequently used (e.g., 20% on a dry basis, based on cement) they will, without use of auxiliary accelerating agents, not retard the setting time of concrete more than permitted by the ASTM specification for Type A mixtures (not over 1 hour).

Tobacco-Molasses Product

T.M. Kelly; R.B. Peppler and J.A. Ray; U.S. Patent 3,485,649; December 23, 1969; assigned to Martin-Marietta Corporation have found that the pyrolytic treatment of tobacco plant material in an aqueous system with molasses yields a product which has in concrete or mortar substantially all the beneficial effects of untreated molasses, but in which the tendency to retard the rate of set is significantly reduced, or even eliminated. In certain applications these products accelerate the rate of set of concrete.

The preferred material is the product resulting from heating some portion of the tobacco plant in any convenient concentration in a water solution together with molasses. The use of the tobacco plant alone or of extracts is described in U.S. Patent 3,432,316.

The tobacco plant material may comprise any portion of the plant but for economic reasons the stem is preferred, and the stem is in its normal condition of having been subjected to some degree of natural or artificial curing. Prior grinding of the stem is desirable since it facilitates the reaction with the molasses, but products of a similar nature are obtained whether stems are previously ground or not.

Products within the scope of the process are secured by heating the tobacco plant material and the molasses in a preferred temperature range from 100° to 200°C. The conversion of the molasses is not very extensive if the reaction is conducted much below 100°C, whereas the organic matter is too severely degraded if the reaction is conducted at 200°C.

However, such products heated or processed in the temperature range from room temperature to 100°C are useful additives and lie within the scope of this process, since some degree of chemical conversion of the constituents is involved. In such products prepared at 25°C, the retardative property is completely removed, but the products are not substantially accelerating.

Mixtures of Hydroxyalkylcelluloses

A process described by *D.S. Weiant; U.S. Patent 3,824,107; July 16, 1974; assigned to AA Quality Construction Material, Inc.* relates to dry mortar-forming compositions containing a minor amount of a mixture of (1) hydroxypropylcellulose and (2) a compound selected from the group consisting of hydroxymethylcellulose and hydroxyethylcellulose. The compositions contain a major proportion of hydraulic cement or a mixture of hydraulic cement and sand.

It is a principal object of the process to provide water retentive agents for hydraulic cement which are lower in cost and produce an instantly usable thixotropic mortar which is effected to a lesser degree by temperature.

Example 1:

	Percent
Hydroxypropylcellulose type HW, viscosity at 2%, 30,000 cp	0.20
Hydroxyethylcellulose, viscosity at 2%, 15,000 cp	0.40
Hydraulic cement	98.2
Inorganic fiber	1.2

Example 2:

	Percent
Hydroxypropylcellulose, viscosity at 2%, 30,000 cp	0.40
Hydroxymethylcellulose, viscosity at 2%, 15,000 cp	0.20
PVA	1.2
Hydraulic cement	48.0
Sand	50.0
Inorganic fiber	0.19
Antifoam	0.01

The above were compared with the following additional compositions.

Example 3:

	Percent
Hydroxypropylcellulose type HW, viscosity at 2%, 30,000 cp	0.60
Hydraulic cement	98.2
Inorganic fiber	1.2

Example 4:

	Percent
Hydroxymethylcellulose, viscosity at 2%, 15,000 cp	0.60
Hydraulic cement	98.2
Inorganic fiber	1.2

In the above examples, the constituents are mixed dry and the dry mixture is mixed with water in amounts to produce a mortar which will remain on a hawk (a plaster's tool). Approximately 1½ gallons of water are mixed with 25 pounds of the dry mixture. A high strength thin-set mortar results. The mortars of this process have the advantage, among others, over mortars heretofore known containing one type of cellulose in that the two types produce a thixotropic agent with increased solubility rate and reduced viscosity change with temperature.

The improved solubility rate is demonstrated by the following test for water retentivity. Water retentivity values were obtained on Portland containing various amounts of the thixotropic agents described. The property was measured by placing a ⅛" layer of the mix previously slurried with the specified amount of water on the porous side of a quartered 4¼" x 4¼" commercial standard 181 glazed wall tile. A thin glass slide was placed over the mortar layer and the assembly positioned under the microscope lens. As the water left the mortar travelling onto the porous bisque of the tile, the mortar layer contracted thereby causing the slide to be displaced downward. This displacement could be accurately measured with a microscope and plotted against the square root of time. Normal procedure is to mix the mortar and let stand for 15 to 20 minutes to allow for solubility of the thixotropic agent. In these tests a premix of 3 minutes was used to demonstrate the increased solubility rate. The slope of the straight line divided into 1,000 yielded the retentivity values listed below.

Example	Water Requirement, %	Retentivity
1	28	35
2	28	30
3	28	20
4	28	5

The extremely poor water retentivity of Example 4 demonstrates the poor rate of solubility of methocel cement systems and the rate improvement derived with hydroxypropylcellulose.

On-the-wall tests demonstrated the remarkable improvement obtained with hydroxypropylcellulose and other hydroxyalkylcellulose mixtures in cement mortar systems.

Those systems not containing the hydroxypropylcellulose cannot be used as dry set tile mortars immediately after mixing because of the slow solubility. Those containing hydroxypropylcellulose can be used immediately and have demonstrated viscosity stability over long periods of use because additional cellulose is not slowly being dissolved.

AIR ENTRAINMENT AND EXPANSION AGENTS

Sulfate Esters

J.A. Komor; U.S. Patent 3,782,983; January 1, 1974; assigned to GAF Corporation describes a composition for imparting air entrainment to cement mortars which comprises salts of sulfate esters derived from: (a) straight or branched chain alcohols containing 0 to 65% by weight of ethylene oxide, where the alcohol portion has 8 to 10 carbon atoms, and (b) alkyl aryl alcohols containing 0 to 65% by weight of ethylene oxide, where the alkyl group has 2 to 7 carbon atoms.

Examples 1 through 28: Selected compositions were evaluated in a standard mortar mix according to the following screening test method, based on field experience which has shown that a surfactant concentration giving 16 to 22% air in cement mortar generally gives 3 to 6% air in concrete mortars, both being optimum air contents. Compositions were evaluated in cement mortars prepared according to the following formula in which all surfactants were 0.01% of the dry Portland cement, by weight:

	Parts by Weight
Ottawa sand, 20 to 30 mesh in accordance with ASTM spec. C 185-59	180
Type I, II or III cement from the Alpha Portland Cement Co.	60
Water and surfactant solution	36

A very exact test procedure was followed to avoid as much random variation as possible. Ingredients were weighed into a 400-ml beaker, partially blended by 2 or 3 swirls with a spatula, and positioned directly below a drill press-driven agitator. The agitator was lowered and positioned as close to the bottom of the beaker as possible without touching. Mixtures were stirred with a 4-slotted circular agitator having a 2-inch diameter stirrer for 3 minutes at 1,360 rpm.

For the first 20 to 30 seconds of agitation, the beaker was moved in a horizontal circular pattern, so that the agitator moved about the periphery of the beaker, in order to effect complete blending of ingredients. Thereafter, the agitator was positioned directly at the center of the beaker, without changing its elevation, for the duration of the 3-minute period. The beaker was then removed from the drill press stand, mixed once with a spatula and the contents transferred to a tared container of known volume.

Excess mortar was removed by multiple passes of a straight edge over the top rim of the container. The sides and bottom of the filled vessel were dried with paper tissue and the weight to the nearest $\frac{1}{10}$ gram was obtained. Percent air was calculated by using the following expression:

$$\text{Percent air} = \frac{(T - t)100}{T}$$

where T = theoretical weight of vessel and mortar and t = observed weight of vessel and mortar. The theoretical weight of the mortar assumes that the sand, cement, water and composition of this process are all packed as efficiently as possible without any voids being present.

Twenty-seven compositions and one standard air-entraining agent, Vinsol NVX, were tested according to this test method, using 0.01% surfactant, based on the weight of dry cement. The experimental results are shown in the following table in which the percentage of ethylene oxide in the nonionic portion of the surfactant is given.

The performance capabilities for the compositions of this process were determined by comparative evaluations under these rigorously standardized conditions, as set forth in the following illustrative examples.

Example number	Composition	Air entrained by using composition in mortar at 0.01% by wt. of dry portland cement, percent	Ethylene oxide, percent by wt. in nonionic intermediate
1	$n\text{-}C_8H_{17}(OCH_2CH_2)_{2.5}SO_4NH_4$	12.01	49
2	$n\text{-}C_{10}H_{21}(OCH_2CH_2)_{2.0}SO_4NH_4$	13.47	39
3	$n\text{-}C_{10}H_{21}(OCH_2CH_2)_{3.0}SO_4NH_4$	13.32	48
4	$n\text{-}C_{12}H_{25}(OCH_2CH_2)_{3.5}SO_4NH_4$	7.20	48
5	$n\text{-}C_{14}H_{29}(OCH_2CH_2)_{4.0}SO_4NH_4$	3.76	47
6	$n\text{-}C_{6-10}H_{13-21}(OCH_2CH_2)_{2.5}SO_4NH_4$	16.31	56
7	$n\text{-}C_{8-10}H_{17-21}(OCH_2CH_2)_{1.0}SO_4NH_4$	15.63	28
8	$n\text{-}C_{8-10}H_{17-21}(OCH_2CH_2)_{2.0}SO_4NH_4$	15.36	44
9	$n\text{-}C_{8-10}H_{17-21}(OCH_2CH_2)_{3.0}SO_4NH_4$	15.13	54
10	$n\text{-}C_{8-10}H_{17-21}(OCH_2CH_2)_{5.5}SO_4NH_4$	10.97	68
11	$n\text{-}C_6H_{13}SO_4Na$	5.7	0
12	$n\text{-}C_{10}H_{21}SO_4Na$	19.6, 18.9	0
13	$n\text{-}C_{10}H_{21}(OCH_2CH_2)_2SO_4NH_4$	13.3	38
14	$n\text{-}C_{8-10}H_{17-21}(OCH_2CH_2)_{2.5}SO_4Na$	17.2	48
15	$n\text{-}C_{8-10}H_{17-21}(OCH_2CH_2)_{2.5}SO_4NH_4$	17.1, 17.3	49
16	$n\text{-}C_{6-7}H_{13-15}$—  —$(OCH_2CH_2)_{10.0}SO_4NH_4$	7.02	73
17	$n\text{-}C_9H_{19}$—  —$(OCH_2CH_2)_{4.0}SO_4NH_4$	5.80 5.80	47 47
18	$n\text{-}C_9H_{19}$—  —$(OCH_2CH_2)_{9.5}SO_4NH_4$	4.80	67
19	$n\text{-}C_9H_{19}$— 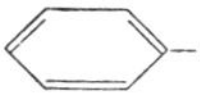 —$(OCH_2CH_2)_{20.0}SO_4NH_4$	5.71	81
20	$n\text{-}C_{6-7}H_{13-15}$— 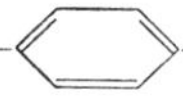 —SO_4NH_4	14.49	0
21	$n\text{-}C_{6-7}H_{13-15}$—  —$(OCH_2CH_2)_{2.16}SO_4NH_4$	17.67 17.67	37 37
22	$n\text{-}C_{6-7}H_{13-15}$— 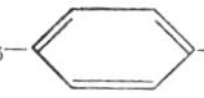 —$(OCH_2CH_2)_{2.84}SO_4NH_4$	17.17	43
23	$n\text{-}C_{6-7}H_{13-15}$— 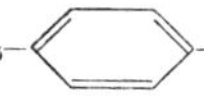 —$(OCH_2CH_2)_{3.0}SO_4NH_4$	14.14	47
24	$n\text{-}C_{6-7}H_{13-15}$— 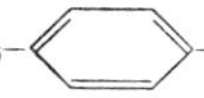 —$(OCH_2CH_2)_{5.80}SO_4NH_4$	11.87	61
25	$n\text{-}C_{6-7}H_{13-15}$—  —$(OCH_2CH_2)_{7.00}SO_4NH_4$	10.46	66
26	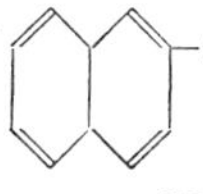 —$(OCH_2CH_2)_2SO_4NH_4$	5.5	37
27	$t\text{-}C_4H_9$—  —$(OCH_2CH_2)_2SO_4NH_4$	14.5	40
28	Vinsol NVX (sodium resinate)	3.50	0

As is evident from the results in this table, the compositions of this process are markedly more effective than the sodium resinate, Vinsol NVX, which is the most widely used air-entraining agent on the market today.

Sodium Tripolyphosphate and Surfactant

G.E. Hardin; U.S. Patent 3,769,051; October 30, 1973 describes a mortar set retarder and air-entrainer composition comprising an aqueous solution of sodium tripolyphosphate, a water-soluble ethylene oxide condensation product and a wetting agent. The retarder may optionally include a polyamino polycarboxylic acid. The following examples illustrate the process.

Example 1:

	By Weight
Water	57%
Sodium tripolyphosphate	2%
Isooctylphenoxypolyethoxyethanol	1%
Sodium dodecylbenzene sulfonate	40%

The compositions were each added to separate batches of Portland cement mortar at the rate of 5 to 7 oz per 5 gallons of water of hydration and wherein the ratio of cement to aggregate was approximately 1:2 by weight, and the water-cement ratio was approximately 1:2 by weight, which is approximately 15% less water than normally utilized in hydrating Portland cement.

Example 2:

	By Weight
Water	69.5%
Sodium dodecylbenzene sulfonate	24.0%
Isooctylphenoxypolyethoxyethanol	3.0%
Sodium xylene sulfonate	2.5%
EDTA	1.0%

Example 3: A masonry mortar consisting of:

Portland cement	1 part by volume
Lime	1.25 parts by volume
Mason's sand	6 parts by volume
Water	(73.8% by weight of cement and lime)
Composition of Example 2	3 liq oz/each 5 gal of water

was admixed and test data compiled as set forth in the following Table 1.

Example 4:

Portland cement	1 part by volume
Lime	1.25 parts by volume
Mason's sand	6 parts by volume
Water	(77.9% by weight of cement and lime)
Composition of Example 2	3 liq oz/each 5 gal of water

TABLE 1

	Mortar of Example 4			Mortar of Example 3		
Wet density, 1 lbs./cu. ft..		126.8			101.4	
Dry density, lbs./cu. ft...		120.8			95.4	
Absorption, percent......		12.5			10.2	
Weight after 28 days, average grams..............		256.3			200.5	
Compressive strength:						
Pounds actual........	7,200	7,000	7,500	1,900	1,780	1,850
Lbs./sq. inch..........	1,800	1,750	1,875	475	445	462
Tensile strength:						
Pounds actual........	320	325	315	200	195	190
Lbs./sq. inch..........	320	325	315	200	195	190

Salts of Styrene-Maleic Anhydride Copolymers

M.A. Stram, J.A. Dieter, R.J. Pratt and D.W. Young; U.S. Patent 3,563,930; February 16, 1971; assigned to Atlantic Richfield Company have found that air entrainment in Portland cements is increased by the addition to the cement prior to curing of minor amounts of water-soluble salts of styrene-maleic anhydride copolymers or the water-soluble half-esters or the water-soluble salts of half-esters of styrene-maleic anhydride copolymers and alkoxy polyalkylene glycols or other alcohols. Increased air entrainment in the cement provides superior durability and resistance. The following example illustrates the process.

Example: A mixture of 350 grams of methoxypolyethylene glycol of molecular weight 350 and 225 grams of styrene-maleic anhydride copolymer having a mol ratio of styrene to maleic anhydride of 1:1 and a molecular weight of 1,600 was heated at 185° to 195°C for 2½ hours under nitrogen atmosphere. The reaction was performed in a resin kettle fitted with a stirrer, thermometer and heating mantle. A portion of the reaction product was taken out and analyzed by alkali titration as having 30% of all carboxyls present converted to ester (i.e., a 60% half-ester). The pot was then neutralized at 160°F to pH 7.0 with an aqueous solution of sodium hydroxide. Approximately 7.0 grams of base were used and the ester completely dissolved.

A fraction of the neutralized ester was diluted with water to 2.0% total solids and tested for air entrainment in a Portland cement mortar according to the ASTM C-185 tests using proportions of 350 grams of Portland Type 1 dry powder cement to 1,400 grams of standard Ottawa sand (20/30). The results using different weight percentages of the polymer in the mortar are shown in the table below and compared with air entrainment for mortar containing no polymer.

Wt. Percent of Copolymer 30% Ester on Wt. of Cement	Percent Air, volume (ASTM C-185)
0	8.09
0.0114	14.83
0.0570	21.33
0.1140	25.80

The table below compares air entrainment in Portland cement using a 25% ester of a styrene-maleic anhydride copolymer having a molecular weight of 1,600 and a mol ratio of styrene to maleic anhydride of 1:1 and a methoxy polyethylene glycol of average molecular weight about 350.

Wt. Percent of Copolymer 25% Ester on Wt. of Cement	Percent Air, volume (ASTM C-185)
0	8.09
0.0114	14.79
0.057	21.76
0.114	24.84

Additional tests were also made on a Portland cement incorporating the following polymers:

Polymer A — Copolymer of styrene-maleic anhydride having a molecular weight of 1,600 and a styrene to maleic anhydride mol ratio of 1:1.

Polymer B — Copolymer of styrene-maleic anhydride having a molecular weight of 1,700 and a styrene to maleic anhydride mol ratio of 2:1.

Polymer C — 70% half-ester of Polymer B and n-propyl alcohol.

Polymer D — 100% half-ester of Polymer A and ethylene glycol butyl ether (butyl Cellosolve).

Polymer E — 50% half-ester of Polymer A and ethylene glycol butyl ether (butyl Cellosolve).

Polymer F — 77% half-ester of Polymer A and methoxy polyethylene glycol having a molecular weight of about 350.

Polymer G — 100% half-ester of Polymer A and methoxy triethylene glycol.

Polymer H — 20% half-ester of Polymer A and methoxy polyethylene glycol having a molecular weight of about 2,000.

Polymer I — 14% half-ester of Polymer A and methoxy polyethylene glycol having a molecular weight of about 5,000.

The table on the following page compares air entrainment values of Polymers A through F in Portland cement compositions. Samples were tested according to ASTM C-185 method using proportions of 350 grams of Portland Type I dry powder cement to 1,400 grams of Standard Ottawa sand (20/30). Air content of the cement used in the tests without the air entrainment polymer was 9.17% by volume. The tests were run at room temperature.

As may be seen from the data, all of the compositions tested gave superior air entrainment in comparison to the cement sample which did not contain an air entrainment additive. Generally the half-esters gave superior air entrainment values compared to corresponding amounts of the unesterified polymers.

Varying proportions and test results	Sample designation A	B	C	D	E	F	G	H	I
Air entraining agent, ml.[1]	2.0	2.0	2.0	2.0	2.0	2.0	2.0	2.0	2.0
Water, ml	232.0	240.0	220.0	230.0	230.0	250.0	240.0	250.0	240.0
Flow, percent	84.0	93.0	80.0	88.0	86.0	92.0	92.0	95.0	86.0
Air content, percent	11.84	10.70	13.73	14.59	13.72	12.07	12.35	11.87	10.86
Air entraining agent, ml.[1]	10.0	10.0	10.0	10.0	10.0	10.0	10.0	10.0	10.0
Water, ml	202.0	220.0	200.0	188.0	200.0	210.0	200.0	220.0	210.0
Flow, percent	80.0	80.0	84.0	88.0	88.0	95.0	95.0	88.0	95.0
Air content, percent	15.94	12.53	18.21	21.76	21.11	18.74	19.50	15.80	17.30
Air entraining agent, ml.[1]	20.0	20.0	20.0	20.0	20.0	20.0	20.0	20.0	20.0
Water, ml	174.0	200.0	166.0	174.0	170.0	160.0	166.0	190.0	170.0
Flow, percent	80.0	80.0	84.0	92.0	84.0	92.0	95.0	95.0	95.0
Air content, percent	19.57	14.99	22.54	26.86	23.37	25.16	26.87	18.41	22.72

[1] 2.0 wt. percent of sodium salt of polymer at pH 7–8.

Styrene Oxide-Sulfonated Lignin Products

V.F. Felicetta and A.E. Markham; U.S. Patent 3,398,006; August 20, 1968; assigned to Georgia-Pacific Corporation describe a Portland cement composition containing a mixture of sulfonated lignin reacted with from 0.01 to 45 weight percent, based upon the dry solids of the sulfonated lignin-containing material, of styrene oxide. The reaction may be carried out by intermixing an aqueous solution of lignosulfonate-containing material with the styrene oxide with or without use of catalysts, followed by heating to 50° to 200°C. The interaction may be effected under acid, neutral or alkaline conditions.

All concrete tests were made with ASTM specifications at a nominal cement factor of 4.75 sacks per cubic yard. Each concrete mix comprised 26.3 lb of Type I Portland cement, 71 to 81 lb of fine aggregate, 116 lb of coarse aggregate, and sufficient water to give a nominal slump of 3½" as measured by a 12" cone. The mixture was added as an aqueous solution as part of the mixing water. The coarse aggregate consisted of an equal weight of each of the following size fractions of rounded gravel: ¾ to 1", ½ to ¾", ⅜ to ½" and 3/16 to ⅜" (pea gravel).

The weight of sand or fine aggregate was adjusted within the limits specified to compensate for the differences in amounts of entrained air. Tests made on the fresh concrete were: slump, air content (Press-Ur-Meter) and time of setting (Proctor penetration resistance needles according to ASTM C-403). Six 6" x 12" concrete cylinders were cast from each mix. Three each were used for compressive strength tests (ASTM C-192).

Example: To illustrate the advantage gained by using a mixture of the reaction product of a sulfonated lignin-containing material and styrene oxide, a series of runs was made in which a mixture was prepared by reacting spent sulfite liquor with styrene oxide. A calcium base was fermented to convert the fermentable sugars to alcohol and the alcohol was removed by distillation. A solution of the fermented spent sulfite liquor containing 50 weight percent of the spent sulfite liquor solids was alkaline treated by heating the solution for 4 hours at 80° to 90°C at a pH of 8 obtained by addition of sodium hydroxide. The alkaline treated spent sulfite liquor was then used for the preparation of the mixture.

The mixture was prepared by mixing the alkaline treated spent sulfite liquor with styrene oxide and sodium hydroxide in an amount of 1.7 to 4 weight percent, respectively, based upon the alkaline treated spent sulfite liquor solids. The mixture was brought to a boil in 1 hour and then refluxed for 4 hours to effect the condensation reaction. The product thus prepared was used as a mixture in concrete and the results obtained compared to concrete containing no mixture and also to concrete prepared using as a control a mixture prepared by further treating a portion of the alkaline spent sulfite liquor in a manner similar to that used in the preparation of the spent sulfite liquor-styrene oxide reaction product except that no styrene oxide was added. The pertinent data and results obtained are shown in the table below.

Mixture	Mixture Added (% of Cement)	Cement Factor (sk/cu yd)	Slump (in)	Water Reduction (percent)	Air Content (% by volume)
Plain concrete	-	4.74	3	-	2.5
Sulfite liquor	0.30	4.74	3¾	14	4.5
Sulfite liquor, 1.7% styrene oxide	0.30	4.77	3¾	21	6.1

Mixture	Time of Setting				Compressive Strength, psi	
	Initial Set		Final Set			
	Hours	Minutes	Hours	Minutes	At 7 Days	At 28 Days
Plain concrete	4	40	6	45	2,380	3,570
Sulfite liquor	6	45	8	40	3,200	4,420
Sulfite liquor, 1.7% styrene oxide	6	50	8	50	3,420	4,510

Polyglycolamides, Amide Oxides and Anionic Detergent

K.L. Johnson; U.S. Patent 3,642,506; February 15, 1972; assigned to Swift & Company describes a method of improving the workability and compressive strength of concrete by the addition of from ½ to 1½ fluid ounces per 94 lb sack of cement or cementitious mixture of a surface-active agent made up of primary polyglycolamides, tertiary amide oxides; an anionic detergent; and a solvent. The following examples illustrate the process.

Example 1: (A) Surface-active agents were prepared from 18 parts lauryl di(2-hydroxy ethyl) amine oxide, 18 parts coconut oil fatty acids, 30 parts polyglycolamine H-163, 30 parts diethanolamine and 39 parts dodecylbenzene sulfonic acid. The following procedure was used. The coconut oil fatty acids were reacted with the polyglycolamine H-163 in a four-neck, 1 liter glass reaction flask under 27 mm Hg vacuum at 150°C for 3 hours, cooled to 70°C and the lauryl di(2-hydroxy ethyl) amine oxide and diethanolamine and dodecylbenzene sulfonic acids were added with mixing. 5% of the detergent prepared as outlined above, 5% isopropyl alcohol and 90% Chicago tap water were combined to form Admixture A.

(B) A second mixture was prepared using the method of (A) and the following ingredients: 18 parts lauryl di(2-hydroxy ethyl) amine oxide, 18 parts coconut oil fatty acids, 30 parts polyglycolamine H-163, 30 parts of diethanolamine and 39 parts tridecylbenzene sulfonic acid. The same procedure was followed as in part (A). 5% of the detergent prepared as above, 5% of a mixture of monopropylene glycol monomethyl ether, dipropylene glycol monomethyl ether and tripropylene glycol monomethyl ether and 90% Chicago tap water were combined to form Admixture B.

(C) A third mixture was prepared from: 36 parts coconut oil fatty acids, 30 parts polyglycolamine H-163, 30 parts diethanolamine and 39 parts tridecylbenzene sulfonic acid. The following procedure was used. The coconut oil fatty acids and polyglycolamine H-163 were reacted in a four-neck, 1 liter flask at a temperature of 150°C for a 3-hour period under 27 mm Hg vacuum. The condensate was then cooled to 70°C and the tridecylbenzene sulfonic acid was then added along with the diethanolamine. 5% of the detergent prepared as outlined above, 5% dipropylene glycol monomethyl ether and 90% Chicago tap water were combined to form Admixture C.

(D) The mixtures prepared as above were incorporated into hydraulic cement mixes which contained the following ingredients and had the following properties:

Cement—Portland Type I, lb	517.00	517.00	517.00
Fine aggregate sand, lb	1,265.00	1,286.00	1,281.00
Coarse aggregate gravel, lb	1,938.00	1,967.00	1,962.00
Admixture A, fl oz	5.25	-	-
Admixture B, fl oz	-	5.50	-
Admixture C, fl oz	-	-	5.50
Water (added), gal	28.33	26.68	26.68
Slump (consistency), inch	3.00	3.00	3.00
Air content, % by volume	5.70	5.40	5.60
Weight of fresh concrete, lb per cu ft	146.52	147.87	147.50
Concrete produced (yield), cu ft	27.00	27.00	27.00
Cement factor, sacks per cu yd	5.50	5.50	5.50
Water-cement ratio by weight	0.46	0.43	0.43

As shown above the mixes had excellent workability with a low water-cement ratio. The resulting concrete had the following average compressive strengths (Standard cured 6 x 12 inch cylinders were tested):

	Pounds per Square Inch	
Mix Containing	**7 Days**	**28 Days**
Admixture A	3,136	4,374
Admixture B	3,101	4,291
Admixture C	3,420	4,415

Example 2: An admixture was formulated by blending under ambient conditions 950 grams Chicago tap water, 10 grams diethanolamine, 15 grams dodecylbenzene sulfonic acid and 25 grams lauryl dimethyl amine oxide. The admixture was incorporated into a cement mix having the following mix characteristics and ingredients:

Cement, Portland Type I, lb	517.00
Fine aggregate, sand, lb	1,272.00
Coarse aggregate gravel, lb	1,949.00
Admixture, fl oz	4.50
Water (added), gal	27.50
Slump (consistency), in	3.00
Air content, % by volume	5.70
Weight of fresh concrete, lb per cu ft	146.93
Concrete produced (yield), cu ft	27.00
Cement factor, sacks per cu yd	5.50
Water-cement ratio by weight	0.44

The average compressive strength of the resulting concrete was 3,449 psi at 7 days and 4,455 psi at 28 days. The 7 day and 28 day compressive strengths are both significantly higher than normal concrete produced from mixes containing no air-entraining agent, and having the same slump, cement factor (sacks per cubic yard), and yield.

The same tests as described above were made substituting lauryl di(2-hydroxyethyl) amine oxide for the lauryl dimethyl amine oxide. The results were substantially identical.

Short Fibers and Air-Entraining Agent

A.C.S. Howe; U.S. Patent 3,679,445; July 25, 1972; assigned to John Laing & Son Limited, England describes a plastic concrete or mortar mix including an incorporated gas such as entrained air, and from 1 to 10 grams of short fibers per pound of cement uniformly distributed throughout the mix.

It has been found that short fiber mixed into the wet mix (cement/water/air paste) at a dosage level from 1 gram per pound cement upwards has the ability to maintain uniform dispersion of solid, liquid and gaseous components within highly air-entrained concrete and mortar mixes during the critical stage of setting when the process of changing from liquid to crystalline solid takes place.

The air-entrained concrete will generally contain a suitable air-entraining agent, as known in the art. A typical example is the neutralized form of the resin obtained by wood distillation. There are, however, three main types usable:

(a) Natural, such as keratin, animal and vegetable fats and fatty acids, natural wood resins, e.g., neutralized by the lime in the cement;

(b) Modified natural, such as alkali salts of wood resins, calcium salts of glue, sodium abietate; and

(c) Synthetic, such as detergents, especially the alkylsulfonates, sodium salts of cycloparaffinic acids, e.g., naphthenates, triethanolamine salts of sulfated aromatic hydrocarbons, aryl polyglycols.

Preferably, the fibers are less than 2 inches long, and usually range in length between about ⅛ inch and 1⅛ inch. They may be of regular or natural shape, or they may include crimped, hammered or tapered profiles.

Although there is no specific limitation on the nature of the fiber used, considerations of toughness, ultimate strength, immunity to attack by any of the ingredients of the mix—at least until the mix is no longer plastic and so on will normally serve to put practical limitations on the choice of fiber for carrying out the process.

Examination of results shows that the extent of air entrainment and even distribution of dimensional stability achieved is directly dependent upon the combined performance of each separate additive present. For instance, where fiber was deliberately left out of formulations and the extent of air entrainment raised to about 15% in concrete-containing aggregates ¾-inch maximum size in A/C (aggregate/cement) ratios between 3:1 and 5:1 aggregate distribution became defective, dimensional stability impaired and replication unreliable.

The results of studies appear to indicate that the population of fibers, when evenly distributed within the highly air-entrained matrix, provides a load transfer system and stress absorption characteristic which maintain the static stability in addition to physical ability to assist in the increase and stabilization of the air-entrainment capability.

The evident ability of fiber to increase the stability and amount of air entrainment in concrete and mortar formulations dosed with air-entraining agents can be related to the fact that the fiber must be associated with and involved in: (a) expansion of the specific surface of the cement/water paste; (b) reducing work required to expand the paste surface; (c) enclosing air and retaining it under the surface; and (d) countering disturbance and drainage which causes bubbles to collapse.

The combined action of the air entraining agent and fiber appears to produce one or more separate effects to an extent which varies with the concentrations of air-entraining agent and of fiber, the A/C and water/cement ratios, fiber dimensions, aggregate size and grading, mixing time and type of mixing action.

Alkaline Earth Lignosulfonate and Alkali Carbonate

S. Brunauer; U.S. Patent 3,689,294; September 5, 1972 describes a method for making a free-flowing expanding Portland cement paste which comprises grinding Portland cement clinker to a specific surface area between 6,000 and 9,000 cm^2/g, mixing the ground cement with at least 0.0025 and preferably 0.005 to 0.010 part by weight of alkali or alkaline earth lignosulfonate, and then with about 0.20 to 0.28 part by weight of water containing at least 0.0025 and preferably 0.005 to 0.015 part by weight of alkali carbonate, all per part of ground cement.

At least 0.0005 and preferably 0.002 to 0.005 part by weight of a grinding aid is added to the clinker to assist in grinding. When alkaline earth lignosulfonate is used, it must be ground or mixed together with the dry cement powder. However, alkali lignosulfonate or other water-soluble sulfonated lignins can be added in the mix water together with the alkali carbonate.

As contrasted to the approximately 4,000 cm^2/g specific surfaces of conventional Portland cement powder (determined by the method of Blaine ASTM Designation C204), the ground clinker used in this process has a specific surface in the range of 6,000 to 9,000 cm^2/g. Clinker ground to such high fineness, particularly when ground without gypsum, would be expected to rapidly hydrate and flash set on mixing with water.

An alkali or alkaline earth metal lignosulfonate, such as the sodium, potassium, ammonium, calcium or magnesium lignosulfonate, or other sulfonated lignin, is added to retard hydration and at the same time maintain the paste workable in spite of its low water content. The water added contains an alkali metal carbonate such as sodium, potassium or ammonium carbonate. This combination of additives obviates the need to add gypsum to the cement powder utilized in the process. The additives work together to obtain the desired result, as will be discussed later.

A grinding aid is added to the cement clinker in order to achieve the high degree of powder fineness required. The grinding aid also participates in maintaining the cement free-flowing and affects the workability and setting properties of the cement paste and the ultimate strength properties of the concrete.

Generally speaking, the most satisfactory grinding aids are surfactants containing both polar and nonpolar groups. The polar groups of the surfactant are attracted to the ions of the surfaces of the cement constituents, leaving the nonpolar portions on the outside. In effect, the cement grains are thus rendered somewhat hydrophobic, thereby inhibiting premature hydration and maintaining the cement free-flowing in spite of its high fineness.

A list of grinding aids suitable for use in the process would include diethyl carbonate, polyglycol derivatives and other nonionic surfactants, sulfonates and other anionic surfactants, alkyl-ammonium salts and other cationic surfactants, and the like.

Alite Crystals Containing Calcium Oxide and Amorphous Calcium Sulfate

T. Kawano, T. Mori and H. Kubota; U.S. Patent 3,947,288; March 30, 1976; assigned to Onoda Cement Co., Ltd., Japan describe an improved finely pulverized expansive cement additive which causes the cement mortar or concrete to retain an increased chemical prestressing property. The clinker of the additive substantially consists of alite crystals containing fine crystals of calcium oxide and a phase substantially consisting of optic-microscopically amorphous calcium sulfate in which the alite crystals are dispersed.

A clinker for the expansive cement additive of this process can be obtained by mixing raw materials of a calcareous substance, siliceous substance and gypsum in a manner allowing a mol ratio of CaO/SiO_2 to range from 4.2 to 9.2 and a mol ratio of $CaSO_4/SiO_2$ from 0.12 to 1.40, pulverizing the mixture to such fineness that a residue on a sieve of 88-micron mesh accounts for about 1 to 40% by weight based on the mixture, and burning the pulverized mixture of raw materials in an oxidizing atmosphere at a temperature of 1350° to 1550°C, the burning being stopped before decomposition of the gypsum mixture reaches 15 to 40% by weight.

The clinker thus obtained substantially consists of alite crystals containing fine crystals of calcium oxide therein and a phase of optic-microscopically amorphous calcium sulfate in which the alite crystals are dispersed. The ranges of crystal sizes of the alite and calcium oxide are 50 to 800 microns, and 5 to 30 microns respectively, and the content of calcium oxide crystals distributed in the alite crystals is at least 20% by weight based on the clinker and at least 50% by weight based on the total weight of calcium oxide crystal contained in the clinker. The content of calcium sulfate phase is in the range of from 5 to 30% by weight based on the clinker. The following example illustrates the process.

Example: Raw materials consisting of 79 weight parts of lime, 10 weight parts of silica stone and 11 weight parts of calcium sulfate anhydrite, whose fineness and chemical composition are shown in Table 1, were mixed together in a manner allowing the CaO/SiO_2 mol ratio to be 8.1 and the $CaSO_4/SiO_2$ mol ratio to be 0.46. After flaking the mixture under about 900 kg/cm^2 pressure, the mass of flake was charged in an electric furnace whose temperature had been maintained at about 1000°C in advance. The charged mass was heated up in increments of 5°C per minute to 1400°C. After being heated at 1400°C for 30 minutes, the charged mass was cooled rapidly to obtain a clinker whose chemical composition is shown in Table 2. The data in Tables 1 and 2 are within the scope of the process.

TABLE 1: FINENESS AND CHEMICAL ANALYSIS OF RAW MATERIALS

Raw material	Fineness Over 88 micron mesh (wt.%)	Ignition loss	CaO	SiO_2	SO_3	Fe_2O_3	Al_2O_3	MgO	Total
		Chemical analysis (wt.%)							
Lime	4.8	5.9	90.9	1.4	0.3	0.3	0.5	0.5	99.8
Silica stone	1.0	1.8	0.4	89.8	tr.	4.3	1.8	0.9	99.0
Anhydrite	5.6	1.5	37.8	1.9	54.4	1.4	2.7	tr.	99.7

TABLE 2: CHEMICAL ANALYSIS AND COMPOSITION OF CLINKER

Chemical analysis (wt.%)							
Ignition loss	CaO	SiO_2	SO_3	Fe_2O_3	Al_2O_3	MgO	Total
0.4	81.3	10.8	4.9	0.8	0.9	0.5	99.6

Composition (wt.%)				
Alite	Total CaO crystals (A)	$CaSO_4$ phase	CaO crystals dispersed in alite crystal (wt.% based) on clinker (B)	$\frac{(B)}{(A)} \times 100$
41	34.8	8.3	31	89

Calcium Sulfoaluminate Clinker

According to a process described by *P.K. Mehta; U.S. Patent 3,857,714; December 31, 1974; assigned to Chemically Prestressed Concrete Corporation* expansive calcium sulfoaluminate clinker containing a high proportion of calcium sulfate is prepared and is included in Portland cement (e.g., by intergrinding calcium sulfoaluminate clinker with Portland cement clinker) to contribute expansive properties to concrete made from the cement, and to contribute false set resistance to the cement by incorporating the calcium sulfate required as anhydrite present in the expansive clinker.

Controlled Expansion Using Calcium Sulfate

J.L. Deets and Z.T. Jugovic; U.S. Patent 3,861,929; January 21, 1975; assigned to United States Steel Corporation have discovered that relatively small amounts of calcium aluminate cement or clinker, about 2 to 17%, together with Portland cement or clinker, and calcium sulfate expressed as percent excess SO_3 over optimum SO_3 determined according to ASTM Standard C563-70 and equal to about 2 to 24%, when properly proportioned, give cements having controlled expansive properties.

The improved expansive cement composition, consisting of Portland cement or clinker, calcium aluminate cement or clinker and calcium sulfate, has the following improved properties:

(1) The workability, placement, compaction and finishing characteristics, all at normal water requirements with no excessive slump loss, are equal to those of a normal Portland cement concrete.

(2) The amount of expansion of the particular formulation can be readily controlled and duplicated making it applicable to special field requirements.

(3) The curing of shrinkage-compensating concrete is subject to only those precautions commonly recognized for curing ordinary concrete in hot or cold weather.

(4) The expansive potential of the shrinkage-compensating concrete is optimized in a relatively short curing period of 3 to 4 days.

(5) Storage requires no limitations in time or facilities other than those recommended for normal Portland, calcium aluminate, or any other hydraulic cement.

(6) Manufacture requires no special expansive components but uses commercially produced hydraulic cements or clinkers (Portland and calcium aluminate) and calcium sulfate.

(7) The strengths of mortars or concrete are equal to or better than those obtained with regular Portland cement mortar or concrete when suitable restraint is used.

(8) The cement can be produced by either intergrinding, blending or a combination of grinding and blending of the components in the proportions recommended for specific range of expansive characteristics.

The following examples illustrate the process.

Example 1: A shrinkage-compensating expansive cement was prepared from 86.3 parts by weight Portland cement clinker, 5.6 parts by weight low iron content calcium aluminate cement clinker and 8.1 parts by weight $CaSO_4{\cdot}2H_2O$ equivalent to 3.6 parts by weight SO_3. The optimum SO_3 level of the Portland cement clinker was determined using ASTM Standard C563-70 as 1.7%. The expansive cement was prepared by intergrinding the components in a laboratory mill to a Wagner surface of 1,900 cm^2/g. The expansion of this cement was measured on 2 x 2 x 10 inch restrained mortar prisms using a length comparator conforming with that described in ASTM Standard C490-70.

The mortar mix consisted of the cement and graded Ottawa sand in a 1:2.75 mix by weight with sufficient water to provide a flow of 100 to 115% comparable with ASTM Standard C109-64. Mixing was done in accordance with ASTM Standard C305-65. The molds were of such dimensions as to provide sufficient space to insert a restraining cage consisting of a 3⁄16 inch diameter threaded mild steel rod having an overall length of 11 9⁄16 inch ± 1⁄16 inch and two mild steel end plates 2 x 2 x 3⁄8 inch. The end plates were positioned on the rod to provide a 10-inch opening between them and were secured in this position with locknuts. The steel cages were assembled and measured for length in the comparator prior to insertion in the mold. Subsequent measurements were referred to this initial measurement.

Molded specimens were placed in the moist cabinet (73° ± 3°F, +90% relative humidity) for 5½ hours, demolded and then placed in saturated lime water (73° ± 3°F) until completion of the test. At the conditions as set forth above, the percent expansion after 3 days was 0.077% and after 7 days, 0.078%. The compressive strength determined in accordance with ASTM Standard C109-64 was 2,950 psi (3 day) and 4,270 psi (7 day).

Example 2: The procedure of Example 1 was followed using a Portland cement prepared from 83.8 parts by weight Portland cement clinker, 5.6 parts by weight low iron content calcium aluminate cement clinker, and 10.6 parts by weight $CaSO_4 \cdot 2H_2O$ (equivalent to 4.7 parts by weight SO_3). The Portland cement had an optimum SO_3 level of 2.6%. The 3-day expansion of this cement composition was 0.091% and the 7-day expansion was 0.087%. Compressive strengths were 3,080 psi (3 day) and 4,230 psi (7 day).

Example 3: A shrinkage-compensating expansive cement was prepared from 85.6 parts by weight Portland cement clinker, 6.0 parts by weight low iron content calcium aluminate cement clinker, 5.3 parts by weight $CaSO_4 \cdot 2H_2O$ equivalent to 2.3 parts by weight SO_3, and 3.1 parts by weight $CaSO_4 \cdot \frac{1}{2}H_2O$ equivalent to 1.7 parts by weight SO_3. The optimum SO_3 level of the Portland cement clinker was 1.8%. The expansive cement was prepared by intergrinding the components in a laboratory mill to a Wagner surface of 1,890 cm^2/g. The 3-day expansion of this cement composition was 0.089% and the 7-day expansion was 0.093%. Compressive strengths were 1,950 psi (3 day) and 3,780 psi (7 day).

The following table is a comparison of cement compositions of Examples 1, 2 and 3. It shows that the percent expansion of the three cement compositions having approximately equal excess SO_3 is approximately equal. It is therefore apparent that the process makes it possible to control both the magnitude and rate of expansion of the expansive cement composition.

	Compressive Strength psi		Expansion, %	
Cement	3 Day	7 Day	3 Day	7 Day
Ex. 1	2950	4270	0.077	0.078
Ex. 2	3080	4230	0.091	0.087
Ex. 3	1950	3780	0.089	0.093

Calcium Sulfoaluminate of Specific Particle Size Distribution

Y. Ono and T. Mizunuma; U.S. Patent 3,819,393; June 25, 1974 describe a calcium sulfoaluminate series cement expanding agent having a specific particle size distribution. This cement expanding agent is not influenced by the variation of curing conditions and can develop a high expandability in the concrete.

The process provides a cement expanding agent which is produced by burning limestone, alumina and gypsum in such a mixing ratio that the resulting sintered body has the following mineral composition: 10 to 40% $3CaO \cdot 3Al_2O_3 \cdot CaSO_4$, 10 to 20% free CaO and 20 to 60% free $CaSO_4$, and pulverizing the sintered body and adjusting the particle size distribution of the pulverized sintered body to the following ranges: 30 to 60% smaller than 44 μ, 10 to 40% 88 to 149 μ and less than 20% larger than 250 μ.

Sintered Mixture of Lime and Ferric Oxide or Calcium Fluoride

H. Ogura, T. Takizawa, Y. Ono and Y. Taketsume; U.S. Patent 3,801,339;

April 2, 1974; assigned to Denki Kagaku Kogyo KK, Japan describe a process for the preparation of an expansive additive for lime system cement which comprises adding at least one member selected from the group consisting of ferric oxide (Fe_2O_3), calcium fluoride (CaF_2) and calcium sulfate ($CaSO_4$) to lime, crushing the mixture and sintering or melting the mixture at a temperature at which the additive or additives are not decomposed.

The clinker thus obtained effectively controls the occurrence of the rapid expansion accompanied by the use of lime alone to cause a mild and prolonged expansion of a cement containing the clinker at the proper time during the progress of hardening of the cement. Hence, it is used as a very effective inflating agent for cement. The following examples illustrate the process.

Example 1: A mixture of 91 parts by weight of limestone having a purity of 98% and 9 parts by weight of mill scale was crushed and baked for 2 hours at 1450°C to provide a clinker for use as an expansive additive. The clinker was crushed and added to a cement in an amount of 5% by weight or 6% by weight to provide cement α and cement β, respectively. For comparison, a conventional cement containing no such expansive additive was prepared.

By using the cement prepared above a 10 x 10 x 50 cm concrete sample was prepared. The composition of the concrete per cubic meter thereof was as follows:

Blend

	Blend α, kg	Blend β, kg
Expansive additive of this process	168	191
Ordinary Portland cement	299	296
Water	218	218
Gravel (2.5 mm in diameter)	955	955
Sand	876	876
Total	2,516	2,536

The expansion coefficient of the concrete during curing in water at 20°C, measured by the above-described method employed with the mortar samples, is shown in the following table.

Expansion Coefficient*

	Age (days)						
	1	2	4	7	14	21	28
Cement α	12	19	31	48	64	78	79
Cement β	18	23	29	63	81	101	101
Conventional	-2	3	-1	2	1	3	2

* These values are x 10^{-3}

Example 2: A powdered mixture of 70 parts by weight of calcium oxide prepared by baking limestone at 1200°C and 30 parts by weight of fluoride was heated for 2 hours at 1400°C to provide a clinker as an expansive additive. The clinker was crushed smaller than 149 μ and compounded in a cement in an amount of 4% by weight to provide cement γ.

Separately, a powdered mixture of 70 parts by weight of calcium oxide, 15 parts by weight of mill scale, and 15 parts by weight of fluoride was heated for 2 hours at 1400°C to provide a clinker as an expansive additive. The clinker was crushed into a size smaller than 149 μ in diameter and compounded in a cement in an amount of 5% by weight to provide cement δ.

The expansion coefficients of the cements prepared above were measured by the same test procedure as used in Example 1 and the results are shown in the table below.

Expansion Coefficient*

	Age (days)						
	1	2	4	9	14	21	28
Cement γ	122	180	195	202	212	223	223
Cement δ	54	81	120	168	187	196	198

* These values are x 10^{-3}

Inorganic Fluoride Melt Produced in Resistance Heating Furnace

K. Nakagawa; U.S. Patent 3,666,515; May 30, 1972; assigned to Denki Kagaku Kogyo KK, Japan describes a process for producing a CaO-$CaSO_4$-Al_2O_3 type cement inflating agent by forming a melt of an inorganic fluoride in a direct resistance heating furnace, gradually adding to the melt a blend of raw materials containing CaO, $CaSO_4$, and Al_2O_3 and at least one inorganic fluoride, and melting the blend by heating the mixture to from 1200° to 1400°C.

Example 1: Bauxite, quick lime (calcium oxide) and anhydrous chemical gypsum were blended to obtain a mol ratio of $CaO/Al_2O_3/CaSO_4$ of 5/1/4. The blend was pulverized to provide the raw material. CaF_2 was added in an amount of 1% by weight to the weight of the raw materials.

A mixture of about 20 kg of CaF_2 and charcoal was charged around the carbon electrodes at the bottom of a Girod-furnace of 450 kw and an electric arc was generated to form a melt having a temperature of 1170°C. To the melt was added gradually the above-prepared raw material and, thereafter, it was confirmed by analysis of the pulsating wave form using an oscilloscope that the system of the electric furnace was converted into a resistance heating furnace.

With the increase of the raw materials, the volume of the pool of the melt increased gradually. In addition, the temperature of the melt reached 1370°C, but neither severe decomposition nor severe scattering of sulfuric anhydride occurred. Thereafter, the melt, about 70% of the pool, was tapped from the furnace and with supplying the raw materials continuously, the tapping operation was repeated continuously for 148 hours. The fluidity of the melt was good and the scattering of sulfuric anhydride did not occur throughout the whole operation and also the expansion of the expansive additive thus prepared was better than one prepared using a rotary kiln system.

The results of the chemical analysis of the raw materials are shown in the table below. In addition, the secondary voltage and the average load in operation were 40 to 122 volts and about 218 kw.

	SiO_2	Al_2O_3	Fe_2O_3	CaO	SO_3	Others	Total, %
Raw material	0.9	12.50	0.7	51.5	33.4	1.0	100
Molten product	1.7	12.00	0.7	52.1	32.0	1.5	100

Example 2: A mixture of 3 parts by weight of BaF_2, 4 parts by weight of CaF_2 and 3 parts by weight of Na_3AlF_6 was charged together with coke particles at the bottom of the electric furnace used as in Example 1 and around the electrodes. By forming an electric arc, a melt of about 1080°C was formed.

Powdered blended raw materials prepared by mixing bauxite, calcium oxide and gypsum, so that the molar blending ratios were 1.33 $CaO/CaSO_4$ and 5.7 CaO/Al_2O_3, were gradually added to the melt, whereby the operation was changed to a resistance operation and thereafter the melt was tapped from the furnace. Subsequently, while supplying the raw material continuously to the furnace, the melt thus formed was tapped from the furnace repeatedly, whereby the expansive additive was produced continuously. The temperature of the melt reached 1320°C, the scattering of sulfuric anhydride was less and an inflating agent having excellent quality was obtained.

In addition, after the operation had been changed to a resistance operation, cryolite was added in an amount of 1% by weight to the weight of the raw materials. The compositions of the raw materials used and that of the inflating agent thus produced are shown in the following table.

	Ignition Loss	Al_2O_3	CaO	SO_3	SiO_2	Fe_2O_3	Others	Total, %
Lime	-	0.5	95.7	0.2	0.4	1.9	0.8	99.5
Bauxite	0.3	86.2	0.3	-	3.9	5.5	3.4	99.6
Gypsum	1.7	0.3	39.4	57.8	0.3	0.1	0.3	99.9
Expansive additives	0.2	9.5	51.8	36.0	1.5	0.8	0.9	100.7

Comparison Example 1: A comparison test was conducted using the same electric furnace and raw materials as used in Example 1. Without forming the preliminary melt as was done in the above examples, the raw materials were charged directly in the furnace around the graphite electrode and by passing an electric current, an electric arc was formed, whereby a melt of the raw materials was formed. In this case, the temperature of the melt was increased to about 1600°C and marked scattering of sulfuric anhydride occurred.

Furthermore, the raw materials supplied afterwards were not mixed smoothly or uniformly, the eutectic mixture was not formed readily and hence continuous operation was impossible. The composition of the melt is shown in the table on the following page.

	Melt Percent
SiO_2	1.4
Al_2O_3	18.4
Fe_2O_3	1.1
CaO	77.2
SO_3	0.6
Others	1.3
Total	100

As is clear from the above table, the gypsum used as the raw material had essentially been decomposed.

Each of the clinkers of the expansive additives prepared in Example 1 and the clinker of the expansive additives prepared by a conventional method was pulverized to 2,800 cm^2/g and 12 parts by weight of the pulverized clinker was blended with 88 parts by weight of ordinary Portland cement to provide an expansive cement. Using the expansive cement thus prepared a specimen of reinforced concrete was prepared.

The specimen was prepared using sand, gravel and with a 40% water/cement ratio. A steel rod (diameter 12 mm) was positioned at the center of a mold, 250 cm in length and 15 cm in height, and the mix was poured into the mold.

After 6 hours, the mold was released and to measure the expansion rate, a test pin was inserted in it. The specimen was cured under a relative humidity of above 90% and a Hargen Bagar type handy strain gauge was used. The effective expansion thus measured periodically and expansion strain data obtained for the period up to 30 days are shown in Figure 2.2. Curve **a** represents the data obtained for the sample prepared using the expansion additive prepared according to this process. Curve **b** represents the data obtained for the sample prepared using a conventionally prepared expansion additive (a calcium sulfo aluminate) and curve **c** represents the data obtained for the sample prepared containing no additive.

FIGURE 2.2: EFFECT OF CEMENT EXPANSIVE ADDITIVES

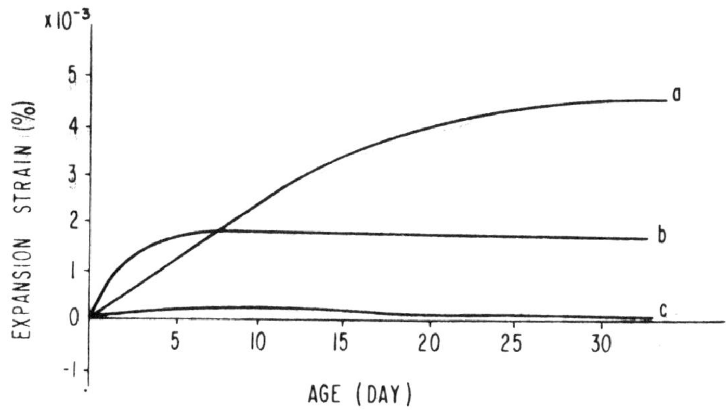

Source: U.S. Patent 3,666,515

The results obtained after 100 days for the strength test (using the three-edge bearing method) are shown in the following table.

		- - - - - Strength Test - - - - -	
Expansive Additive	Effective Expansion Force, ton	Initially Cracked Load, ton	Load, tons
Process	1.35	0.42	2.75
Conventional	0.50	0.37	2.32
None	0.0	0.23	1.5

As is clear from the above table, when the expansive additive of this process was applied to a chemical prestress process, better properties were obtained than with expansive additives conventionally prepared.

In addition, the effective expansion force data above were obtained by the method described in the report of Muguroma et al in Fifth International Cement Chemical Symposium, Part IV, Session 4 using 243,000 kg/cm^2 as the concrete elasticity and 1,980,000 kg/cm^2 as the steel rod elasticity.

Coated Expansion Agents

A process described by *T.M. Houseknecht; U.S. Patent 3,649,317; March 14, 1972; assigned to Fuller Company* relates to a shrinkage compensating hydraulic cement. Portland cement normally shrinks as it is cured, and it would be desirable to overcome this shrinking characteristic.

By this process it has been found that such shrinkage can be compensated for by the addition of an expansive agent to the Portland cement. Since water is added to the cement and aggregate in the making of concrete, the expansive agent is preferably reactive upon contact with water. Upon reaction with water, the expansive agent should not add a foreign substance to the concrete.

One such expansive material is quicklime (CaO). When water is added to quicklime, it expands and increases in volume by about 20%. The quicklime plus water forms calcium hydroxide [$CaO + H_2O \longrightarrow Ca(OH)_2$] which is a normal hydration product of calcium silicates, which comprise up to 80% of commercial Portland cement. Thus, the addition of the quicklime to the Portland cement will not result in the addition of any foreign substances to the finished concrete. The addition of the quicklime may be beneficial to the composition as the calcium hydroxide will react with certain forms of silica to form cementitious calcium silicate. This may add durability to the finished concrete.

A second material which may be used as an expansive agent is plaster of Paris. When mixed with water, plaster of Paris will expand and produce gypsum:

$$CaSO_4 \cdot \tfrac{1}{2}H_2O + 1\tfrac{1}{2}H_2O \longrightarrow CaSO_4 \cdot 2H_2O$$

Since gypsum is often added to Portland cement, the addition of plaster of Paris to Portland cement will not add any foreign substances to the finished concrete. However, the plaster of Paris will expand only about one-half the amount that the same volume of quicklime will expand. This factor must be considered in determining the amount of expansive agent to be added to the Portland cement. Both quicklime and plaster of Paris are inexpensive when compared to prior expansive cement additives.

In order for an expansive cement to be useful, it should not expand as soon as water and aggregate are added. However, most of the expansion should occur within 24 to 48 hours after the concrete has been mixed. After such initial 24 to 48 hour period, the concrete should be dimensionally stable and substantially no further expansion or subsequent shrinkage should occur.

Because the expansive agents noted above expand upon contact with water, it is necessary to delay contact between the expansive agent and water in order to insure that the greater part of the expansion takes place within the initial 24 to 48 hour curing period. The expansive material is coated with a water-insoluble material. Such coating will prevent the water from contacting the expansive agent until the coating has been dissolved and thereby delay the expansion of the cement.

Since alkalis are naturally formed when water is added to Portland cement, it has been found to be desirable to make the coating alkali-soluble. When water is added to the cement, the alkalis formed will dissolve the coating. As the coating is dissolved, the water in the cement will contact the expansive agent expanding the cement.

A coating material which was found to be useful is a wood resin commercially available under the name Vinsol. This material meets the qualifications of being alkali-soluble and water-insoluble and has the further advantage of being able to be placed in solution with organic liquids. The ability to be placed in solution with an organic liquid permits easy application of the coating to the expansive material.

It has been found that the coating may be applied by placing it in solution with an organic liquid, such as denatured alcohol, and the reactive compound being stirred in the solution. The coated reactive compound is then dried. Other techniques for applying the coating to the reactive compound may include spraying the coating solution onto the reactive compound. In laboratory applications, the immersion of the reactive compound in the solution of coating material and organic liquid followed by filtration and drying proved to be the most successful technique for applying the coating material to the reactive compound. The amount of coating applied to the reactive compound may be varied by increasing or decreasing the amount of coating placed in solution with the organic liquid.

Experimentation was carried out by stirring the expansive compound in a solution of denatured alcohol and Vinsol for 5 minutes followed by suction filtration and drying at 105°C. The coated reactive compound was blended with Portland cement in amounts up to 20% by weight of the finished product.

Test specimens were of neat cement paste, i.e., no aggregate was added. Water content of the test specimens was kept as close to 25% as possible. After the specimens were molded, they were moist-air cured for 1 day, then the mold forms removed and length measurements taken. Subsequent length measurements were made at one-day intervals. Identical specimens were made without the expansive agent added for purposes of comparison.

Three factors affected the expansion of the cement. Those factors were (1) the amount of reactive agent added to the Portland cement, (2) the thickness of coating and (3) whether moist-air cured or water-submersed cured.

The amount of the coated reactive compound blended with the Portland cement can be used to control the amount of expansion of the cured concrete. It has been found that up to 20% by weight of the finished composition may be coated reactive compound. Using the same coating thickness, a composition of Portland cement and 10% by weight of finished product coated reactive compound expanded about 0.19% after 1 day, about 0.20% after 2 days and about 0.23% after 12 days while a composition of Portland cement and 5% by weight of finished product coated reactive compound expanded about 0.07% after 1 day, about 0.08% after 2 days and about 0.13% after 12 days.

Each of these test specimens was cured in a similar manner. No reversal of the expansion was noted after 20 days when measurements were discontinued. It became apparent that a higher percentage of expansive agent produced a greater expansion, most of which took place over a shorter period of time.

The thickness of the coating also affects the amount of expansion of the cement. In experimentation, the coating thickness was varied by dissolving greater or lesser amounts of the resin in alcohol. The range of coating in solution ran from 1 to 50%. In all cases, as the strength of the coating solution decreased, and hence the thickness of the coating decreased, the ultimate linear expansion of the cured cement increased.

In addition, the thicker the coating, the slower the rate of expansion. This is believed to be attributable to the length of time it takes to dissolve the coating before the water contacts the reactive compound.

Using an equal amount of coated reactive compound blended with Portland cement, a comparison of coating thickness was made. A solution of 10% Vinsol and 90% denatured alchol resulted in an expansion of about 0.13% after 1 day, about 0.15% in 2 days and about 0.19% in 12 days. A solution of 1% Vinsol and 99% denatured alcohol resulted in an expansion of about 0.26% in 1 day, about 0.27% in 2 days and about 0.32% in 12 days. Again, no tendency to shrink was observed after 20 days.

GRINDING AIDS

Melamine-Formaldehyde Condensation Product

According to a process described by *A. Aignesberger and H.-G. Rosenbauer; U.S. Patent 3,856,542; December 24, 1974; assigned to Suddeutsche Kalkstickstoff-Werke AG, Germany* an aqueous solution of melamine-formaldehyde condensation product is sprayed over dry cement product in the fine grinding compartment of a cement grinding mill at a uniform rate essentially throughout the grinding process in concentration and amount chosen to evaporate all the aqueous solvent during the normal cement grinding process and introduce an aggregate of 0.1 to 2.0 weight percent of condensation product solid in relation to the dry cement product being ground.

Example: In a two-compartment mill (air-separating mill with coarse and fine compartment; throughput capacity 16 to 18 tons of cement per hour with a specific surface area of the cement of about 4,800 cm^2/g measured by Blaine air permeability apparatus used to measure the surface area of a finely ground substance), 50 kg of an aqueous melamine-formaldehyde condensation product modified with sulfide or sulfonic acid and having a solids content of 20%, produced according to Austrian Patent 263,607, is sprayed into the fine grinding compartment per 1 ton of cement, at a uniform rate over the cement production period. The finished cement contains the calculated quantity of 1% condensation product with respect to total weight of cement and it is distributed uniformly therein.

Morpholine or Amides

V.H. Dodson and F.G. Serafin; U.S. Patent 3,443,976; May 13, 1969; assigned to W.R. Grace & Company have found that in the grinding of minerals (e.g., Portland cement) an additive which can be morpholine, N-methylmorpholine or a by-product from the continuous commercial production of heterocyclic amines consisting predominantly of 4-(2-aminoethoxy) ethylmorpholine, 2-(4-morpholinylethoxy) ethanol, and bis-2-(4-morpholinyl) ethyl ether is added to increase the efficiency of the grinding operation.

The additive is employed effectively over a relatively wide range. The preferred range is about 0.001 to 1%, and more preferably about 0.005 to about 0.1% based on the weight of the mineral, i.e., weight of additive solids based on weight of mineral solids (herein referred to as solids on solids). In a particular example about 0.05% of the additive based on the weight of the mineral is employed.

In the table below the effectiveness of the additives as grinding aids is reported. The data reported were obtained using a Type I cement clinker and gypsum ground in a laboratory mill at a temperature of 210°F for 7,411 mill revolutions.

Additive	Amount of Additive (% Solid on Solid)	Blaine Surface Area, cm^2/g
Blank	-	3,441
Aniline	0.01	3,503
Aniline	0.05	3,660
Morpholine	0.01	3,480
N-methylmorpholine	0.01	3,500
p-Toluidine	0.01	3,570
Pyrrolidine	0.01	3,542
Pyrrolidine	0.05	3,515
Piperidine	0.01	3,499
Diphenylamine	0.01	3,550

The pack set data are reported below on cement interground with the additives of this process.

Additive	Addition Rate (% Solid on Solid)	Pack Set Index
Blank	-	12.6
Aniline	0.05	10.0
Morpholine	0.01	11.3
Morpholine	0.05	2.0
N-methylmorpholine	0.01	11.3
p-Toluidine	0.01	12.3
p-Toluidine	0.05	4.0
Pyrrolidine	0.01	8.0
Pyrrolidine	0.05	1.0
Piperidine	0.01	9.0
Diphenylamine	0.05	7.3

In a process described by *F.G. Serafin; U.S. Patent 3,459,570; August 5, 1969; assigned to W.R. Grace & Company* minerals and Portland cement are interground with a compound of the formula $R—CONH_2$ where R is an alkyl or a phenyl group (e.g., acetamide) to enhance the efficiency of the grinding operation.

Urea

F.G. Serafin; U.S. Patent 3,420,687; January 7, 1969; assigned to W.R. Grace & Company has found that in the grinding of a mineral or cement (e.g., Portland cement), a urea compound may be interground with the mineral or cement to increase the efficiency of the grinding operation. Pack set of the resulting ground material is also inhibited. The urea compound can be urea, acetyl urea or diacetyl urea.

The following table shows the results of intergrinding Type I Portland cement with urea at various concentrations. The grinding was carried out in a laboratory steel ball mill at a temperature of 220°F for 2,000 revolutions.

Additive	Amount of additive (solids on solids)	Blaine surface area (sq. cm./g.)
Blank		3,015
Commercial grinding aid	0.02	3,140
Urea	0.01	3,160
Do	0.015	3,220

The results show that urea is an effective grinding aid and is even more effective than a commercial grinding aid when used at a level lower than that of the commercial grinding aid. The table below shows air entrainment data of the additive of this process compared with a blank. The method used to determine air entrainment was ASTM C185.

Additive	Amount of Additive (solids on solids)	W/C ratio	Percent air entrained
Blank		0.70	8.7
Urea	0.01	0.67	8.8
Do	0.015	0.67	8.8

The results indicate that grinding cement with urea does not result in an increase in entrained air. The data below show that the use of urea as an additive in cement does not appreciably increase shrinkage of the cements. The shrinkage test used was ASTM C157. The W/C (water/cement) ratio was 0.40.

Additive	Amount of additive (solids on solids)	Length, inches 7 days	28 days	3 months	Percent change*
Blank		11.6601	11.6532	11.6495	0.091
Commercial grinding aid	0.02	11.6300	11.6211	11.6170	0.112
Urea	0.01	11.6360	11.6292	11.6252	0.093

$$*\text{Percent change}=100\times\frac{(7\text{ day length}-3\text{ month length})}{7\text{ day length}}.$$

Ethanolamine Salts

V.H. Dodson, Jr. and F.G. Serafin; U.S. Patent 3,329,517; July 4, 1967; assigned to W.R. Grace & Company describe an additive for Portland cement which serves a dual function as a grinding aid and pack set inhibitor. The additive is selected from the group consisting of an alkanolamine acetate and an acetylated alkanolamine acetate. Only a small amount of this additive need be used to achieve the desirable results, such as 0.005 to 0.050% based on the weight of the cement.

The additive is derived from the residue obtained in preparing ethanolamines. The residue product may be derived from a number of well-known methods which are employed to synthesize ethanolamines. It may be obtained from such reactions as the ammonolysis or amination of ethylene oxide, the reduction of nitro alcohols, the reduction of amino aldehydes, ketones and esters, and the reaction of halohydrins with ammonia or amines. The exact composition of the residue product varies within certain limits and, therefore, the term "ethanolamines" as used refers to one or more mono-, di- or triethanolamines.

In general the residue product is predominantly triethanolamine, preferably, between 40 to 84% by volume triethanolamine. A specific residue product which is employed in a particularly preferred embodiment of this process is a mixture of mono-, di- and triethanolamine which is available commercially and has the following chemical and physical properties:

Triethanolamine	45 to 55% by volume
Equivalent weight	129 to 139
Tertiary amine	6.2 to 7.0 meq/g
Water	0.5% by weight, maximum
Density	9.49 lb/gal

Example: The ethanolamine acetate was produced by neutralizing 100 grams of the above-identified residue product with 45 grams of glacial acetic acid. The reaction was exothermic and heat of neutralization was produced.

The table below shows the results of intergrinding the ethanolamine acetate of the example with a Type I Portland cement clinker containing 2.5% gypsum. The additive was incorporated at a rate of 0.01% solids per solids cement. An untreated cement is also included for purposes of comparison.

Additive	Grinding Time (minutes)	Blaine Surface Area ($cm.^2/g.$)	Pack Set Index
Blank	110	3,220	9
Ethanolamine acetate	100	3,260	4

It is noted that the cement which was interground with the ethanolamine acetate had a pack set index of 125% lower than the untreated cement. The lower index shows that the additive lessens the tendency of the cement to compact. The increase in surface area of the treated cement using a shorter grinding time indicates that improved grinding efficiency is achieved with the use of the additive of this process. The table on the following page reflects the comparative compressive strengths of the above table. The strengths were determined on ASTM C-109 2-inch mortar cubes.

	Compressive Strength, psi		
Additive	1 Day	7 Day	28 Day
Blank	1,308	3,588	5,100
Ethanolamine Acetate	1,328	3,525	5,438

The improvement in the 28-day strength of the treated cement is quite marked. This is significant because it serves to indicate the ultimate structural strength. Results similar to those reported above were found when the same procedure was applied to Type III Portland cement. The additive was interground with the cement at a rate of 0.0167% solids additive per solids cement.

Aromatic Acetates

In a process described by *F.G. Serafin; U.S. Patent 3,420,686; January 7, 1969; assigned to W.R. Grace & Company* aromatic acetates are used as grinding aids and pack set inhibitors in minerals and cement. Typical additives within the scope of the process are the following:

- Phenyl propionate
- Phenyl butyrate
- 1,2-Dimethylphenyl acetate
- 2,4-Dimethylphenyl acetate
- p-Nitrophenyl acetate
- Methylphenyl acetate
- α-Naphthyl acetate
- α-Naphthyl propionate
- p-Chlorophenyl acetate

Furan and Thiophene

F.G. Serafin; U.S. Patent 3,492,138; January 27, 1970; assigned to W.R. Grace & Company has found that furan and thiophene increase the grinding efficiency and pack set characteristics of minerals especially cement clinker.

The additive is employed effectively over a relatively wide range. A preferred range is about 0.001 to 1% based on the weight of the mineral, i.e., the weight of the additive solids based on the weight of mineral solids (herein referred to as solids on solids). In a particularly preferred example the amount of additive employed is about 0.004 to 0.04%, more preferably 0.02%.

Example: Type I Portland cement clinker was ground in a laboratory steel ball mill for 5,490 revolutions at room temperature. The following table reports the results of the grinding test.

Additive	Amount of additive, percent solids on solids	Blaine surface area, sq. cm./g.	Percent improvement over blank
Blank		2,966	
Furan	0.039	3,098	4.44
Thiophene	0.048	3,132	5.59
Tetrahydrofuran	0.41	3,260	9.91

The following table reports the results of Portland cement clinker ground in a laboratory steel ball mill for 5,490 revolutions at 220°F.

Additive	Amount of additive, percent solids on solids	Blaine surface area, sq. cm./g.	Percent improvement over blank
Blank		3,142	
2-furaldehyde	0.012	3,290	4.70

The following table reports the results of grinding Portland cement clinker in a laboratory steel ball mill for 5,655 revolutions at room temperature.

Additive	Amount of additive, percent solids on solids	Blaine surface area, sq. cm./g.	Percent improvement over blank
Blank		3,049	
Tetrahydrofuran	0.041	3,354	10.0
Tetrahydrothiophene	0.050	3,172	4.03

EFFLORESCENCE AND COLOR

Alkylene Polyamine-Epichlorohydrin Condensates

According to a process described by *J.P. Brants and J. van der Meulen; U.S. Patent 3,674,522; July 4, 1972; assigned to Akzo NV, Netherlands* efflorescence of masonry structures composed of cold-hardened masonry materials, such as concrete, sand-lime bricks, asbestos cement and cement plasterwork, or of stone or fired bricks, is suppressed by the condensation product of at least one alkylene polyamine and/or polyalkylene polyamine, and epichlorohydrin.

The condensation product in the form of an aqueous or more volatile solution may be applied to the surfaces of the masonry structure or, in the case of such structures composed of cold-hardened materials, is preferably incorporated in such material prior to the hardening or setting thereof. The condensation product is further found to increase the coloring power and permanence of pigments when added with the latter to cold-hardening materials prior to the hardening or setting.

Examples of polyamines which may be used for the preparation of the condensation products are tetraethylene pentamine (TEPA), pentaethylene hexamine (PEHA), triethylene tetramine (TETA), diethylene triamine (DETA) and ethylene diamine (EDA). A mixture of two or more of these amines can also be used. TEPA or technical TEPA, which is a mixture of polyalkylene polyamines with an average stoichiometric composition equivalent to TEPA, is preferred.

The molar ratio of the alkylene polyamine and/or polyalkylene polyamine to the epichlorohydrin may be in the range from 0.3:1 to 2.1:1 and is preferably in the range from 0.5:1 to 1.5:1.

The condensation products for incorporation in, or application to the surfaces of masonry materials in accordance with this process may be easily prepared, for example, as follows. The alkylene polyamine and/or the polyalkylene polyamine is dissolved in water and then the desired quantity of epichlorohydrin is added slowly, while stirring, at a temperature of from 10° to 70°C. The mixture is allowed to go on reacting for a time while stirring is continued.

The clear aqueous solution thus obtained, which may, for example, contain about 10% by weight of condensation product, may then be evaporated on a steam bath or in vacuo at 70°C to give a more highly concentrated solution. Alternatively, the quantity of water used in the process can be so chosen that, on completion of the reaction, a solution having the desired final concentration of condensation product is obtained. Specific illustrative condensation products for use in accordance with this process were produced as follows.

Condensate A: 80 kg of TEPA were dissolved in 90 kg of water. At a temperature between 55° and 60°C, over a period of 30 minutes, 40 kg of epichlorohydrin were added to this mixture, while stirring. After the addition of the epichlorohydrin, stirring was continued at 55°C for a further 30 minutes. On cooling to ambient temperature, there was obtained a clear watery liquid containing about 55% by weight of condensate A having a tetraethylene pentamine/epichlorohydrin molar ratio of approximatelv 1:1.

Condensate B: 8 kg of TEPA were dissolved in 184 kg of water. At a temperature of 15° to 20°C, over a period of 1 hour, 8 kg of epichlorohydrin were added to this mixture, while stirring vigorously. After the addition of epichlorohydrin, stirring was continued for a few hours. The solution obtained, containing 8% by weight of condensate B, was evaporated in vacuo at 70°C until it contained 30% by weight of condensate B. If desired, evaporation can be continued until a solution containing 50 to 80% by weight of condensate B is obtained. The obtained condensate B has a tetraethylene pentamine/epichlorohydrin molar ratio of approximately 0.5:1.

Condensate C: The same procedure was followed as is described in connection with the production of condensate A, except that 27 kg of EDA were used instead of 80 kg of TEPA. The resulting condensate C has an ethylene diamine/epichlorohydrin molar ratio of approximately 1:1.

Condensate D: 96 kg of TEPA were dissolved in 135 kg of water. This solution was caused to react with 40 kg of epichlorohydrin in the same manner and under the same conditions as described above in the preparation of condensate A. There was obtained a solution containing about 50% by weight of condensate D having a tetraethylene pentamine/epichlorohydrin molar ratio of approximately 1.2:1.

Condensate E: 103 kg of DETA were dissolved in 240 kg of water. This solution was reacted with 155 kg of epichlorohydrin in the same manner and under the same conditions as described in the preparation of condensate A. There was obtained a solution containing about 50% by weight of condensate E having a diethylene triamine/epichlorohydrin molar ratio of approximately 0.6:1. The following examples illustrate the process.

Example 1: A mortar mixture (Mixture 1) comprising 1 part of Portland cement,

3 parts of sand and 0.4 part of water (parts by weight) was prepared. A mixture (Mixture 2) of the same proportions of cement, sand and water was also prepared with, in addition, a sufficient amount of the aqueous solution containing 55% by weight of condensate A to provide the resulting mixture with 0.6%, by weight, of the condensation product, based on the weight of dry cement.

Two mixtures (Mixtures 3 and 4) were prepared to be similar to Mixture 2, but in which, instead of condensate A, the condensates C and D, respectively, were incorporated in respective quantities of 1.0% and 2.0%, by weight, based on the weight of dry cement. A number of iron molds were filled with these four Mixtures 1 through 4 and were then vibrated. After the mortars had hardened, the specimens were removed from the molds and placed in the open air (maritime climate of Holland). After 6 months, the moldings made of Mixture 1 were found to be clearly affected by efflorescence, whereas those made of the Mixtures 2, 3 and 4 showed no efflorescence. At the time of placing the mortar in the molds, it was found that the Mixtures 2, 3 and 4 had a better plasticity and therefore flowed better than Mixture 1.

Example 2: A wall of sand-lime bricks having a sand/lime weight ratio of 10:1, and which showed a strong tendency to efflorescence, was brushed clean with very dilute hydrochloric acid and then washed down with very dilute ammonia and finally rinsed with water. One-half of the wall thus freed of efflorescence was coated with the aqueous solution containing 30% by weight of condensate B. After several months' exposure to the weather, the untreated part of the wall was found to have again become affected by efflorescence, whereas no efflorescence was detectable on the treated part.

Example 3: 25 kg of carbon black were placed in the rotatable stainless steel pan of an Eirich mixer. During operation of the mixer, 11.5 kg of an aqueous solution containing about 55% by weight of condensate A were added over a period of 25 minutes. Operation of the mixer was then continued for a further 5 minutes. An apparently dry, loosely granular product was obtained, in which the ratio of carbon black to condensate was 4:1, by weight.

A mortar mixture was prepared, comprising 1 part of Portland cement, 3 parts of sand and 0.4 part of water (parts by weight). Quantities of the loosely-granular carbon black/condensate mixture were mixed with four portions of this mortar such that the mortar mixtures, hereinafter referred to as mixtures of a, b, c and d, respectively contained 1, 2, 3 and 4% by weight of carbon black, based on the amount of dry cement, and therefore respectively contained 0.25, 0.5, 0.75 and 1.0% by weight of condensate A.

Four additional mixtures e, f, g and h were respectively prepared from the described mortar mixture and 1, 2, 3 and 4% by weight of carbon black based on the amount of dry cement. A number of iron molds were filled with the 8 mortar mixtures a through h and then vibrated. After hardening, the specimens were demolded. Immediately after demolding, the specimens were visually inspected and compared with one another as to color. They were then placed in the open air (maritime climate of Holland), and thereafter visually inspected and compared with one another at regular intervals.

The observations directly after demolding, and after 6 months, are summarized in the table on the following page. For clarification, the percentage quantities of carbon black and condensation products in each mixture are also included in the table.

Mixture	Percent carbon black (by weight)	Percent condensation product (by weight)	Visual Inspection for color after demoulding	After 6 months
a	1	0.25	Grey to black	Grey.
b	2	0.5	Intense black	Black.
c	3	0.75	do	Do.
d	4	1.0	do	Do.
e	1	None	Grey	Effl. [1]
f	2	None	Grey to black	Effl.
g	3	None	Black	Effl.
h	4	None	do	Effl.

Effl.=efflorescence.

After 6 months' exposure to the weather, the specimens e through h showed a blotchy gray efflorescence. Even after a week's exposure in the open air, the differences in color between the specimens made from mixtures a through d and those made from mixtures e through h had become noticeably greater, with the mixtures e through h having become distinctly grayer in color.

Hardly any difference in blackness was found between the mixtures b, c and d. There was hardly any difference in blackness between the mixtures g and h either, but immediately after demolding, these mixtures g and h were distinctly less black than the mixtures b, c and d, and the difference in color became greater and greater during exposure in the open air.

Sodium Stearate and Stearic Acid

According to a process described by *R.T. Jordan; U.S. Patents 3,720,528 and 3,720,529; March 13, 1973* colored cements having improved color and reduced efflorescence are prepared by adhering to a portion of the surface of a majority of the unset cement particles a surface-tension reducing agent and/or a dispersant which effects dispersion by establishing a common charge on each of the cement particles without substantial reduction in the size of the particles. Preferably, the organic matter is adhered to the cement particles by impinging a stream of one of the particles on a stream of the other and preferably both a surface-tension reducing agent and a dispersant are utilized.

While only one additive agent will provide an improved color effect, the use of a plurality of species is preferred. Generally, there is no appreciable color enhancement on addition of more than about 4 to 5% pigment to the cement where a plurality of additive agents, for example, a sulfonic acid-formaldehyde condensation product and sodium stearate, are used.

The additives useful in reducing surface tension are organic and have both polar and nonpolar units in the molecule. Preferably, these additives are soluble in water or form micelles therein. The most preferred additives are the alkali metal fully saturated fatty acid soaps having 12 to 22 carbon atoms. The most preferred surface active organic additives are sodium stearate and stearic acid.

The Day Flash Mixer is preferred for adhering additive agent to the cement and/or coloring agent. The Flash Mixer utilizes two horizontal discs rotating in opposite directions at 3,600 rpm to propel streams of particles against one another. Materials to be mixed are introduced into the top of the unit and fall onto the counter rotating discs which impart a rotational component to the movement of the particles to impinge one particle stream on another before the mixed coated material passes downwardly through an exit chute for bagging or bulk storage.

The mixer is, with slight modification, also quite useful for grinding particles of, for example, pigments or cement. Three tons of improved cement can be readily coated per hour by all Models C, CD and CS (18-350) Flash Mixer or Centri Flo Mixers.

Example 1: Aliquots of white, gray and pozzolanic cements are mixed with 3% by weight iron oxide pigment in the absence of an additive. Additional aliquots and these cements are mixed with a mixture of, by weight of the cement, 3% pigment and 2%, by weight of the cement-pigment mixture, of equal parts of stearic acid and Tamol SN. Coating of 357 grams of the various mixtures is accomplished in a Tyler Portable Sieve Shaker in a 9½" diameter by 2" deep pan containing steel balls of various sizes and shapes. Samples of the coated and uncoated mixtures are periodically removed and compared, by color reflectance meter. Maximum color development occurs in about 15 minutes.

Prior to hydration and setting, the cement containing additive shows, under the microscope, pigment particles uniformly dispersed in random fashion on the surface of and in fissures in the cement particles. Few or no particles are found in the interstitial spaces between cement particles. Further, the cement and pigment do not segregate on handling during bagging, shipment, etc.

Cement and pigment mixtures containing no additive show nonuniform pigmentation, deposit of some pigment in the interstitial spaces, severe efflorescence in the set cement, and segregation in the unset cement. The x-ray patterns and UV patterns for the optimally-coated cement with and without additives also differ.

Example 2: To test the effect of mixing and coating on the pigment-additive cement mixtures, Portland cement is mixed with 1% stearic acid and 1% Tamol SN and amounts and types of pigments are indicated in the table below. Color development is then compared with neat raw pigment as a standard.

Color	Pigment-cement ratio	Color development as percent of standard	
		Impingement	Paddle mixer
Red (Fe_2O_3)	3:97	45	15
	5:95	51	18
	7:93	52	23
	10:90	52	30
Green (Chrome green)	3:97	54	18
	5:95	60	22
	7:93	68	29
	10:90	68	34
Yellow (Fe_2O_3)	3:97	38	9
	5:95	44	15
	7:93	46	17
	10:90	51	20

The preceding table sets out the differences incident to coating in a mixer where the two streams of particles are centrifugally forced into fine streams which impinge to cause adhering at a rate of 100 lb of coated mixture in 1 minute and in a paddle mixer (Hobart Model H-50) wherein 10 lb of mix was stirred for 5 minutes. The table sets out the effectiveness of the processes as a percent of color development, the standard being 100%.

Example 2: Table 1 shows the amount of pigment required for comparable tinting where coating is accomplished by impingement on the one hand and mixing by paddle mixing on the other, where both mixtures contain 1% Tamol SN and 1% stearic acid. There is little color enhancement above 4 or 5%, by weight of cement, pigment where coating is by impingement. Color is measured using the Lovibond Tintometer, Photovolt Reflection Meter and visual comparison (Munsell Book of Color). The methods of ASTM D 307-44 and Fed. Spec. TT-P-141G Method 425.1 are used in the comparative status.

TABLE 1

Blue		Green		Yellow		Red	
Impingement	Paddle	Impingement	Paddle	Impingement	Paddle	Impingement	Paddle
1	2.5	1	3	1	3	1	4.5
2	4	2	5.5	2	6	2	6
3	6	3	8	3	8	3	9
4	10	4	10	4	10+	4	12

Table 2 sets out the percentage increase in efflorescence inhibition occurring when the listed additives are used alone and in combination in Portland cement with pigment. The standard for comparison is ASTM efflorescing cement. 1% by weight of additive was added in each case.

Efflorescence is measured by mixing a test mortar in accordance with ASTM-C 12 Subcommittee 11, Experimental Procedure for Pilot Study for the Efflorescence Test Procedure. The mortar mix contains 1,080 grams of 230 grade sand, 1,080 grams graded Ottawa sand and 630 grams cement containing regulated amounts of additive. Minimum water is added to achieve plasticity. The mixed mortar is placed between the ends of test bricks with a one-half inch excess at the joint. After curing for 3 days, the cemented bricks are placed in a shallow pan of water. Efflorescing material migrates onto the brick during the 14-day soaking in 1 inch of water. The test sample is compared against standard photographs at the end of the test.

TABLE 2

Additive	Adipic	Arachidic	Glutaric	Oleic	Palmitic	Stearic	Undecylenic
Adipic	7	–	74	78	–	95	–
Arachidic	–	3	–	–	36	85	–
Glutaric	–	–	5	–	–	–	–
Oleic	–	–	–	48	–	96	–
Palmitic	–	–	–	–	6	92	–
Stearic	–	–	–	–	–	65	–
Undecylenic	–	–	–	–	–	–	1

Example 4: A cement of comparable whiteness to white cement is prepared by mixing 4%, by weight of cement, of rutile and 1% each of stearic acid and Tamol 721 diluted with 1% by weight Tamol active solids at 6,000 lb/hr in a Flash Mixer. The Tamols are alkali metal salts of condensation products of formaldehyde and organic acids.

Polyvinyl Alcohol and Barium Hydroxide

A.R. Ronzio and B.T. Pull; U.S. Patent 3,645,763; February 29, 1972; assigned to Bio-Organic Chemicals, Inc. describe a composition for substantially lessening or eliminating efflorescence in Portland cement products. The composition prevents bleeding out of salts present in Portland cement which lead to efflorescence, and produces improved surface smoothness, enhancement of color in products made of cement to which the composition has been added, and resistance to moisture penetration.

The composition is prepared by grinding Portland cement, either clinker or previously powdered, polyvinyl alcohol and barium hydroxide or barium oxide in a ball mill. Water and sand are added to the mixture to make concrete. A typical formula used as the base for the composition is the following:

Polyvinyl alcohol	1 g
Barium hydroxide	0.2 g
Portland cement	100 g
Sand	300 g
Water	44 cc

The polyvinyl alcohol in this formula is 1% of the Portland cement. Experiments were conducted in which the polyvinyl alcohol content was varied between ¼ and 5%; the desired effect began to appear at ¾% (0.75 gram of polyvinyl alcohol to 100 grams Portland cement).

In these experiments a light shade of Portland cement in the form of fine gray powder known as Type I or Grade A was used. Polyvinyl alcohols vary in molecular weight and viscosity and hence in solubility in water. Experiments showed that the best effect is obtained by the use of very sparingly soluble and insoluble alcohols, having high viscosity and high molecular weight. Elvanol 72-60 is an example of polyvinyl alcohol well suited for the process.

Coloring Agents Added Prior to Burning for Colored Cement

T.F. Tanner; U.S. Patent 3,667,976; June 6, 1972; assigned to General Portland Cement Company has found that Portland cement can be colored to a variety of different hues by mixing selected metals with the raw materials from which the cement is made prior to or during burning of the raw materials. The metals can be introduced in elemental or combined form. Generally, the metals are introduced in amounts from 0.2 to 2.0% by weight expressed as the oxide of the metal based on the total amount of dry raw materials which are introduced into the kiln. Burning temperatures at which the raw materials and coloring agent are reacted are conventional.

All formulations in the following examples are intended to produce a Type I general purpose Portland cement which has the color, tone and hue characteristic of the metal which has been added, although the coloring mechanism will be equally effective with other cement formulations.

Example 1: A cement raw material is prepared using a mixture of limestone, clay and iron ore with the following formulation.

Composition	Percent by Weight
SiO_2	14.0
Al_2O_3	3.6
Fe_2O_3	1.7
CaO	41.2
Loss on ignition	37.0

This mixture is altered by varying the addition of limestone and ilmenite, a mineral complex of the formula $TiO_2 \cdot FeO$. The ilmenite utilized in this particular example is obtained from Australia in a reprocessed ore form containing about 50% TiO_2 by weight. 1,000 grams of 7 different raw mixes are prepared in this manner and are shown in Table 1.

TABLE 1: RAW MIXES WITH Ti VARIED

	Samples (weight percent)						
	1	2	3	4	5	6	7
Composition:							
SiO_2	14.0	14.0	14.0	13.9	13.8	14.0	13.7
Al_2O_3	3.6	3.6	3.6	3.6	3.6	3.5	3.5
Fe_2O_3	1.7	1.7	1.7	1.7	1.7	1.7	1.7
CaO	41.2	41.2	41.1	41.0	41.4	41.0	41.1
Loss on ignition	37.0	37.0	37.0	36.5	36.7	37.1	37.0
Composition TiO_2	0.0	0.1	0.5	0.75	1.0	2.0	3.0

About 100 grams of each of the raw mixes is pellitized. 20-gram portions from each of these 100-gram samples are separately burned in an electric furnace for 20 minutes at temperatures varying from 2500° to 3000°F. After burning for 20 minutes, the samples are removed at once from the furnace and allowed to cool by normal radiation and convection to room temperature. The samples are ground with mortar and pestle and visually compared for color. The results are set forth in Table 2, using the following color code: G, gray; GY, gray with yellow trace; B, good buff hue; B/M, buff tone which is slightly muddy.

TABLE 2: COLOR RESULTS FROM EXAMPLE 1

	Sample Number						
Temperature, °F.	1	2	3	4	5	6	7
2,500	G	GY	GY	GY	GY	GY	GY
2,600	G	GY	GY	GY	GY	GY	GY
2,700	G	GY	B	B	B	B/M	B/M
2,750	G	GY	B	B	B	B/M	B/M
2,800	G	GY	B	B	B	B/M	B/M
2,900	G	GY	B	B	B	B/M	B/M
3,000	G	GY	B	B	B	B/M	B/M

This is repeated using rutile and purified titanium dioxide, rather than ilmenite. The results are substantially the same.

Example 2: A cement raw mix weighing 175 pounds was prepared using a mixture of limestone, clay, iron ore and ilmenite ore. The chemical composition of the cement raw mix is set forth below.

Composition	Percent by Weight
SiO_2	13.4
Al_2O_3	3.7
Fe_2O_3	2.2
CaO	42.3
MgO	1.4
Loss on ignition	35.2
TiO_2	1.2

This raw mix is pellitized and burned in a 5" x 14" x 10' long laboratory rotary kiln. Samples of the clinker are taken and visually checked for color. The color is recorded at 30-minute intervals. In addition, the temperature of the kiln burning zone is taken and recorded at 1-hour intervals. The run was continued for about 7 hours. The results are set forth in Table 3.

TABLE 3: COLOR RESULTS OF EXAMPLE 2

	Clinker		
	Production rate, lb./hr.	Temperature leaving burning zone, ° F.	Color
Hour:			
1		2,700	Buff.
2	4.60	2,630	Do.
3	6.44	2,590	Yellow-gray.
4	4.27	2,640	Buff.
5	4.60	2,640	Do.
6	5.10	2,645	Do.
7	4.84	2,620	Do.

As can be seen from the foregoing data, when the burning temperature reaches a point below 2600°F, the clinker retains some of its natural gray tone. When the temperature is above 2600°F, a good quality buff clinker is produced.

Burning Technique Using Metal Oxides

T.F. Tanner; U.S. Patent 3,799,785; March 26, 1974; assigned to General Portland Cement Company describes an improved method for producing a colored cement which comprises initially burning argillaceous and calcareous materials with a sufficient amount of a metal oxide coloring agent in the burning zone of the kiln to produce a cement clinker and thereafter maintaining it at a temperature above about 2600°F prior to rapidly quenching the same. Preferably, the material passing from the burning zone in the kiln is heated by an auxiliary burner adjacent the outlet thereof to maintain the material at a temperature above about 2600°F as it is passed to a quenching zone. Preferably, the material is rapidly quenched with water.

The metals which have been discovered to result in a desirable and predictable hue and color in a cement clinker are those which appear in Groups IV-b, V-b and VII-b of the Periodic Table. Preferred among these metals are titanium, zirconium, manganese and vanadium. Addition of titanium to the raw materials will produce a cement clinker which has a buff color, that is, it will produce a yellowish hue in the final cement clinker. Addition of manganese to the raw materials will produce a clinker with a definite blue tone. Addition of vanadium will produce a silver-gray bordering on a green tone. Zirconium will again produce a buff-colored cement clinker.

It has been found that the color and hue is somewhat dependent upon the reaction or burning temperatures. For example, when titanium is mixed with raw materials, a reaction temperature of between 2500° and 2700°F will produce a grayish buff or grayish yellow cement clinker. However, when the reaction temperature ranges between 2600° and 3000°F, a definite buff tone and very desirable hue will be present. Preferred percentages of titanium to be added to the raw materials, based on the dry weight of the raw materials, range between 0.3% by weight (expressed as titanium dioxide) to about 2.0% by weight. The most preferred burning zone temperatures or reaction temperatures range between about 2650° to 3000°F.

Additionally, it has been found that quenching the clinker produced by adding the metals set forth above will additionally fix and produce a brighter hue of the same tone as is produced by merely adding the metal alone. Quenching can be accomplished by an air blast upon the hot clinker as it leaves the kiln, or can be accomplished by other quenching media such as steam or water. Normally, the clinker will cool to ambient temperature in about 10 to more than 20 minutes. If the clinker is quenched from its burning temperature of around 2700°F to near room temperature in about half that time, that is, from 5 to 12 or so minutes, the clinker will take on a very definite bright, esthetically desirable hue and color characteristic of the metal which has been added to the raw materials.

As a general guideline, the desirable quality contributed by quenching will occur if the clinker is cooled about 1.5 or more times faster than occurs in conventional production. Generally, quenching in a time span of about 25 to 75% of that which the clinker normally takes to cool to room temperature will produce the desirable increase in hue and color fixation.

WATER REPELLANT

Tertiary Aliphatic Monocarboxylic Acids

W.K. Striebel and W. Loch; U.S. Patent 3,656,979; April 18, 1972; assigned to Dyckerhoff Zementwerke AG, Germany describe a hydrophobic cement especially suited for soil stabilization which comprises a certain amount of a mixture of tertiary, saturated, aliphatic monocarboxylic acids of certain carbon atom content.

Soil consolidation or stabilization increases the load-carrying capacity of a soil so that the soil is capable of withstanding the stress which is due to the traffic load and weather. The load-carrying capacity of a soil can be increased either by admixture of other suitable soil types which have the desired characteristics or by mixing with cement. After mixing the soil must be consolidated or compacted while maintaining the optimum water content determined by physical testing of soil.

One of the most important tests used for this purpose is what is known as the Proctor test which determines the relationships between the water content and the dry density of a disturbed soil during constant consolidation work. The highest dry density which can be achieved in this test is referred to as the Proctor density and the associated water content is referred to as the most favorable or optimum water content.

It is known to add water repellent materials to cement in cement production to improve the stability of finely ground cement in storage so that it can be stored unprotected in bags at the construction site without deteriorating. Various carboxylic acids or salts also in mixture with fats, phenols, etc., have been proposed as agents providing water repellency. See German Patent DAS 1,239,605. When using these hydrophobic cements for soil stabilization, the effect of the additives is not always satisfactory chiefly because natural fatty acids or derivatives thereof are degraded in the soil by bacteria.

This process provides a hydrophobic cement which comprises a highly branched and predominantly tertiary, saturated, aliphatic monocarboxylic acid having from 5 to 20, and especially from 9 to 11 carbon atoms in the molecule. The content of acid is in an amount of from 0.05 to 2% by weight, preferably 0.1 to 0.5% by weight.

It has been found that these monocarboxylic acids which are predominantly tertiary combine a very satisfactory water repellent effect with outstanding properties when used as additive in soil stabilization. They are not attacked by bacteria in the soil. Stability is developed rapidly and the stabilized soil retains its stability and frost resistance even over extended periods of time.

The advantages of the cement which is ideally suited for soil stabilization are illustrated by the following example. A PZ 375 cement and a cement according to the process having added 0.2% of a mixture of tertiary monocarboxylic acids having from 9 to 11 carbon atoms, sold under the name Versatic 911, were tested. When tested according to DIN 1164, these cements had the following properties:

Comparative Properties of Cements

	PZ 375	PZ with Additive
Specific surface area, sq. cm./g.:	3,350	3,330
Setting, hours:		
beginning	2 1/3	2 2/3
end	3 1/4	3 1/3
Flexural tensile strength/compressive strength, kp./sq. cm.:		
3 days	7/267	61/275
7 days	75/387	83/398
28 days	86/474	82/473

It is apparent from the above data that the treated cement has highly satisfactory rapid setting properties, and the requisite strength properties. Likewise, when 0.05, 0.5, 1 and 2% by weight of such mixture of acids is incorporated into Portland cements PZ 275 and 475, the treated cement has satisfactory properties.

The above prepared two cements were used to stabilize two types of soil, viz., a sandy soil having a nonuniformity coefficient of 2.5 and a cohesive soil. The amount of cement added was 10% by weight, in each case corresponding to a cement consumption of 25 kg/m^2 with a layer depth of 15 cm.

Compressive Strengths of the Stabilized Sandy Soil

Cement	7 Days, kp./sq. cm.	28 Days, kp./sq. cm.
PZ 375	47	58
PZ with additive	59	74

Compressive Strengths of the Stabilized Cohesive Soil

Cement	7 Days, kp./sq. cm.	28 Days, kp./sq. cm.
PZ 375	37	52
PZ with additive	45	67

The data show that with the cements according to the process, the compressive strength of both soils is increased by about 30%, when adding the same amount of cement. Freeze-thaw tests carried out according to ASTM D-560 on 7 cm cubes showed the following results after 25 freeze-thaw cycles.

	Loss in Weight, %		Compressive Strengths, kp./sq. cm.	
Cement	Sandy Soil	Cohesive Soil	Sandy Soil	Cohesive Soil
PZ 375	18	26	22	14
PZ with additive	2	3	68	62

After the freeze-thaw treatment, the decrease in compressive strength with the use of PZ 375 was 62 and 73%, respectively. With the cement according to the process, the loss in strength was only 8 and 7%, respectively. The data also show greater stability of the soils, as evidenced by smaller loss in weights, even for the cohesive soil.

Animal Fat and Fatty Alcohol

R.L. Johnson; U.S. Patent 3,547,665; December 15, 1970; assigned to Fats and Proteins Research Foundation, Inc. describes compositions for adding to concrete, which impart water repellency without any loss of strength, which develop plasticity without the cost of an additional ingredient, and which promote a general increase in the strength of the cured product. A typical dispersion includes the following ingredients:

	Parts by Weight
Inedible animal fat	2.0
Water	18.4
Calcium chloride	2.2
Calcium hydroxide	2.0
Silica flour	11.2
Calcium carbonate	1.0

Such a composition has been usefully employed when added to concrete and mortar mixtures at levels which have been calculated to provide from ¼ to 5 lb of the fatty substance per each 100 lb of dry Portland cement.

W.A. Cordon; U.S. Patent 3,486,916; December 30, 1969 describes a mixture for hydraulic cement mixes which comprises an emulsion in water of a fatty alcohol having from 14 to 20 carbon atoms. The emulsion is added to a mix before pouring and concentrates in the bleed water of the mix after pouring to form an evaporation-inhibiting film.

Alkyl Polysiloxane and Magnesium Carbonate

A process described by *H. Weissbach; U.S. Patent 3,318,839; May 9, 1967; assigned to Farbenfabriken Bayer AG, Germany* involves the manufacture of water repellent products of lime or cement mortar. Such products containing alkyl polysiloxane in powdered or liquid form are known. This siloxane content leads to a very effective hydrophobic property, in most cases already in quantities of less than 1% by weight of dry building material composition, but it has been found in the meantime that under unfavorable hardening conditions, for example, in cold wet weather, there are formed on the surface of the finished products whitish films of calcium carbonate which may substantially impair their appearance. The removal of such so-called efflorescences, for example, from a facade plaster, is hardly possible without damaging it and involves in any case substantial expenditure.

It has been found that the occurrence of these efflorescences can be obviated by adding to the building material composition of known type, besides the hydrophobic alkyl polysiloxanes, a small amount of commercial, finely powdered magnesium carbonate. The necessary amount varies according to the composition of the building material mixture, especially its content of alkyl polysiloxane. Normally 5% by weight at most are required, in most cases an amount between 1 and 3% by weight, referred to the dry composition, is sufficient. The addition of these quantities not only prevents the efflorescence, but also improves the water repellent effect, without noticeably changing the strength and other properties.

WORKABILITY

Dispersed Cellulose Fibers

In concrete compositions based on hydraulic cement, improved properties in respect to workability, especially moldability, capacity for being impelled along a pipe as by pumping, and frost resistance in the concrete, are obtained by incorporating in the composition a minor proportion of dispersed cellulose fibers such as papermaking fibers, according to *W.O. Nutt; U.S. Patent 3,753,749; August 21, 1973; assigned to The Cement Marketing Company Ltd., England.*

In particular the incorporation of cellulose fibers in a concrete mix provides a lubricating effect for the surface of the particles of large aggregate material and enables such modified concrete to be pumped without the necessary increase of either the sand, cement or water proportion. Such a concrete composition containing cellulose fibers offers improved workability properties, particularly an improved capacity for molding and is moreover found to perform at least comparably in resistance to freezing and thawing cycles, to air-entrained concrete.

The incorporation of fiber increases the overall specific surface area of the concrete mix and this in turn limits the quantity of fiber it is practical to include in the mix. This is because with increasing specific surface area the concrete mix demands higher ratios of water to cement in order to achieve practical workability. Each fiber appears to have specific properties in this respect and the following examples illustrate the effects of fiber additions.

Mix 1 (Control Mix):

	Parts by Weight
Coarse river flint aggregate	4
River sand	2
Portland cement	1
Water to provide a water/cement ratio of 0.6	

Mix 2:

As Mix 1 but with a commercially available air-entraining agent containing ionic and nonionic surfactants added at the rate of 1% by weight based on the cement. This mix typifies normal air-entrained concrete.

Mix 3:

As Mix 1 but including the addition of 2% by weight of the cement content of a fiber prepared from coniferous wood pulp as manufactured for white papermaking and comminuted dry in a hammer mill to provide an average fiber length of 2 mm. This pulp was added dry to the dry components of the mix and distributed throughout the concrete composition prior to the addition of water.

Mix 4:

As Mix 3 but with additional water to provide a comparable workability to Mix 1.

The following workability measurements were established according to BS 1881: 1970 testing procedures.

	Mix 1	Mix 2	Mix 3	Mix 4
Compacting factor	0.92	0.95	0.75	0.89
Slump	¾ inch	3½ inch	Nil	¾ inch
Vebe consistometer	4 sec	1.5 sec	40 sec	6 sec

The following compressive strength results were obtained by crushing 4-inch cubes (average of 3 per test) cured in accordance with the conditions of BS 12: 1958.

	Mix 1	Mix 2	Mix 3	Mix 4
7 days age, lbf/in^2	4,850	3,800	4,250	4,200
28 days age, lbf/in^2	5,650	4,950	5,750	3,300

From these results it will be observed that the inclusion of air in Mix 2 reduces its strength over Mix 1 whereas the inclusion of fiber in Mix 3 does not significantly reduce its strength over Mix 1 and materially improves its strength over Mix 2; the strengths quoted for Mix 4 illustrate the effect of high water/cement ratio in comparison with Mix 3. However, the above cube strengths should be read in conjunction with the density of the concretes produced which were tested to give the following results.

	Mix 1	Mix 2	Mix 3	Mix 4
Density lb/ft^3	148.75	144.5	148.25	144.25

From these it will be observed that Mixes 1 and 3 are comparable whereas Mix 2 by inclusion of its air has a lower density than the control mix.

Tests were conducted to establish the air content using an air meter, following BS 1881:1970 testing procedures.

	Mix 1	Mix 2	Mix 3	Mix 4
Air content %	0.8	4.0	1.9	1.1

Although the cited procedure does not contemplate fiber inclusion, these results clearly indicate the higher air content of Mix 2 which is not of the same order indicated for Mix 3 containing fiber.

Cubes molded from the above mixes were subjected to a series of freezing and thawing cycles in an attempt to establish the frost resistance of the concrete. The tests commenced when the concrete cubes had been cured under water for 7 days. Initially the method of test entailed freezing at -10°C for 24 hours cubes which were immersed in trays of water to half their depth. The frozen trays were then thawed at room temperature (18°C) for 24 hours; this procedure constituted one cycle.

After 15 cycles of the above test no signs of deterioration of the surface of the specimens were apparent. The specimens were therefore treated more severely by being thawed in an oven at 80°C for 24 hours instead of thawing at room temperature. After 12 such cycles specimens from concrete Mix 4 began to show signs of deterioration to an extent which was judged to represent severe spalling of the surface of a concrete pavement.

After a further 30 cycles specimens from concrete Mixes 1, 2 and 3 were examined; by this time these specimens had received 15 cycles of the first described test and 42 cycles of the more severe second test. Specimens from Mix 1 showed spalling at the edges and corners to a degree greater than specimens from Mixes 2 and 3; particularly specimens from Mix 3 showed no sign of deterioration. The most reliable indications are that concrete made from Mix 3 would not provide worse performance in practice than concrete made from Mix 2. Concrete made from Mix 2 is known to perform satisfactorily in general application once allowance has been made in its design use for the reduced strength and density resultant upon its air entrainment.

The conclusion is therefore that fiber inclusion in pavement quality concrete enables full concrete density and strength to be achieved while still affording protection against freezing and thawing. This aspect has material significance in the design of adequately durable concrete roads, aircraft runways and similar paved areas. The following example illustrates the process.

Example: A particular concrete specification which called for a nominal volume mix design of 1:4:8, of cement:sand:stone, could only be placed according to the prior art when formed from the following mix composition:

Cement	215 lb
Sand containing 7% moisture	643 lb
Crushed stone graded below ¾ inch and containing 1% moisture by weight	985 lb
Water	85 lb

This provided a nominal mix volume proportion of 1:2.5:4.5, which is uneconomical in comparison with calculations based on the nominal 1:4:8 composition. However, a mix suitable for placement by pumping was composed as follows, representing a nominal mix volume proportion close to the original specification (1:4:7:5):

Cement	130 lb
Sand containing 7% moisture	643 lb
Crushed stone graded below ¾ inch and containing 1% moisture by weight	985 lb
Tissue paper	1½ lb
Water	70 lb

It has further been found that the addition of cellulose fiber to concrete mixes which are currently considered to be capable of pumping improves their pumping transport characteristics. In a typical site installation by which a standard 1:2:4 concrete mix was being pumped, the addition of tissue paper fiber as described reduced the power required by the pump to transport the concrete through the pipeline, by 10 to 20%. Similarly sand/cement mortars and comparable compositions used for screeds, plasters and mortars are more easily pumped through narrow bore distribution lines when paper fibers are incorporated as described above. It will be appreciated that the process thus provides for substantial reduction of the power required to transfer a concrete mix through a pipeline, whether in the sense of thereby making such transfer possible or merely rendering it easier.

In a broad aspect, therefore, the process provides a method of transporting a concrete mix from a mixing site to a placement site which comprises dispersing in the mix a minor proportion of a cellulosic material in segregated fiber form whereby the power required to transfer the mix through a pipeline is reduced and thereafter conveying the fiber-modified mix through a pipe.

Hydroxyethylcellulose

J.B. Batdorf and S.M. Weiss; U.S. Patent 3,788,869; January 29, 1974; assigned to Hercules Incorporated describe the use of hydroxyethylcellulose (HEC) as an additive for concrete compositions or mixes, substantially dry concrete compositions comprising conventional ingredients plus HEC, and concrete compositions comprising conventional ingredients (including water) plus HEC. The HEC renders the concrete compositions (when water is added) readily pumpable and gives percent total air values within American Concrete Institute specifications.

A typical, readily-pumpable concrete mix comprises the amounts of ingredients per cubic yard as follows: (a) ⅛ to 1½ lb of hydroxyethylcellulose having an MS range of about 1.8 to 3 and a viscosity from about 5,000 cp at 1% concentration in water to about 6,500 cp at 2% concentration, (b) 800 to 6,500 lb of coarse aggregate, (c) 700 to 2,200 lb of fine aggregate, (d) 380 to 895 lb of Portland cement and (e) 30 to 48 gallons of water.

Ball Clay and Sodium Carbonate

B.E. Dunworth and W.T. Bakker; U.S. Patent 3,817,770; June 18, 1974; assigned to General Refractories Company describe a technique and composition which afford substantial improvement in reducing rebound of gunnable hydraulic concrete compositions. Rebound is reduced for concretes containing calcium aluminate cements from on the order of 25 to 50 weight percent to about 10 to 15 weight percent. Corresponding and similar improvements are attained with other hydraulically settable concretes. The technique and composition comprise adding about 0.05 to 0.3 weight percent of an inorganic salt of an alkali metal ionizable in water and 0 to 2.0 weight percent ball clay to the concrete formulation.

Example: A plurality of hydraulic concrete formulations, based on a mixture of 17.5 weight percent of high alumina calcium aluminate cement and 82.5 weight percent aluminum silicate aggregate was projected by a dry pneumatic gun and mixed with water in an approximate weight ratio of 2.25:1, based on the weight ratio of cement-to-water, and the rebound loss was measured. The formulations were varied as illustrated in the following table, where the rebound losses are also shown. All percentages in the table are by weight, with the amounts of additives based on the total weight of the concrete formulation.

Run	Additive	Amount	Rebound loss
1	None		25
2	Ball clay	2.0	20
3	Na_2CO_3/ball clay	0.2/2.0	10
4	do	0.15/2.0	13
5	Na_2CO_3	0.175	15
6	$NaNO_2$	0.20	11
7	$NaNO_2$/ball clay	0.20/2.0	11
8	$NaNO_2$	0.15	14
9	$NaNO_2$/ball clay	0.15/1.0	11
10	do	0.15/2.0	15

It is readily apparent from the table that substantial improvement is effected by the process, where rebound losses are reduced by 40 to 60%. It is likewise apparent that substantial improvement can be effected by the inclusion of the inorganic alkali metal salts alone, but the inclusion of up to about 2.0 weight percent ball clay can also be of benefit and is accordingly a preferred ingredient.

While the foregoing operations in the example are confined to calcium aluminate, the process is equally applicable to other hydraulic cements as well. Similarly, still other alkali metal inorganic salts could be employed, such as, for example, sodium nitrate, sodium phosphate, sodium silicate, sodium chloride, and the corresponding salts of other alkali metals, notably potassium. Sodium and potassium salts are preferred, principally for economic reasons.

BINDERS, FILLERS AND OTHER ADDITIVES

Clay-Containing Binder for Furnace Lining

According to a process described by *J. Rafine; U.S. Patent 3,915,719; October 28, 1975; assigned to ECOTEC, SARL, France* concrete comprises a binder containing 3 to 12% of an aluminous cement having little free lime, from 4.56 to 9.73% of a phyllite clay having a base-exchange capacity of at most 10 meq per 100 grams and a flake-like structure consisting of at most two layers so that when water is added the mixture flows according to Newton's laws instead of Binghame's equation. The percentages given are based on the total weight of the concrete.

The process relates particularly to the use of a binder for use in making refractory concretes for lining furnaces and imparts to these concretes advantageous properties with respect to their resistance to compression, both before and after firing, as well as improved mechanical properties which are necessarily different from those of certain refractory coatings intended to be simply sprayed onto the subjacent refractory linings of the furnaces.

Two examples of a refractory concrete according to the process are described below. One uses as the aggregate a burnt expanded clay of German origin and a binder containing Rollandshutte standard aluminous cement, while the other utilizes corundum base aggregates and a binder containing aluminous cement known as Secar 250. The composition of the aluminous cement is as follows:

	Percent
$Al_2O_3 + TiO_2$	46 to 53
CaO	37 to 42
SiO_2	6 to 99
FeO	0.4 to 1
MnO	0.3
MgO	1.5 to 2
S	1
SO_3	0.4

This aluminous cement is of the fused type and is light in color. It is made in a blast furnace in a reducing atmosphere. Its fusion point is 1380°C and its apparent density 1.22 g/cm^3. Such a cement is useful for making concretes which may be worked at temperatures of as low as -15°C. The Secar 250 cement is of the calcium aluminate type, of high purity and white color. Its composition is as follows: 70% alumina; 29% lime; and 1% impurities, such as iron, silicon, etc. It will not collapse under a pressure of 2 kg/cm^2 until heated to 1300°C. Its MP is 1650°C. Its apparent density is 1.08 g/cm^3.

Example 1: A refractory concrete having an expanded burnt clay base has the following composition:

Aggregates* –	Granules, 1-2 mm diameter	550 kg
	Granules, <1 mm diameter	280 kg
Binder –	Rollandshutte standard cement	100 kg
	Kaolin clay	70 kg
Additives –	Sodium hexametaphosphate	620 g
	Sodium pyrophosphate	620 g
Mixing water –		18-22 l/100 kg

*Expanded burnt clay (German), 0-2 mm (containing >35% Al_2O_3)

A concrete with a binder according to the process as in the above example has the following characteristics:

Apparent density	1.10-1.20 g/cm^3
Resistance to compression while cold	50-70 kg/cm^2

(These two characteristics are those of a raw, dry concrete.) After firing at 1300°C, this concrete has the following properties:

Apparent density	1.10-1.20 g/cm^3
Resistance to compression	135-190 kg/cm^3
Dimensional variations	-1 to -2%

It should be noted that in this example the raw concrete has been tamped in place.

Aggregates	-	Corundum particles, 1-7 mm diameter	750 kg
		Corumdum particles, <1 mm diameter	100 kg
Binder	-	Cement (Secar 250 Lafarge)	50 kg
		Kaolin clay	50 kg
Additives	-	Sodium hexamethaphosphate	500 g
		Sodium pyrophosphate	500 g
Mixing water	-		6-8 l/100 kg

The binder, when used according to the principles and the proportions of the process imparts the following characteristics to the concrete:

	Raw and Dried	After Baking at 1650°C
Apparent density	3-3.10 g/cm^3	3.10-3.20 g/cm^3
Resistance to compression	600-800 kg/cm^2	1,400-1,800 kg/cm^2
Porosity	12-15%	12-15%
Dimensional variations	–	-1 to -2%

In this example the raw concrete has been put in place with a vibrator. It will nevertheless be noted that when a dense concrete is to be produced, the fact that perfect liquefaction of the clay can be obtained with a minimum amount of water makes it possible to produce concretes of greater density. Moreover, in the case of light concretes, the choice of clay used may make it possible to decrease the quantities of cement which must be incorporated in the binder.

Phosphorus Furnace Slag

A. Schneider-Arnoldi, H. Gäbler and J. Kandler; U.S. Patent 3,825,433; July 23, 1974; assigned to Knapsack AG, Germany describe the production of a cement-based hydraulic binder containing from 10 to 80% by weight slag ground to cement fineness and originating from the electrothermal production of phosphorus, and use of the hydraulic binder for making massive concrete.

The electrothermal production of phosphorus from phosphates is known to entail the formation of slag, which is called phosphorus furnace slag and has a chemical composition different from that of Portland cement or blast-furnace slag, with respect to its individual constituents. Phosphorus furnace slag has long been held to be unsuitable for the production of hydraulic binders, particularly in view of the very slight proportions of aluminum oxide contained therein. Cement, more particularly Potland cement, containing from 10 to 18% by weight, preferably from 15 to 70% by weight, of slag, ground to cement fineness and originating from the electrothermal production of phosphorus, has unexpectedly been found to be a hydraulic binder very suitable for use in massive concrete. In concrete produced with the hydraulic binder of the process, the increase in strength occurs more slowly than in concrete made with pure Portland cement or blast-furnace cement, but this is accompanied by an extremely long after-hardening period for the concrete, which is substantially longer than that of concrete produced with standard cements and incomplete even after 360 days.

It has been found that concrete made with the binder of the process has a strength greater than that of concrete produced with pure Portland cement, after 360 days. However, the most important property of the hydraulic binder resides in the fact that concrete produced therewith has a heat-evolution rate considerably lower, particularly during the first days, than that of concrete produced with standard cements. In other words, the concrete mixtures so made are of considerable interest for the production of concrete bodies having large dimensions.

The phosphorus furnace slag can be mixed with the Portland cement either by grinding the two binder components separately and then mixing them, or by grinding the phosphorus furnace slag together with the Portland cement clinker. No difference in the strength of the concrete has been observed. It has already been proposed to use foamed phosphorus furnace slag as an additive in the production of building materials, particularly lightweight concrete. However, this is granular phosphorus furnace slag which has no hydraulic properties. With this in mind, it is all the more an unexpected result that ground phosphorus furnace slag does possess such hydraulic properties.

Example 1: Phosphorus furnace slag having the following composition was ground so as to have a Blaine index of 3,470 cm^2/g and then mixed with Portland cement PZ 275, which had a Blaine index of 2,490 cm^2/g in the ratio of one part phosphorus furnace slag to two parts Portland cement. The symbol PZ 275 stands for a Portland cement with a crushing strength of at least 275 kg/cm^2 after a hardening period of 28 days.

	Percent
CaO	48
SiO_2	42
Al_2O_3	2.8
MgO	0.3
Fe_2O_3	0.3
F (bound)	3.0
S (bound)	0.4
P_2O_5	1.5

The following test data were obtained. The test data for PZ 275 are indicated for the purpose of comparison. The crushing strength was determined in this and the following examples on prisms with edges 4 x 4 x 16 cm long, in accordance with testing method DIN 1164.

	PZ 275	Mixture (1:2)
Crushing strength, kg/cm^2		
After 28 days	413	200
After 365 days	599	817
Heat of hydration, cal/gram		
After 1 day	59	8.8
After 7 days	79.6	38.4

Example 2: The ground phosphorus furnace slag of Example 1 was mixed with the Portland cement PZ 275 of Example 1, in the ratio of 1:1. The following data were obtained.

	PZ 275	Mixture (1:1)
Crushing strength, kg/cm^2		
After 28 days	413	280
After 365 days	599	741
Heat of hydration, cal/gram		
After 1 day	59	21.8
After 7 days	79.6	41.1

Example 3: The phosphorus furnace slag of Example 1 was mixed with Portland cement PZ 375, which had a Blaine index of 4,140 cm^2/g, in the ratio of 1:1. The following test data were obtained. The symbol PZ 375 stands for a Portland cement with a crushing strength of at least 375 kg/cm^2 after a hardening period of 28 days.

	PZ 375	Mixture (1:1)
Crushing strength, kg/cm^2		
After 28 days	543	389
After 365 days	657	765
Heat of hydration, cal/gram		
After 1 day	79.5	38.7
After 7 days	93.7	58.0

Boron Modification Clinker

According to a process described by *T.C. Slater and F.C. Hamilton, Jr.; U.S. Patent 3,861,928; January 21,1975; assigned to The Flintkote Company* ground mixtures of conventional Portland cement-making components are mixed with relatively small amounts of boron-containing components, and the mixtures are burned at kiln temperatures substantially below those normally utilized to form clinker. The resultant clinker is more easily grindable than that produced in conventional processes of Portland cement production, and the cement resulting from the grinding of the clinker yields cementitious products possessing substantially higher compressive strength than is exhibited by Portland cement made from the same components in the absence of the boron-containing component. The boron-containing additive used in the process is an oxide of boron, a compound containing an oxide of boron or a compound that yields an oxide of boron at the temperatures prevailing in a cement kiln. Examples of such boron-containing additives are boric acid anhydride, acids of boric acid anhydride, salts of boric acid anhydride, polyboric acids and polyborates such as diborates, triborates and tetraborates of ammonium, sodium, potassium, calcium, barium, strontium and magnesium, and boron-containing refined and unrefined minerals or ores, including colemanite, ulexite, danburite, pinnoite, ascherite and rasorite.

Boron trioxide (B_2O_3), raw colemanite and refined colemanite are preferred. The quantity of the boron-containing additives incorporated in the pulverized mixture of raw ingredients fed to the kiln is relatively small. An amount sufficient to provide a B_2O_3 content in the range of about 1 to 3% by weight of the pulverized mixture, preferably about 1.5 to 2.5%, should be used, which amounts approximately can then be found in the clinker. In an actual test, a modified typical Type II Portland cement pulverized mix was made up with the following components and designated A Modified:

A Modified	Percent
Cataract limestone	72.58
Cataract shale	16.60
Winship clay	3.67
Pyrite cinders	2.39
Special Additive No. 1	4.76

Special Additive No. 1 contained:

	Percent
SiO_2	8.00
Fe_2O_3	0.19
Al_2O_3	0.35
CaO	14.66
MgO	2.67
Na_2O	5.19
K_2O	0.23
B_2O_3	36.13
Ignition loss	32.57

Thus, the mix had the following calculated composition:

	Percent
SiO_2	22.55
Al_2O_3	5.82
Fe_2O_3	4.45
CaO	64.85
B_2O_3	1.86

Batches of the foregoing A Modified mix were burned at various temperatures for a period of time equal to a typical Portland cement kiln retention time at normal operation temperature of about 2800°F. The clinker prepared from each such batch contained free lime in the amount shown in the table below. The data clearly show that clinker having a desirably low level of free lime can be prepared at burn temperatures in the range of about 2350° to 2550°F, a range of temperature markedly lower than the temperatures usually employed to make

Portland cement clinker having satisfactory low free lime content.

Burn Temperature °F.	% Free Lime
2150	11.11
2250	4.24
2350	1.76
2450	0.77
2550	1.98
2650	2.26
2750	1.60
2850	1.27

The production of a satisfactory cement clinker thus is possible at reduced kiln temperatures, and substantial savings in fuel cost as well as other benefits can be obtained. Fuel savings that amount to the order of about 300,000 Btu per barrel of cement clinker produced have been calculated. Also, by virtue of the lower temperatures utilized in the kiln operation, longer kiln shelf life and less frequent replacement of the refractory lining of the kiln are benefits that result in substantially reduced kiln maintenance cost. This cost reduction has been estimated to be on the order of about 10%.

In addition to and perhaps even more important than the foregoing advantages accruing from the use of the described boron-containing additives is the improvement thereby obtained with respect to the compressive strength of the resultant cement. As a control, a typical Type II Portland cement mix without modification by an additive component of this process was burned at about 2800°F for about 30 minutes. The mix had the following raw materials and calculated compositions:

A Control	Percent
Cataract limestone	76.21
Cataract shale	17.43
Winship clay	3.85
Pyrite cinders	2.51
Mix Composition	**Percent**
SiO_2	22.50
Al_2O_3	5.89
Fe_2O_3	4.52
CaO	65.17

The free lime content of the clinker was 0.72%. Compressive strength tests (ASTM Test C 109-64) were conducted on mortar cubes 2 x 2 x 2 inches made of the A Control cement described above and the A Modified cement according to this process from clinker burned at 2450°F. The cubes were kept in a cabinet in which the atmosphere was controlled at 70°F and 100% relative humidity. The compressive strength of the cubes was tested after 7 and 28 days.

The table shown on the following page presents the results obtained and various factors involved in making the tests for the cements described above and also for another cement according to this process (B Modified) which is described on the following page.

	A Modified	A Control	B Modified
Clinkered temp (°F)	2450	2800	2550
B_2O_3	1.86	-	1.54
Free lime (%)	0.77	0.72	2.17
Blaine fineness	3591	4195	3585
SO_3 added (%)*	1.52	1.94	0.59
Water (cc)	230	250	230
Initial set (hr:min)	4:20	1:10	2:30
Final set (hr:min)	7:00	3:40	7+
Compressive strength:			
7 days (psi)	4,558	3,133	4,308
28 days (psi)	9,375	5,267	9,258

*As gypsum to make cement.

The B Modified cement was prepared from a raw material mix containing 95.24 parts by weight of the A Control cement raw material mix and 4.76 parts by weight of Special Additive No. 1, and had a calculated composition as follows:

	Percent
SiO_2	21.40
Fe_2O_3	4.30
Al_2O_3	5.60
CaO	63.57
B_2O_3	1.54
Na_2O	0.26
Insolubles	0.82
H_2O	0.82

Clinker was formed by burning the raw mix at 2550°F for about 30 minutes, and had a free lime content of about 2.17%.

Chromium Oxide

A process by *T. Uno and M. Mochizuki; U.S. Patent 3,649,316; March 14, 1972; assigned to Onoda Cement Company, Ltd., Japan* provides Portland cement having one-day compressive strength upwards of 150 kg/cm². The mineral composition of early strength Portland cement clinker consists of alite and interstitial materials which consist of glass phase, C_4AF phase and C_3A phase (C, A and F represent CaO, Al_2O_3, and Fe_2O_3 respectively), and it is generally known that what contributes to the strength of Portland cement clinker is mainly alite. Alite is the solid solution of C_3S (S represents SiO_2), containing solid solution of such impurities as MgO, Al_2O_3 and Fe_2O_3 up to approximately 1 to 2% by weight. The strength of clinker varies notably according to kinds of such impurities in solid solution.

Accordingly, studies have been conducted with a view to producing extra high one-day compressive strength Portland cement, by including solid solution of Cr_2O_3 in addition to the solid solution of the above-mentioned impurities in the composition of alite, so as to improve its hydraulic characteristic.

Studies to produce early compressive strength Portland cement from clinker which is produced by adding Cr_2O_3 to raw materials for clinker and burning the mixture have already been reported, but those studies point out that the maximum three-day compressive strength is attained where Cr_2O_3 content in clinker is in the neighborhood of 1% by weight, and that early compressive strength of Portland cement cannot be obtained, if Cr_2O_3 content in clinker is increased or decreased from the above-mentioned percentage (1%).

After systematic studies in regard to the chemical composition of early strength Portland cement clinker it has been found that the maximum solid solubility of Cr_2O_3 is the highest in interstitial materials followed by belite, while alite is the lowest in interstitial materials and that where the content of Cr_2O_3 in alite is increased, one-day compressive strength of early strength Portland cement is augmented, and furthermore that where the content of Cr_2O_3 in clinker is 1.2 to 2.6%, one-day compressive strength, not to speak of three-day compressive strength, increased sharply. Figures 2.3a and 2.3b show the changes in the one-day compressive strength in accordance with variations in the Cr_2O_3 content, (Al_2O_3 + Fe_2O_3) content and particular lime saturation degree. The table below shows chemical analysis values of raw materials used to produce extra high early strength Portland cement.

	Percent										
Raw Materials	**Ig Loss**	**SiO_2**	**Al_2O_3**	**Fe_2O_3**	**Cr_2O_3**	**CaO**	**MgO**	**SO_3**	**Na_2O**	**K_2O**	**Total**
Limestone	43.7	0.3	0.3	0.1	-	55.2	0.6	-	-	-	100.2
Clay	7.7	60.5	16.5	17.1	-	1.5	3.2	-	1.88	1.42	99.8
Siliceous material	3.6	83.2	8.3	3.2	-	0.4	0.9	-	0.00	0.16	99.4
Copper slag	-4.4	35.8	8.1	47.4	-	6.3	2.4	-	0.60	0.83	97.0
Chromite	0.0	4.4	26.4	15.2	33.8	0.8	17.1	-	-	-	97.7
Gypsum	21.2	------0.2------			-	32.5	-	46.1	-	-	100.0
Serpentinite	12.2	39.1	32.8	11.1	-	2.0	33.3	-	-	-	100.5

Where the component materials shown above were used, Cr_2O_3 content was altered to meet the conditions noted. The mixed materials were pulverized so as to pass 99% by weight through an 88 μ sieve.

Condition 1: In order to make the composition of clinker substantially a mixture of C_3S, C_4AF and C_3A (note: C_2S is not included), the particular lime saturation degree (LSD) was worked out as follows:

(1) $$\frac{CaO + 1.39\ MgO}{2.80(SiO_2 + 0.79\ Cr_2O_3) + 1.65\ Al_2O_3 + 0.35\ Fe_2O_3} = 1.00$$

The particular LSD was worked out by the following calculation. On the supposition that the composition of cement minerals consists of C_3S-C_3A-C_4AF system and LSD of the composition is calculated by Bogue's method so as to be equal to 1.00, the LSD may be shown as follows:

(2) $$LSD = \frac{C}{3S + 4F + 3(A - F)} = \frac{C}{3S + 3A + F}$$

FIGURE 2.3: HIGH EARLY STRENGTH PORTLAND CEMENT

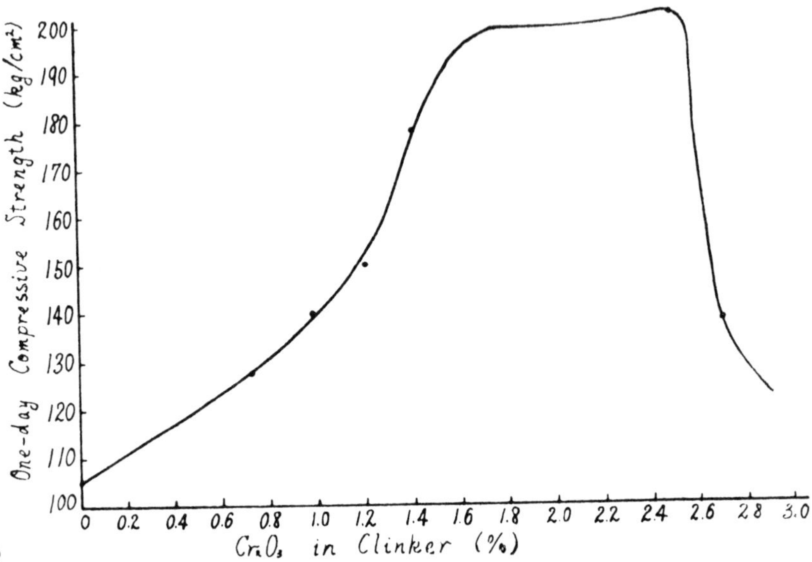

a.

Strength as a Function of Cr_2O_3 Content

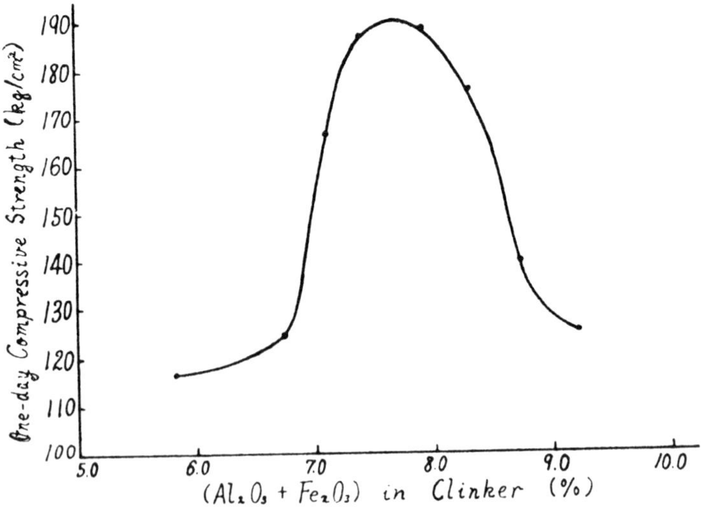

b.

Strength as a Function of Al_2O_3 + Fe_2O_3 Content

Source: U.S. Patent 3,649,316

In equation (2) C is CaO, S is SiO_2, A is Al_2O_3, and F is Fe_2O_3, expressed as mols. When Cr (i.e., Cr_2O_3) is added, it will form a solid solution in the form of $6CaO{\cdot}Cr_2O_3$, i.e., C_6Cr. By replacing C (i.e., CaO) in an equal mol ratio with M (i.e., MgO), then, accordingly, when M and Cr are introduced in the equation (2), the next equation (3) is worked out. The equation (3) is named particular LSD.

$$\text{(3)} \qquad \text{Particular LSD} = \frac{C + M}{3S + 6Cr + 3A + F}$$

where M is MgO and Cr is Cr_2O_3. When C, S, A, F, M, and Cr are represented by weight percent, the above equation (1) is induced from the equation (3).

Condition 2:

$$Al_2O_3 + Fe_2O_3 = 7.4\%$$

$$\text{Iron Modulus} = \frac{Al_2O_3}{Fe_2O_3} = 1.65$$

Mixed materials thus prepared were burned in an electric furnace at 1500°C for one hour, and to the clinker thus obtained gypsum (2.5% as SO_3) as shown in the table above was added. The mixture was then pulverized so that the cement had the specific area of 4,600 cm^2/g by the Blaine method. Strength test was conducted on the cement thus produced by the Japanese Industrial Standards, JISR-5201 (1964). The size of tested sample specimen was 4 x 4 x 4 cm. The results of strength tests are shown in the table below. Also changes in one-day compressive strength in accordance with the varied content of Cr_2O_3 are also shown in Figure 2.3a.

Sample Number	Percent SiO_2	Al_2O_3	Fe_2O_3	Cr_2O_3	CaO	MgO	Na_2O	K_2O	Total	SM	IM	Particular LSD	F-CaO, percent
1	22.0	4.6	2.8	0.0	68.3	1.5	0.44	0.36	100.0	3.0	1.6	1.00	0.6
2	21.4	4.6	2.8	0.7	68.2	1.5	0.37	0.29	99.9	2.9	1.6	1.00	0.5
3	21.2	4.6	2.8	1.0	68.3	1.5	0.29	0.24	99.9	2.9	1.6	1.00	0.6
4	21.1	4.6	2.8	1.2	68.5	1.5	0.26	0.18	100.1	2.8	1.6	1.01	0.6
5	20.9	4.6	2.8	1.4	68.5	1.5	0.23	0.17	100.1	2.8	1.6	1.01	0.5
6	20.7	4.6	2.8	1.6	68.6	1.5	0.21	0.15	100.2	2.8	1.6	1.01	0.6
7	20.5	4.6	2.8	2.0	68.2	1.5	0.15	0.10	99.8	2.8	1.6	1.00	0.9
8	20.2	4.6	2.8	2.5	68.7	1.5	0.08	0.03	100.4	2.7	1.6	1.00	0.7
9	20.0	4.6	2.8	2.7	68.5	1.5	0.05	0.01	100.2	2.7	1.6	1.00	2.5
10	19.8	4.6	2.8	2.9	68.6	1.5	0.02	0.00	100.2	2.7	1.6	1.00	2.4

Sample Number	Specific Surface (Blaine) cm^3/g	Compressive Strength, kg/cm^2 1 Day	3 Days	4 Weeks	Observation Through a Microscope Color of Alite
1	4.690	105	204	422	Colorless*
2	4,550	126	235	431	Very pale green*
3	4,570	140	256	445	Very pale green*
4	4,530	150	245	477	Pale green*
5	4,520	177	260	484	Pale green*
6	4,510	199	284	489	Green*
7	4,520	197	292	472	Green*
8	4,700	201	275	481	Green*
9	4,550	138	230	411	Green**
10	4,530	122	210	440	Green**

*Belite (C_2S) was not included.
**Belite and F-CaO was included.
Note: In Sample No. 1, when mixed material was burned for 30 minutes at 1500°C, Belite and F-CaO was included in clinker.

It is clear from the above data that clinkering was best facilitated, and Portland cement having one-day compressive strength upwards of 150 kg/cm^2 was obtained where particular LSD = 1.00, (Al_2O_3 + Fe_2O_3) content = 7.4%, Al_2O_3/Fe_2O_3 ratio = 1.65 and Cr_2O_3 content = 1.2 to 2.5%, more preferably 1.6 to 2.5%.

Where Cr_2O_3 content = less than 1.2%, the content of Cr_2O_3 solid solution in alite was insufficient and, therefore, one-day compressive strength did not increase satisfactorily, and when Cr_2O_3 content = more than 2.6%, alite decomposed to belite and CaO and one-day compressive strength of cement dropped sharply, because Cr_2O_3 content was over the limit of solid solution of alite and of interstitial materials.

Next, mixed materials were prepared so the content of R_2O_3 = Al_2O_3 + Fe_2O_3 in clinker consisting of particular LSD 1.00 and 2% Cr_2O_3 was altered, and clinkers were made in the same manner described in the above, to which gypsum was added and the mixture was pulverized. The strength tests were conducted on the cement thus produced. The results are shown in the table below, and the change in one-day compressive strength in accordance with the varied contents of R_2O_3 is shown in Figure 2.3b.

Sample No.	Percent											
	SiO_2	Al_2O_3	Fe_2O_3	Cr_2O_3	CaO	MgO	Na_2O	K_2O	Total	R_2O_3	SM	IM
1	21.6	3.6	2.2	1.9	69.0	1.6	0	0.01	99.9	5.8	3.7	1.6
2	21.1	4.2	2.6	2.0	68.5	1.5	0.06	0.01	100.0	6.8	3.1	1.6
3	20.8	4.4	2.7	2.0	68.5	1.5	0.11	0.06	100.1	7.1	2.9	1.6
4	20.5	4.6	2.8	1.9	68.4	1.5	0.15	0.09	99.9	7.4	2.8	1.7
5	20.4	4.7	2.8	1.9	68.4	1.5	0.15	0.10	100.1	7.5	2.7	1.7
6	20.1	4.9	3.0	1.9	68.2	1.5	0.22	0.15	100.0	7.9	2.5	1.6
7	20.1	5.0	3.0	1.9	67.9	1.5	0.23	0.20	99.8	8.0	2.1	1.7
8	19.8	5.2	3.1	1.9	67.8	1.5	0.26	0.29	99.8	8.3	2.4	1.7
9	19.6	5.4	3.3	1.9	67.7	1.5	0.31	0.35	100.1	8.7	2.2	1.6

Sample No.	Particular L.S.D.	F. CaO, percent	Specific surface (Blaine), cm.^{2}/g.	One-day compressive strength kg./cm.^{2}
1	1.00	1.0	4,580	117
2	1.00	0.7	4,520	125
3	1.00	0.8	4,660	167
4	1.00	0.9	4,560	187
5	1.00	0.8	4,680	195
6	1.01	0.7	4,650	190
7	1.00	0.5	4,570	185
8	1.01	0.6	4,610	175
9	1.01	0.3	4,630	141

It is evident from these data that extra high early strength Portland cement having one-day compressive strength upwards of 150 kg/cm^2 was obtained where particular LSD = 1.00, R_2O_3 content = 7.1 to 8.3%, preferably 7.4 to 7.8%.

In other experiments, it was found that where the Cr_2O_3 content in clinker varied in the range of 1.2 to 2.6%, the result of the same tendency was observed, and the same was true when the ratio of Al_2O_3/Fe_2O_3 varied 1.65 ± 0.50. Where R_2O_3 content was less than 7.0%, solid phase reaction took place partially or uneven chemical reaction took place, due to the shortage of interstitial materials content, and therefore, Cr_2O_3 was unevenly melted in alite, and one-day compressive strength of cement dropped. On the other hand, where R_2O_3 content was more than 8.6%, interstitial materials content which contained more Cr_2O_3

than alite increased, causing the Cr_2O_3 content in alite to decrease, and also the relative content of alite in clinker decreased, and accordingly, one-day compressive strength dropped.

Aluminum Flake Powder

L. Kampf; U.S. Patent 3,442,672; May 6, 1969 has found that when a cement mortar material has incorporated 0.005 to 0.010% of aluminum by weight of the material and where the aluminum used is in a powder form which results from stamping flake powder, the concrete that is formed from such aluminized cement material undergoes no shrinkage but a slight expansion during its forming period, i.e., from the time that the material is mixed until it is set. It has also been found that such expansion occurs sufficiently rapidly to take place before hardening occurs but not so rapidly that it goes to completion prior to the completion of the period in which shrinkage would normally take place in the absence of aluminum.

Such results ensue because the aluminum powder, being in flake form, presents a comparatively large surface to permit a satisfactorily rapid reaction, i.e., the evolution of hydrogen gas resulting from the reaction between the cement mortar material and the aluminum and yet is in a controlled concentration which does not permit an expansion great enough to degrade the strength and other required physical characteristics of the concrete. In this connection, it has been further found that where the aluminum powder is in a form which results from the stamping of the flake powder and is in a concentration of 0.005 to 0.010% by weight of the cement mortar material, the strength characteristics of the concrete such as its compression strength are enhanced.

Special Anhydrous Thermal Treatment of High Siliceous Materials

A process described by *A. Rio, M. Cerrone and A. Saini; U.S. Patent 3,885,979; May 27, 1975; assigned to Societá Italiana per Azioni per la Produzione di Calci e Cementi di Segni, Italy* consists essentially of subjecting products manufactured from Portland cements or from mixtures of same with reactive siliceous materials, after a preliminary aging at ambient temperature, to an initial treatment with steam under pressure at temperatures ranging from 150° to 250°C, for 1 to 6 hours and then to a second thermal treatment, in anhydrous environment, at temperatures ranging from 120° to 350°C, for 1 to 12 hours.

It has been observed in fact, that when a hydrothermal treatment in autoclave is followed by a thermal treatment in anhydrous environment at the temperature intervals and for the time indicated, the mechanical resistance, already intrinsically high, displayed a further substantial increase reaching values ranging from 950 to 1,500 kg/cm^2, according to the binder composition and to the temperature and duration of the thermal treatment.

For the manufacture of the products, according to the process, a particular suitability is displayed by Portland cements with a high siliceous modulus, preferably higher than 3, characterized by the presence of a high percentage (higher than 78 to 80%) of calcium silicates and by a moderate content (lower than 15 to 18%) of the ferric-aluminic phases. In Portland cement-reactive siliceous material mixtures, the respective proportions can vary from 50 and 90% to 50 and 10%, depending on the characteristics of the material available and on the properties required for the final product. Some reactive siliceous materials, very rich in silica (above 90%), obtained from processing of trachytic rocks, materials which display also considerable pozzolana properties, are particularly advantageous for these purposes. Excellent results are obtained equally, using directly pozzolana cements prepared with the above materials already available on the market.

The possibility of obtaining, in a very short time, even less than 12 hours, cement products endowed with very high mechanical characteristics is particularly useful in prefabrication, as it allows, under equal stresses, the use of lighter structures with reduced dimensions and it is clearly linked not only with the thermal treatment, but also with the nature of the pozzolana or siliceous addition used.

The mechanical resistance obtainable, increases with the passage from the use of normal Portland cements to Portland cements with a high silicic modulus and from the latter to mixtures of Portland cements with reactive pozzolana or siliceous additions, in dependence on the quantity and quality of the addition itself. The following examples illustrate the process.

Example 1: Mortars manufactured with ARC pozzolana cement and sand standardized according to the Italian specifications, proportioned in the 1:3 ratio, with a water/cement ratio = 0.5, were submitted to the following aging modalities:

(a) 12 hours of preaging at ambient temperature;

(b) 3 hours treatment in autoclave at 215°C (20 atm); and

(c) 4 hours thermal treatment at 1 atm at 200°C.

The following mechanical resistances were displayed: 855 kg/cm^2 at the end of treatment (b); and 1,356 kg/cm^2 at the end of treatment (c).

Example 2: Products manufactured according to Example 1, using as a binder a mixture of Portland cement with a high silicic modulus (3.10) and with a low modulus of melting materials (1.5) and sand with a high (90%) SiO_2 content in the proportions of 70:30, aged according to Example 1, displayed the following mechanical resistances: 750 kg/cm^2 at the end of treatment (b); and 1,112 kg/cm^2 at the end of treatment (c).

Example 3: Products manufactured according to Example 1, using as a binder normal Portland cement with a high silicic modulus (3.10) and with a low modulus of melting materials (1.5), aged according to Example 1, displayed the following mechanical resistances: 580 kg/cm^2 at the end of treatment (b); and 920 kg/cm^2 at the end of treatment (c).

Example 4: Products manufactured according to Example 1, using as a binder normal Portland cement, aged according to Example 1, displayed the following mechanical resistances: 560 kg/cm^2 at the end of treatment (b); and 836 kg/cm^2 at the end of treatment (c).

Example 5: Products manufactured according to Example 1 and aged according to the following modalities:

(a) 6 hours preaging at ambient temperature;

(b) 6 hours treatment in autoclave at 180°C (12 atm); and

(c) 2 hours thermal treatment at 1 atm at 300°C.

The following mechanical resistances were displayed: 830 kg/cm^2 at the end of treatment (b); and 1,150 kg/cm^2 at the end of treatment (c).

Magnesium Oxychloride-Oxysulfate

A.N. Vassilevsky and A. Renke; U.S. Patent 3,667,978; June 6, 1972; assigned to V.R.B. Associates, Inc. describe an excellent hydraulic cement binder for both organic and inorganic fillers constituted of magnesium oxide, magnesium sulfate and an alkaline earth metal chloride (e.g., calcium chloride, barium chloride and strontium chloride), when the magnesium oxide is present in an amount by weight exceeding that of the magnesium sulfate and the alkaline earth metal chloride is present in at least 10% by weight of the cement but in a molar quantity less than that of the magnesium sulfate.

It has been found that excellent binders can be prepared without the danger that hygroscopicity will cause deterioration of the hydraulic cement before it is combined with water and a filler, and which yields a hardenable composition with inorganic and organic fillers of substantially all types with characteristics approaching those of the highest-quality Portland cement compositions.

The process is based upon the fact that the binder, when combined with water and permitted to set, appears to be a highly complex magnesium oxychloride-oxysulfate. This complex appears to have a binding strength, resistance to abrasion, durability and stability far exceeding the corresponding characteristics of both magnesium sulfate cement and Sorel-type magnesium chloride cement.

The alkaline earth metal chloride forms, upon mixing of the binder with water, a more or less insoluble sulfate by double displacement with the magnesium sulfate, the resulting calcium, barium or strontium sulfate forming a filler bound in the complex. Unlike prior Sorel cements, however, only a fraction of the magnesium sulfate component is converted into magnesium chloride while the balance remains in the sulfate form. The hardened composition is thus an oxychloride-oxysulfate of unique characteristics. While the hardened cement binder of the process is referred to as essentially a magnesium oxychloride-sulfate, it will be noted that it is likely to contain the following species:

$MgO \cdot X_1H_2O$	$Mg_mOSO_4 \cdot X_5H_2O$
$MgCl_2 \cdot X_2H_2O$	$Mg_kClSO_4 \cdot X_6H_2O$
$MgSO_4 \cdot X_3H_2O$	$Ca_jClSO_y \cdot X_7H_2O$
$Mg_nOCl \cdot X_4H_2O$	$CaSO_4 \cdot X_8H_2O$

where X_1 through X_8 represent water of hydration and the alkaline earth metal is, for purposes of illustration, indicated as calcium. The subscripts j, k, m, y and n represent integral or fractional values depending upon the actual composition of the particular portion of the complex. Each of these species appears to be present in chemical combination with the balance in the hardened binder, and further chemical bonds connect the binder with the filler. Thus the hardened binder can be empirically described as having the empirical formula:

$$(MgCl_2 \cdot wH_2O)_a \cdot (MgO \cdot xH_2O)_b \cdot (MgSO_4 \cdot yH_2O)_c \cdot (MaeSO_4 \cdot zH_2O)_d$$

In the formula Mae is an alkaline earth metal selected from the group consisting of calcium, barium and strontium; w, x, y and z are units representing molecules of water of hydration; a ranges between 1 and 3; b ranges between 12.5 and 15.5; c ranges between 1.5 and 2.2; and d ranges between 1 and 3.

The above molecular relationships are based upon the discovery that best results are obtained when the hydraulic cement, which is designed to be mixed with water and inorganic or organic fillers, consists in its powdered form of anhydrous magnesium sulfate present in an amount ranging between 2.8 and 3.3 parts by weight, anhydrous magnesium oxide present in an amount ranging between 4.5 and 5.5 parts by weight and the alkaline earth metal chloride present in anhydrous form and in an amount ranging between 1 and 3 parts by weight.

The resulting cement has an excellent shelf life, a surprising fact in view of the tendency of calcium chloride, for example, to interact with magnesium sulfate in a double-displacement reaction. Furthermore, it has been discovered that the cement, when prepared in this fashion, combines with water and probably the filler in a highly exothermic reaction which yields a setting rate far superior to any obtainable with Sorel-type cement and increasing both the strength of the bond to organic and inorganic fillers and the rate of formation of a hard abrasion-resistant and nonhygroscopic body.

It has been found to be possible to decrease further the hygroscopicity of the improved hydraulic cement binder by adding minor proportions of silicates, especially calcium or sodium silicate, which have been found to yield a composition which is substantially waterproof and less brittle than the products using the improved hydraulic cement binder of the process in the absence of such silicates.

More particularly, between 6 and 15% by weight of calcium or sodium silicate may be combined with the hydraulic cement binder prior to its use in the cementitious composition. Moreover, it has been discovered that, when sodium silicates are used in conjunction with sodium silicofluoride, the improvement from the point of view of waterproofing characteristic and reduced brittleness is synergistic in nature.

Cement No. 1: 5 parts by weight of aluminum silicate or silica, 40 parts by weight of previously calcined magnesium oxide, or an equivalent amount of previously calcined magnesite, 20 parts by weight of barium chloride and 20 parts by weight of calcined magnesium sulfate are mixed, ground to a fine powder, and stored in polyethylene bags or other suitable airtight containers. The calcination is carried out at 750° to 780°C.

Cement No. 2: 2 parts by weight of aluminum silicate or silica, previously calcined at 250° to 300°C, are mixed with 5 parts by weight of calcined magnesia or magnesite and the mixture is ground (Powder A). Separately, 2 parts by weight of previously calcined barium chloride and 1 part by weight of strontium chloride are mixed and ground together (Powder B). The Powders A and B are mixed in a weight ratio of 2:1, and 3 parts by weight of magnesium sulfate are added to the mixture which is then stored as in the case of Cement No. 1.

Cement No. 3: The ingredients enumerated below are separately subjected to a preliminary calcination at 350°C and then, after cooling, are mixed in the proportions shown on the following page.

	Parts by Weight
Silica	5
Calcined magnesite	48
Calcined magnesium sulfate	29
Calcined calcium chloride	10
Calcined strontium chloride	8

The mixture is ground to a fine powder and, if necessary, subjected to additional drying. It is stored as in the case of Cement No. 1. The following examples are given to illustrate the manner in which the above cements are used in the preparation of hardened compositions.

Example 1: 80 parts by weight of Cement No. 1, 2 or 3 is mixed with 50 parts by weight of water to a consistency of heavy cream. Then there is added 30 parts by weight of sawdust, and the slurry is stirred to uniform consistency. It is then poured into molds and allowed to set in the open air without heating.

Example 2: 20 parts by weight of Cement No. 1, 2 or 3 is mixed with 12 parts by weight of water to a consistency of heavy cream. Then there is added 7 parts by weight of sawdust and 5 parts by weight of sand. The resulting slurry is placed in a mold with application of pressure as the mixture is fairly dry and does not flow freely.

Example 3: 20 parts by weight of Cement No. 1, 2 or 3 is mixed with 12 parts by weight of water, and to this mixture 10 parts by weight of stone powder is added. The mixture is stirred until a thick slurry is obtained, and then treated as in Example 1.

Example 4: A hydraulic binder is prepared by individually calcining calcium chloride, magnesium oxide and magnesium sulfate at a temperature of 750° to 780°C. Each of the calcined components is ground and combined with the others in the following proportions: 3 parts by weight of magnesia, MgO, 3 parts by weight of anhydrous magnesium sulfate, $MgSO_4$ and 1 to 2 parts by weight of anhydrous calcium chloride, $CaCl_2$. The powder is placed in a polyethylene bag and stored in a humid environment for several weeks. No deterioration of the cement was observed.

A comparison was made between a Portland cement composition prepared in accordance with ASTM procedure with 1 part Portland cement and 4 parts of clean washed and sieved sand. This composition was compared with a soil cement prepared by mixing 1 part of the powder stored as above with 4 parts by weight of ordinary garden earth containing substantial proportions of organic material in addition to the usual mineral matter. The results and characteristics of the tests are set forth below.

(A) Flexural Strength – Tests were conducted in accordance with ASTM C348. Test specimens for each type of mortar were aged for 9 days at room temperature.

	Flexural Strength, psi	
Test Specimen	Soil Mortar (this process)	Portland Cement Mortar
1A	309	382
1B	281	418

(continued)

Test Specimen	Flexural Strength, psi: Soil Mortar (this process)	Flexural Strength, psi: Portland Cement Mortar
2A	281	400
2B	315	437
3A	304	406
3B	270	448
Average	293	415

(B) Resistance-to-Abrasion Wear — The abrasion-wear tests were conducted after test specimens had cured for 9 days at room temperature. The Tinius Olsen Wearometer, as specified in Military Specification, Mil-D-3134F, was used. The wear after 1,000 cycles for each type mortar, given as loss in thickness in inches, is shown below.

Specimen	Soil Mortar	Portland Cement Mortar
No. 1	0.101	0.094
No. 2	0.096	0.089
Average	0.099	0.091

The above results demonstrate that an expensive Portland cement composition using high quality sand can effectively be replaced by an inexpensive composition such as the soil cement of the process using ordinary earth available at all necessary locations. Best results are obtained when 20 to 40 parts by weight of earth or clay are admixed with 10 parts by weight of the cement composition of the process.

Moreover, it was found that the magnesium oxide (anhydrous) content should be confined between the critical limits of 4.5 to 5.5 parts by weight, the anhydrous magnesium sulfate should be confined to an amount between 2.8 and 3.3 parts by weight while the calcium chloride (anhydrous) should be present in an amount of 1 to 3 parts by weight.

Soda Lime Glass

A.G. Jansen; U.S. Patent 3,823,021; July 9, 1974 describes a cementitious composition for use, for example, as overlay layers for concrete bases of improved adhesion property which comprises from about 20 to 66 parts by weight of cement, from about 80 to 34 parts by weight of silica sand and from about 0.1 to 1.5% by weight, based upon the weight of the cement, of glass particles having a particle size in the range of 2.6 to 233 mesh and being a soda lime type glass. The following examples which illustrate the process were prepared with approximately the same water-cement ratios.

Example 1: A concrete batch was made up using 9 pounds of Portland cement, 30 pounds of washed silica sand, and 1 ounce of soda lime glass beads of size number 801-206. The concrete batch was mixed with approximately 0.5 gallon of water. The wet mortar mixture is applied by free hand forming to various substrates.

Example 2: A concrete batch was made up using 9 pounds of Portland cement, 30 pounds of No. 4098 silica sand, and 2 ounces of soda lime glass beads of size number 801-206. The concrete batch was mixed with about 0.5 gallon of water.

Example 3: A concrete batch was made up using 9 pounds of Portland cement, 9 pounds of No. 2 torpedo sand, 9 pounds of floured silica sand and 2 ounces of soda lime glass beads of size number 801-206. The concrete batch was mixed with approximately 0.35 gallon of water.

Example 4: A concrete batch was made up using 224 parts by weight of Portland cement, 176 parts by weight of No. 2 torpedo sand, 96 parts by weight of No. 4 size silica sand, 5 parts by weight of fly ash additive and 0.67 part by weight of soda lime glass beads of size number 801-204. The concrete batch was mixed with approximately 6.4 gallons of water.

Concrete bases, the surfaces of which are to be overlayed are thoroughly cleaned by one or more of the procedures known as sandblasting, washing with acid solution, detergent solution or ammonium solutions. After mixing the products of Examples 1 through 4, to obtain a plastic mixture, the mixtures were applied to cleaned concrete bases, using the appropriate tools of the trade with the thickness of the coating visually regulated using screed rings as a guide. After aging, for about a week, the layers of Examples 1 through 4 showed strong adhesion and no tendency to flake or crack.

Sparingly Soluble Barium Compounds

According to a process described by *H. Jaklin; U.S. Patent 3,950,178; April 13, 1976* sparingly soluble barium compounds are added to compositions comprising cement or mortar, calcium sulfate and water, the barium compounds being unable to react with the calcium sulfate setting regulator during the normal setting of the concrete. In the finished concrete the sulfate to a large extent is bound up as ettringite, so-called trisulfate, or also as monosulfate, which does not use up barium compounds in forming barium sulfate.

Sparingly soluble barium compounds include barium oxalate, barium fluoride and/or barium silicate hydrate having a $BaO:SiO_2$ mol ratio of about 0.5 to 1.2, the barium compound being added in at least about 0.5% by weight of the cement. This procedure improves the resistance of concrete to attack by corrosive liquids.

Calcium Sulfoaluminate Hydrate-Forming Mineral Powders

A process described by *T. Kitsuta, I. Mino and K. Nakagawa; U.S. Patent 3,901,722; August 26, 1975; assigned to Japanese National Railways and Denki Kagaku Kogyo KK, Japan* comprises the production of a high strength concrete under such a condition that the concrete is allowed to stand under atmosphere and a sufficient curing cannot be effected, which is characterized in that calcium sulfoaluminate hydrate-forming mineral powders are added to concrete materials in such an amount that the concrete does not expand.

It has been known that the presence of calcium sulfoaluminate hydrate-forming minerals compensates the shrinkage of concrete by utilizing the expansion energy generated in the hydration, but in this process it is impossible to increase the compression at an age of 28 days to more than 800 kg/cm^2. This process enables one to increase the strength of concrete and improve the freezing and thawing resistance by making the calcium sulfoaluminate hydrate-forming minerals present in concrete in such an amount that the concrete does not expand and therefore the process is different in the technical idea from the conventional method where the shrinkage of concrete is compensated by utilizing the expansion energy. The calcium sulfoaluminate hydrate-forming minerals to be used in the process include the following substances and the item (1) is the most effective.

(1) A mixture of a crystalline or amorphous calcium aluminate, such as CA, C_3A, CA_2, $C_{12}A_7$, $C_{11}A_7{\cdot}CaF_2$ and $C_3A_3{\cdot}CaF_2$ and gypsum or a product obtained by simultaneously burning the calcium aluminate-forming materials and gypsum can be used. A mixture of an amorphous product of $C_{12}A_7$ or $C_{11}A_7{\cdot}CaF_2$ and gypsum is preferred. The weight ratio of calcium aluminate to gypsum is 1:0.5-10, preferably 1:0.8-5.

(2) C_4A_3Y-gypsum. The weight ratio of C_4A_3Y to gypsum is 1:0.2-2, preferably 1:0.4-1.0.

(3) Alumina containing slag-gypsum.

In any case, when free CaO is not contained, the improvement of strength is high and effective. In the above descriptions, C means CaO, A means Al_2O_3 and Y means SO_3. The amount of these minerals is the range within which the concrete does not expand when the expansion percentage is determined but for cement the amount is 2 to 13% by weight, preferably 4 to 8% by weight. The fineness in Blaine value is not less than 3,000 cm^2/g, preferably 4,000 to 8,000 cm^2/g.

When less than 3,000 cm^2/g, the function for promoting the hydration of cement considerably lowers and the unreacted product remains for a long period of time and the stability is poor, while when the Blaine value exceeds 8,000 cm^2/g, the hydration is too rapid in some minerals and the false setting occurs. In the process, the calcium sulfoaluminate hydrate-forming mineral powders are compounded in cement in an amount of 2 to 13% by weight, the unit cement amount (an amount of cement required for the formation of 1 m^3 concrete) is 500 to 700 kg, a water/cement ratio is 18 to 35%, and 0.3 to 5% by weight based on cement, of a surfactant is added while blending concrete.

The unit cement amount of 500 to 700 kg is essential for the production of tho high strength concrete and further is a necessary requirement for developing a high freezing and thawing resistance. When the unit cement amount is less than 500 kg, even if the calcium sulfoaluminate hydrate-forming minerals are added in such a range that the concrete does not expand, it is impossible to increase the compression strength at an age of 28 days to more than 800 kg/cm^2 and the freezing and thawing resistance cannot be improved.

When the unit cement amount exceeds 700 kg, the strength is not improved in proportion to the increased amount and instead Young's modulus decreases and the heat of hydration is too great. The cement to be used in the process includes Portland series cement, mixed cement and the like.

When the water/cement ratio exceeds 35%, it is impossible to increase the compression strength at an age of 28 days to more than 800 kg/cm^2, while when the ratio is less than 18%, even if a surfactant is used, it is impossible to blend a concrete having a workability suitable for concreting. In order to provide the workability, such as placing and compaction, the concrete is blended at a slump of 10 to 20 cm.

Calcium Fluoroaluminate

In a process described by *N.R. Greening, L.E. Copeland and G.J. Verbeck; U.S. Patent 3,628,973; December 21, 1971; assigned to Portland Cement Association* a modified Portland cement is provided which contains as one ingredient a substantial amount of a ternary compound which is essentially a calcium haloaluminate having the chemical formula

$$11CaO \cdot 7Al_2O_3 \cdot CaX_2$$

where X is a halogen, i.e., fluorine, chlorine, bromine or iodine. The ternary calcium haloaluminates having the above indicated molecular formula are per se known to the art and have been described by Brisi et al, *Annoli di Chimica,* 56, 224 (1966) and by Jeevaratnam et al, *Jour. Amer. Ceram. Soc.,* 17, 105 (1964). It has been found that if between 1 and 30% by weight of the ternary compound is contained in a conventional Portland cement, the cement will develop a high early set strength. The preferred ternary calcium haloaluminate to be incorporated into the cement is calcium fluoroaluminate having the formula $11CaO \cdot 7Al_2O_3 \cdot CaF_2$.

The modified Portland cement may be used in a conventional way in applications where a short initial setting time and high early set strength are desired. For example, the cement may be used with conventional aggregates to produce a patching mixture for highways and airplane runways. Such normal density compositions will develop an initial set strength in one hour to support a compressive force of 500 to 2,500 psi. Also, the cement finds utility in mixing with lightweight aggregates, such as pumice, expanded vermiculate or expanded perlite, for the pouring of roof decks in a conventional manner.

Employing appropriate formulations, such a low-density composition upon setting will develop a sufficient initial set strength to support a compressive force of 30 to 150 psi, thereby supporting the weight of a man, within 30 to 120 minutes after the deck has been poured.

Calcium Aluminate and Nepheline Syenite

J.J. Capellman, J.C. Stultz and J.E. Cooper; U.S. Patent 3,824,105; July 16, 1974; assigned to PPG Industries, Inc. describe a hydraulic-setting bonding agent consisting essentially of a hydraulic-setting calcium aluminate and nepheline syenite. The nepheline syenite is present in a concentration of 5 to 50% by weight based on the weight of the hydraulic-setting calcium aluminate and nepheline syenite. The nepheline syenite increases the fired cold flexural strength provided by the hydraulic-setting calcium aluminate. The bonding agent can be combined with an aggregate such as fused silica to provide a refractory such as the following, on a weight basis:

	Percent
Fused silica	60 to 80
Hydraulic-setting calcium aluminate	15 to 38
Nepheline syenite	2 to 15

This refractory composition when combined with 9 to 18% by weight water forms a fluid castable refractory composition which can be cast and hardened to form a refractory body such as a gas hearth module.

Acidulated Fly Ash

J.W. Bainton; U.S. Patent 3,953,222; April 27, 1976; assigned to Tekology Corp. has found that acidulation of a pozzolan generally, and fly ash specifically, by using a selected mineral acid or a combination thereof, greatly improves their pozzolanic properties useful for subsequent addition to concrete or mortar mixes with or without grinding or calcining. The improvement in the strength of concrete structures resulting from the addition of acidulated pozzolan can exceed by more than 100% the strength of the same products or structures made with identical materials except that the pozzolans were not acidulated.

Basically, fly ash itself is glass particles ranging anywhere from one micron to perhaps 100 microns in size. They average about 40 microns, and they are in the form of hollow spheres made up of a lattice work of SiO_2 molecules that have, interspersed between them, the oxides of various other metallic elements that were present in the fuel burned to produce the ash. These would be principally oxides of calcium, magnesium, aluminum, and iron. The exact makeup would depend upon the source of the fuel and the burning conditions.

When this fly ash is treated with a suitable strong mineral acid it becomes possible to remove these oxides from the SiO_2 lattice since they form other salts which can be washed out, but in addition their removal leaves the silica glass spheres pockmarked, as can be seen in pictures made with a scanning electron microscope which demonstrate the surface roughening very clearly. When such acidulated pozzolan is added to cement and aggregate to form concrete, the chemical reaction that results is the same as it would be if an unacidulated pozzolan had been used. However, this action is much more rapid and complete than it is with unacidulated pozzolan.

This improvement results from the creation of a great deal more surface area on the glass spheres which include evacuated pockets, or so-called hot spots, in the spheres which comprise exposed clean surfaces that are very active in their chemical combination with the lime which may have been added to the mix and/or with free lime liberated by the hydration of Portland cement. The following examples illustrate the process and show the advantage gained by acidulating pozzolans before adding them to certain products, as compared with adding unacidulated pozzolans, or adding no pozzolans at all, other factors being the same.

Example 1: Pressed concrete cubes were made and air-cured at ambient temperature, certain of the cubes containing a pozzolan comprising fly ash acidulated with hydrochloric acid and then drained but not washed prior to inclusion in the mix, and others of the cubes containing untreated fly ash, and still others of the cubes comprising control cubes containing no fly ash but otherwise made at the same time with identical ingredients and aged under the same conditions. The ingredients in the cubes are as follows, by weight of the dry mix:

	CONTROL CUBES	UNACIDULATED ASH CUBES	ACIDULATED ASH CUBES
Portland Cement	10%	10%	10%
Sand & Fine Aggregate	90%	80%	80%
Water	8%	8%	8%
Fly Ash Untreated	—	10%	—
Acidulated Fly Ash	—	—	10%

(continued)

	CONTROL CUBES	UNACIDULATED ASH CUBES	ACIDULATED ASH CUBES
Measured Compressive Strength (24 SAMPLES after 3 days, shown on a relative scale where the control sample's strength is adjusted to equal 100.)	100	120	229

Example 2: This example is similar to Example 1 with two exceptions, i.e., that the fly ash was acidulated with sulfuric acid (unwashed), and that the cubes were cured at a temperature elevated somewhat above ambient (140°F) for 16 hours:

	CONTROL CUBES	UNACIDULATED ASH CUBES	ACIDULATED ASH CUBES
MEASURED COMPRESSIVE STRENGTH (24 samples after 48 hours)	100	177	324

The fact that the differential in compressive strength appears greater in Example 2 than in Example 1 is attributable to the use of an elevated curing temperature which always increases the pozzolanic activity both with and without acidulation. Moreover, it is well known that a sulfate improves steam curing, and in Example 2 the treated pozzolan was not washed to remove the sulfate after the ash was drained. If the pozzolan had been washed before adding it to the mix, the relative strength figure of 324 would have been decreased.

Example 3: The example is similar to Example 1 with two exceptions, i.e., that the fly ash was thoroughly washed to remove remaining soluble salts and acid before adding the fly ash to the mix, and that the resulting cubes were autoclaved at 350°F for 8 hours.

	CONTROL CUBES	UNACIDULATED ASH CUBES	ACIDULATED ASH CUBES
MEASURED COMPRESSIVE STRENGTH (12 samples)	100	245	303

The showing of Example 3 emphasizes the fact that autoclaving and acidulation of the pozzolan tend to provide similar benefits. Thus, the unacidulated figure of 245 is approaching the acidulated figure of 303, and if the autoclaving were continued for a longer time the differential between the two would be reduced still further. However, autoclaving is expensive since it uses energy at a high rate.

Blast Furnace Slag and Gypsum

According to a process described by *T. Mori, T. Iwai, A. Yoda and M. Oshima; U.S. Patent 3,565,648; February 23, 1971; assigned to Kajima Construction Co., Ltd., Japan* blast furnace quenched slag or fly ash is blended with gypsum, and after adding water, the slurry is put in a rotary ball mill, a vibrating ball mill, or the like, to be sufficiently ground so as to cause the alumina and calcium sulfate to come in contact and react with each other. The ground slurry thus obtained is added to a cement mix composed of a mixture of cement, aggregate and blending water to form the hydrate of calcium sulfoaluminate by means of the reaction with calcium hydroxide released on hydration of the cement. The amount of blast furnace slag or fly ash utilized is generally within a range of from 50 to 90% of the total weight of the blast furnace slag or fly ash mixture, and is preferably within a range of 60 to 80% by weight.

The amount of gypsum utilized is generally within a range of from 5 to 50% of the total weight of the blast furnace slag or fly ash mixture, and is preferably within a range of 10 to 40% by weight. After mixing the blast furnace slag or fly ash with the desired additional materials to obtain the mixture, water is added in an amount generally within a range of from 50 to 500% of the total weight of the blast furnace slag or fly ash mixture. Preferably, the amount of water utilized is about 60 to 200% by weight. Preferably, in the case of mortar, the weight ratio of cement, water and aggregate is 1:0.4-0.8:2-4, and in the case of concrete, the weight ratio of cement, water, fine aggregate and coarse aggregate is 1:0.4-0.8:2-4:3-6.

The ground slurry is added to the cement mix in such proportion that the weight ratio of the cement to the mixture consisting of blast furnace slag or fly ash and any other materials, is generally within a range of from 50:50 to 95:5, and is preferably about 80:20. The following example illustrates the process.

Example: Blast furnace quenched slag (or fly ash) and other materials such as anhydrous gypsum (or crystalline gypsum), calcium hydroxide, sodium sulfate, and calcium carbonate were blended in the ratios shown in the table below, to form a mixture weighing 4 kg. 2.6 kg (65% by weight of the mixture) of water were added to the mixture, and the resultant mixture was put in a rotary ball mill and ground for one hour. The ball mill used was a cylinder type having a 35 cm diameter and a 35 cm height; iron balls of 1 cm diameter (the total weight of the balls used being 20 kg) were employed.

After grinding, 188.76 grams of the obtained ground slurry (consisting of 114.4 grams of the mixture and 74.36 grams of water) were taken out, and blended with 405.6 grams of cement, 263.64 grams of water and 1,040 grams of sand. The cement mortar was then molded. The total weight of the cement and the mixture was 520 grams and the total weight of water was 338 grams; they met the specification of JISR-5201. Specimen No. 1 was the control which had not been treated according to the method of the process.

Ratios of Cement and Admixtures, %

Number	Rapid-Hardening Cement	Blast Furnace Slag	Fly Ash	Anhydrous Gypsum	Crystalline Gypsum	Sodium Sulfate	Calcium Carbonate	Calcium Hydroxide
1	100	0	0	0	0	0	0	0
2	78	16	0	5	0	0	0	1
3	78	16	0	4	0	1	0	1
4	78	16	0	3	0	2	0	1
5	78	0	5	15	0	1	0	1
6	78	0	10	10	0	1	0	1
7	78	0	15	5	0	1	0	1
8	78	16	0	0	5	0	0	1
9	78	16	0	0	4	0	1	1
10	78	16	0	0	3	0	2	1

The results in the next table show that the initial strength of this cement is greatly increased by this treatment.

Number	1 Day	3 Days	7 Days	28 Days
	- - - - -Compressive Strength (kg/cm^2)- - - - -			
1	91	205	322	442
2	175	299	352	498
3	197	331	376	494
4	202	363	410	518
5	118	220	333	455
6	159	254	345	470
7	189	326	377	496
8	158	231	335	457
9	172	263	348	467
10	170	282	346	472

S. Bastian and M.H. Gruener; U.S. Patent 3,522,068; July 28, 1970 describe a process of preparing a cement composition which comprises dissolving a water-soluble carbonate of a mono to trivalent metal, a water-soluble sulfate of the same metal and, if needed, a sodium or aluminum fluosilicate, sodium chloride, a surface active substance, and a polyvinyl polymer in water.

The aqueous solution is kept at 25° to 50°C for 8 to 16 hours and then ripened for at least 72 hours with free access of air. Then, just prior to the intended use for the cement composition, to the ripened aqueous solution is added a hydraulic additive such as a siliceous earth, a granulated blast furnace slag, an aerial binding agent such as quicklime or slaked lime, Portland cement, and possibly a filler to the solution. There may also be added with the cement the aerial binding agent and the hydraulic additive, a small quantity of a solid gas generating substance.

Fluid Coke

A.B. Small and H.N. Babcock; U.S. Patent Reissue 26,597; June 3, 1969; assigned to Esso Research and Engineering Company have found that the addition of a relatively small amount of fluid coke, that is to say coke recovered from the fluid coking of petroleum, to cement mixes will cause the system to expand during the setting rather than shrink. Expansion occurs between the initial and final set. By controlling the amount of coke used, the cement system can be regulated to give a cement mass that neither shrinks nor expands and hence a nonshrinking Portland cement mixture can be produced.

The shrinkage of the cementitious system is controlled by replacing a portion of the sand or fine aggregate with fluid coke. About 15 to 50% by volume of coke replacing sand in a system containing 0.1 to 2 weights Portland cement, to 2 to 4 weights fine aggregate to 2 to 5 weights of coarse aggregate will yield a system that is essentially of constant volume during cure. All or part of the fine aggregate may be replaced by fluid coke. The amount of fluid coke added to cement mixes, such as Portland cement, coarse aggregate and fine aggregate may be between 0 to 100 weight percent of the fine aggregate depending on the degree of expansion desired.

Still further advantages result from the lower density of the cement system or mixture. Coke has a bulk density of about 60 lb/ft^3 and a real density of about 92 lb/ft^3. By contrast, sand has a bulk density of about 100 lb/ft^3 and a real density of about 165 lb/ft^3. This ability to expand Portland cement in a controlled manner is limited exclusively to fluid coke; no other type of coke or carbon has this effect. It should be noted that coal clinkers may also cause expansion but in a harmful and uncontrolled amount. In the clinker case, the expansion is related to the sulfur content. The behavior of fluid coke in cementitious systems is believed to be due to its inherent ability to adsorb small amounts of inert gases which can be released by the heat generated by the setting concrete.

Polystyrene Particles

B.J. Harvey; U.S. Patent 3,883,359; May 13, 1975; assigned to The Carborundum Company describes a pumpable refractory insulating composition comprising from about 20 to 50% of hydraulic setting cement, from about 10 to 25% of finely divided refractory material, from about 1 to 5% of a particulate synthetic organic polymer, from about 0.15 to 0.3% of a surface active agent and from about 20 to 70% water. The refractory material may be particulate, fibrous or a mixture of these, and the particulate organic polymer may comprise materials such as polystyrene beads. In addition, the mixture may include up to about 4% of colloidal silica. The following examples illustrate the process.

Example 1:

Composition	Percent by Weight
Alumina-silicate ceramic fiber	10.0
Calcium aluminate hydraulic setting cement	29.6
Colloidal silica	4.0
Synthetic organic polymer (spherical, cellular polystyrene particles)	1.7
Surface active agent (Lissapol N)	0.1
Water	54.6

A batch weighing 68 kg sufficient for a dried insulation volume of 0.085 m^3 was prepared by mixing the dry ingredients first and then mixing with the water in a small concrete mixer for 2 minutes, after which the mix was transferred to the feed hopper of a pump having a double internal helix stator with helical rotor, specially designed for plastic viscous materials.

The mix had a cream consistency and could be pumped to a height of 9 meters. It was found that the composition, as pumped, would bridge gaps of up to 6.5 millimeters, provided that the hydrostatic head imposed while still in the fluid condition was not more than 1½ meters. Once the initial hydraulic set had taken place (about 20 minutes), additional amounts of mix could be added. A 100 millimeter thickness of insulation pumped into position had a linear shrinkage, wet to fired, of zero at 600°C and 0.5% at 800°C. Final density was 0.4 g/cc.

Example 2:

Composition	Percent by Weight
Refractory aggregate (420 micrometers and finer)	61.8
Calcium aluminate hydraulic setting cement	15.5
Synthetic organic polymer (spherical, cellular polystyrene particles)	3.2
Surface active agent (Lissapol N)	0.1
Water	19.4

The above mix was prepared and pumped as in Example 1. A 100 mm thickness of insulation pumped into position had a wet to fired shrinkage of zero at 600°C and 0.2% at 1000°C. Final density was 0.48 g/cc. The advantages of the pumpable compositions may be summarized as savings in labor costs, reduction of down time of an installation under repair, improved insulation effectiveness by virtue of the lower density of the insulation, greater resistance to thermal shock, and the feasibility of pumping insulation into normally inaccessible positions. The compositions are easily prepared and applied without the use of special equipment, the resulting insulation setting up and being ready for use within 24 to 36 hours after application.

Acrylamide-Formaldehyde Polymer

B. Bonnel; U.S. Patent 3,692,728; September 19, 1972; assigned to Progil, France describes compositions for mortars and cements which harden rapidly and give great resistance including a hydraulic binder of high alumina cement and a polymer of acrylamide and formaldehyde and a polymerization catalyst.

The applications of the compositions are numerous. They may be used advantageously by thick injections, for filling the fissures in the worked concrete. With these compositions may be made thin covers and ground coatings which harden very quickly. An application giving especially good results is the use of mortars in sealing operations, especially in the sealing of anchor rods used in mine ground supporting or further for the positioning of cables by pumping mortar a long distance. The compositions can also be used for joint making, consolidation of soils, etc. The following examples illustrate the process.

Example 1: The following two types of formulations, which had been stored at 190° ± 1°C were used:

Formulations	Grams
Pulverulent Formula	
Cement (Fondu Lafarge)	320
Dry Fontainebleau sand	70
Ammonium persulfate	2
Liquid Formula	
Aqueous solution of monomeric acrylamide and formaldehyde at 44% of the dry extract	96
Additional water	19.8
Diethylaminopropionitrile (solution of 5% by weight)	4.2
Potassium ferricyanide	0.0132

A mortar paste was prepared by pouring the liquid all at one time into the powder and by stirring intensely for some seconds. A very fluid paste is obtained which can be easily conducted by a pump. The paste was molded in the test-tubes described in the standard AFNOR (P 15,401) and traction tests were made according to prescriptions of the standard. The results obtained are shown below hereafter where each figure represents an average of three specimens.

Elapsed Time, days	Traction Resistance, kg/cm^2
1	73.2
2	89
3	92.5
6	117.3
7	>120

The setting time measured by means of a Vicat needle, according to the standard AFNOR (P 15,431) was 15 to 17 minutes at a temperature of 20° ± 1°C.

As a comparison, it should be noted that a mortar with 80% by weight of high alumina type cement as Fondu Lafarge, without the acrylic mixture, has a traction resistance at the end of 2 days of 48 kg/cm^2 and after 6 days, a resistance reaching hardly 50 kg/cm^2.

Tests for resistance to flexion, conducted according to the standard AFNOR (P 15,451) have furnished excellent results as shown in the table below.

Elapsed Time, days	Resistance to Flexion, kg/cm^2
1	99
3	>150
7	>150

Example 2: Under the same conditions as in Example 1, mortar samples were prepared according to the process, starting with a paste obtained by mixing the following two formulations stored at 19° ± 1°C.

Formulations	Grams
Pulverulent Formula	
Cement (Fondu LaFarge)	280
Dry Fontainebleau sand	118
Ammonium persulfate	2
Liquid Formula	
Aqueous solutions of monomeric acrylamide and formaldehyde at 50% of dry extract	96
Additional water	19.8
Diethylaminopropionitrile (aqueous solution of 5% by weight)	4.2
Potassium ferricyanide	0.036

Setting time measured under the same conditions as in Example 1 was 28 to 30 minutes at a temperature of 20° ± 1°C. Flexion and traction resistance tests were conducted according to the standards AFNOR (P 15,451) and (P 15,401). The results obtained are shown in the table below.

Elapsed time until measurement	Resistance to flexion in kg./cm.^{2}	Resistance to traction in kg./cm.^{2}
Days:		
1	108	68.3
2		90
3	147	
5		109
7	>150	>120

Fatty Esters of Polyglycols and Sulfonated Naphthalene Derivatives

W.A. Proell; U.S. Patent 3,537,869; November 3, 1970 describes an organic additive which, when incorporated in cementitious mixtures in a ratio of 0.1 to 3.0% of the mass of cement in the mixture, will produce strength increases on the order of 25 to 125% without significant deleterious effect upon the handling characteristics of the mixture. The additive may be mixed, in a dry state, with the cement to be used in the cementitious mixture. It may also be added to the water used in preparing the mixture, or it may be introduced into the cementitious mixture in any other manner so long as it is reasonably uniformly dispersed in the final mixture before setting occurs. The optimum form of the additive should comprise at least 3 ingredients, including at least one selected from each of the following catagories:

Category 1

The partial fatty acid esters of glycerol.
The partial fatty acid esters of sorbitan.
The partial fatty acid esters of sorbitol.
The ethoxylates of the partial fatty acid esters of sorbitan.
The ethoxylates of the partial fatty acid esters of sorbitol.
The partial fatty acid esters of choline glycerophosphates such as are termed lecithins.

Category 2

The sulfates of the partial fatty acid esters of polyglycols.
The sulfate salts of the partial fatty acid esters of polyglycols.
The sulfates of the partial fatty alcohol ethers of polyglycols.
The sulfate salts of the partial fatty alcohol ethers of polyglycols.
The sulfates of the partial fatty acid esters of glycerol.
The sulfate salts of the partial fatty acid esters of glycerol.

Category 3

The sulfonated condensation products of formaldehyde and a naphthalene.
The salts of the sulfonated condensation products of formaldehyde and a naphthalene.

For optimum results, it has been found that a significant proportion of an ingredient from each of the above categories must be present in the additive. The proportions of the several ingredients should preferably be on the order of: 25 to 50% Category 1; 10 to 35% Category 2; and 20 to 80% Category 3.

Alkyl Oxydibenzene Disulfonate and Fatty Ester

W.A. Proell; U.S. Patent 3,468,684; September 23, 1969 describes an additive for cementitious mixtures capable of reducing the water-cement ratio, the thixotropy and the air entrainment of such a mixture. The additive consists of an alkyl oxydibenzene disulfonate and the fatty acid ester of a sorbitan or the fatty acid mono ester of glycerol. The additive mixture is used in a ratio, by weight, of 0.5 to 5.0 parts additive to 100 parts cement.

Carbohydrate Fermentation Products

According to a process described by *A. Klein; U.S. Patent 3,536,507; October 27, 1970* one or more characteristics of hydraulic cementitious mixture are improved by including a mixture which comprises between 0.02 and 1.0 percent by weight, reckoned on a dry basis on the cement content of the mixture, of the part of a fermentation liquor, produced by fermenting an aqueous carbohydrate medium aerobically in the presence of an assimilable nitrogen compound under conditions to produce glutamic acid or its precursors, that remains after the separation of glutamic acid.

Example 1: A plurality of fresh concrete batches were prepared from portions of the same Portland cement of ASTM Type II and portions of a stock of graded mineral aggregate from 0 to ¾ inch in size and having a fineness modulus of 5.0, in the ratio of 1:6.52, cement to total aggregate by weight.

The first batch was the control, the others contained the mixture indicated by capital letters in the table below, which denote: FEL, a fermentation end liquor having a solids content of 39% by weight and produced by submerged aerobic fermentation of beet sugar molasses with an organism of the genus Corynebacterium in the presence of ammonia, from which glutamic acid was separated; A and B, different commercial mixtures purchased in the open market, both classified as modified hydroxylated carboxylic type, used to reduce the water required and thereby promote early strength. The solids content of mixtures A and B is not known but is believed to be higher than that of FEL.

Admixture		Water to cement ratio by weight	Compressive strengths							
			1 day		3 days		7 days		28 days	
Type	Dosage, fl. oz. per sack of cement		P.s.i.	Percent	P.s.i.	Percent	P.s.i.	Percent	P.s.i.	Percent
None		0.59	470	100	1,080	100	1,670	100	3,060	100
FEL	[1] 6.00	0.54	620	132	1,550	143	2,290	137	3,640	119
A	2.75	0.54	680	145	1,210	112	2,050	123	3,410	111
B	2.50	0.55	640	136	1,310	121	1,910	114	3,250	106

[1] 2.34 oz. on basis of 100 percent solids.

The quantities of the mixture shown in the table above, were reckoned as fluid ounces per sack (94 pounds) of cement, and were selected to achieve a nominal slump of 3½ inches (about the average consistency used in construction practice) with approximately the same water to cement ratio; the same slump was attained in the control batch by the use of more water. The concretes were molded in 3 x 6 inch cylinders and cured at standard conditions in a moist atmosphere. Strengths were determined at ages 1, 3, 7 and 28 days; each value tabulated is the average of three specimens of concrete.

It will be noted that compressive strengths are reported both in pounds per square inch and as the percentages of the strengths of the control batch. The data show that FEL was superior at all ages except at one day, but as is indicated hereafter, even the one day strength was easily raised with a slight increase in the dosage of FEL. Mixtures A and B were used at dosages about those considered optimum in construction practice. The effect of large increases in dosage of mixtures A and B may cause a drastic reduction of early-age strengths.

Example 2: To demonstrate the efficacy of the mixture when used in greatly varying dosages, one control and four test batches containing various amounts of FEL were prepared, molded, cured and tested as described for Example 1, but using another cement, also ASTM Type II, and adjusting the water to a nominal slump of 3 inches. The results are shown in the table below.

The data show that the mixture was progressively effective at increasing dosages, and no significant deterioration in strength (save at one day) was noted even when 16 fluid ounces per sack of cement (94 pounds) were used. Using the data for FEL from the first table for a different control concrete, it is demonstrated that while the optimum dosage is 6 to 8 fluid ounces of liquid, or 2.3 to 3.1 ounces on the basis of solids, there is no significant effect on strengths after one day, even though the dosage was doubled.

Admixture			Water to cement ratio by weight	Compressive strengths							
	Dosage per sack of cement			1 day		3 days		7 days		28 days	
Type	Fl. oz.	Oz. of solids		P.s.i.	Percent	P.s.i.	Percent	P.s.i.	Percent	P.s.i.	Percent
None			0.62	530	100	1,150	100	1,840	100	3,290	100
FEL	2	0.78	0.60	580	109	1,350	119	1,890	103	3,270	99
FEL	4	1.56	0.57	680	128	1,430	124	2,120	115	3,560	180
FEL *	6	2.34			132		143		137		119
FEL	8	3.12	0.55	720	136	1,650	143	2,410	131	3,680	112
FEL	16	6.24	0.54	550	104	1,650	143	2,540	138	3,700	113

*Data from the first table.

Saccharide Polymers

T.M. Kelly, R.C. Mielenz and R.B. Peppler; U.S. Patent 3,432,317; March 11, 1969; assigned to Martin-Marietta Corporation describe a hydraulic cement additive which, unlike simple sugars, does not severely retard the hydration or hardening of Portland cement and is greatly superior to the dextrins in the amount they increase the strength of concrete and mortar.

The process provides a multiple-component additive for hydraulic cement mixes, which comprises from 1 to 40 parts of saccharide polymers composed of glucose units with the saccharide polymers having a size range from 3 glucose units to on the order of 25 glucose units, from 0.5 to 90 parts of a water-soluble chloride, and from 0.2 to 10 parts of a water-soluble amine.

To illustrate the process, a plain concrete mix was prepared and compared with similar concrete mixes to which had been added increasing amounts or dosages of a glucosaccharide having a polymer size in the range of from 3 to 25 glucose units. In all the similar concrete mixes the proportions and kinds of cement, sand, and coarse aggregates in the concrete were substantially the same. A sufficient amount of water was added to each mix to effect hydraulic setting of the cement and to produce concrete mixes of essentially the same slump. The results are shown in the table below.

	Percentage Addition by Weight of Cement	Compressive Strength of Concrete, psi 7 Day	28 Day
Plain concrete	None	2,665	4,265
Glucosaccharide (3 to 25 glucose units)	0.05	2,925	4,595
Glucosaccharide (3 to 25 glucose units)	0.10	3,315	5,080
Glucosaccharide (3 to 25 glucose units)	0.15	3,360	5,160
Glucosaccharide (3 to 25 glucose units)	0.20	3,555	5,655
Glucosaccharide (3 to 25 glucose units)	0.30	3,690	6,260
Glucosaccharide (3 to 25 glucose units)	0.40	2,850	5,680

It will thus be seen that varying amounts of the glucosaccharide produce advantages in the compressive strength of cement mixes.

Polymeric Acrylic Resin

D.J. Peters and R.J. Frazier; U.S. Patent 3,538,036; November 3, 1970; assigned to Harry T. Campbell Sons Corporation describe a dry cement composition comprising 20 to 40 weight percent Portland cement, 60 to 80 weight percent aggregate, 1.3 to 5.5 weight percent powdered emulsifiable copolymeric methyl methacrylate-ethyl methacrylate resin, 0.08 to 0.55 weight percent emulsifying agent and 0.08 to 0.55 weight percent setting accelerator.

The emulsifiable acrylic resin employed is a copolymeric methyl methacrylate-ethyl methacrylate, known as Resin E-460 (Rohm and Haas Co.) having the following properties: appearance, white powder; pH (reemulsified, 10% solids), 9.5 to 10.0; bulk density, 31.2 lb/ft^3; mean particle size after reemulsification, 0.11 microns; particle size of powder, 50.0 to 75.0 microns. It has been found that to successfully use the powdered, emulsifiable acrylic resin, sodium citrate must be employed as an emulsifying agent. It has also been found that the presence of sodium citrate in the defined amounts acts as a setting retarder and that to overcome this disadvantage a setting accelerator, preferably sodium carbonate, must be used to achieve the advantages of this process.

Example 1: A topping or patching composition is prepared by intimately mixing the following dry ingredients:

	Percent by Weight
Sand	68.963
Portland cement	27.50
RE-460 (copolymeric methyl methacrylate-ethyl methacrylate)	2.75
Sodium citrate	0.137
Sodium carbonate (anhydrous)	0.275
Colloids 513-DD (defoamer)	0.070
Trimethanol ethane (plasticizer)	0.270
Methylcellulose	0.035
Water	$<$0.1%

To 100 parts by weight of the above dry mix composition there is added in a cement mixer with agitation 18 parts by weight of H_2O. The topping composition is then trowelled sufficiently to force the mix into the pores of an old concrete surface thus providing a neat looking fresh concrete surface.

A similar hydraulic composition using instead 20 parts by weight of water per 100 parts of dry mix ingredients is also suitable for patching or topping irregular or cracked concrete surfaces and provides equally advantageous results. The freshly applied cement composition of this process sets in one hour and can be exposed to light traffic within 24 to 48 hours or heavy traffic after about three days. The fresh surface can be applied in thicknesses up to 0.5 inch or greater.

Example 2: 100 parts by weight of the dry mix cement composition of Example 1 was mixed with 40 parts by weight of water to produce slurry having a thick paint-like or thick pancake batter-like consistency. A worn driveway previously cleaned of oil and grease spots by application thereto of a suitable solvent such as mineral spirits, gasoline, conventional cleaning fluids or the like followed by washing with hot water and a household detergent was resurfaced by spreading the above hydraulic cement composition thereon. Generally, the cement composition can be applied in amounts sufficient to provide a fresh surface having a thickness of $^1/_{16}$ inch.

A similar hydraulic cement composition is useful as a stucco finish, as a bonding agent between layers of fresh and hardened concrete or as a sealing agent for swimming pools and patio decks. The composition can be applied by brushing, spraying or trowelling.

Lime-Sugar Additive Blend

C.F. Booth and B.M. Whitehurst; U.S. Patent 3,262,798; July 26, 1966; assigned to Mobil Oil Corporation describe a process of forming high-strength cements from phosphate ores having a high fluorine content. This process also relates to a method for increasing the strength of high fluorine-containing calcium aluminate cements and, in another aspect, relates to a particular additive, per se, which has been found to increase the strength of such cements by a factor of up to 15.

Calcium aluminate or so-called aluminous cement and methods for its production are well known. Aluminous cement is characterized in that it contains roughly equal amounts of CaO and Al_2O_3, e.g., in the neighborhood of 40% of each and a relatively low content of SiO_2, e.g., not more than 12%. The remainder of the cement is constituted by what may be regarded as being impurities such as compounds of alkali metals, F, Fe, Mg, Ti, S, Zr, etc. Aluminous cement is further characterized by its high early strength, as compared to Portland cement, and by other useful properties such as resistance to acids, alkalies, sea water and aqueous solutions of sulfates.

A preferred additive to aluminous cement is a mixture of lime and sugar in a ratio of from 1:1 to 6:1 by weight. While the mixture can be ground to the size for the desired cement, the very fine particles of sugar become hygroscopic and, unless a carrier is added to the lime-sugar mixture soon after grinding, water picked up by the sugar tends to prevent an even distribution throughout the cement. Accordingly, the amount of carrier for the lime-sugar mixture should be sufficient to prevent the sugar from picking up water, but not enough to have a deleterious effect on the cement. Carriers are those of nonporous, nonhydratable materials which do not adversely affect the properties of the cement, and include silica, mullite, granite and bauxite.

The amount of additive, including the carrier for the sugar-lime mixture, is from 1.5 to 3% by weight of the aluminous cement slag composition. To impart the improved high-strength properties to the aluminous cement, the sugar should be present in the final cement in amounts of from 0.1 to 0.3% by weight of the cement, with excellent results being achieved when the amount of sugar is within the range of 0.13 to 0.17%. About 0.15% sugar in the cement is usually preferred. An aluminous cement slag composition was prepared having the following analyzed components.

Ingredients	Slag Analysis, wt %
CaO	38.29
Al_2O_3	47.19
SiO_2	5.57
P_2O_5	2.50
F	2.43
Fe_2O_3	0.51
Insoluble	1.06
Al_2O_3/CaO	1.23

In the slag composition were uniformly mixed the additives set forth in the following table and the various cements were tested for compressive strengths at the end of 1, 3, 7 and 28 days.

Weight Lime, g*	Weight Sugar, g	Lime:Sugar Ratio	Carrier	Weight Carrier, g	Strength, psi 1 Day	3 Days	7 Days	28 Days
158	50	3:1	SiO_2	450	6,192	8,058	9,033	9,375
632	150	4:1	SiO_2**	1,200	5,567	6,308	7,258	8,433
158	50	3:1	Al_2O_3	442	6,700	9,008	9,458	11,033

*95% $Ca(OH)_2$.

**RASC Bauxite.

Enough of each additive composition was used to give a final concentration of sugar in the cement of about 0.154% by weight, i.e., 2% by weight of the total additive. A preferred lime-sugar additive is a mixture of sugar, hydrated lime and powdered silica preferably in the proportions of 7.7% sucrose, 15.4% $Ca(OH)_2$ and 76.9% SiO_2. This composition is preferred because sucrose alone is quite critical and the amount which may be added to a particular cement must be controlled within quite narrow limits whereas the lime appears to moderate the effect of the sugar and to make the amount that may be used less critical.

Although lime or hydrated lime may be used with sugar it is not an effective additive alone, due to the tendency of the sugar, after the material is ground, to absorb water from the air. The silica component of the above-described additive composition is inert and acts merely as a carrier of diluent for the sugar. The following examples illustrate the process.

Example 1: A cement was prepared in a phosphorus reduction furnace by the continuous feeding of naturally occurring, fluorine-containing raw materials. The burden composition consisted of 100 parts of Florida phosphate rock, 64 parts of bauxite and 19.25 parts of coke, each having a typical analysis. This composition was preselected so that the compensated composition of the cement portion of the slag after smelting (exclusive of the CaO equivalent of fluorine, titania (TiO_2) and P_2O_5, would have a compensated alumina to lime ratio of 1.34 and would lie inside of the defined **ABCD** area of the phase diagram shown in Figure 2.3.

The molten slag thus formed was trapped at regular intervals and cooled in carbon crucibles to give pigs of 50 to 135 pounds, with no attention being paid to the rate of cooling of the slag in air. The slag cooled from 1600°C to room temperature in about 4 hours. The slag contained by gross analysis 42.38% CaO;

FIGURE 2.3: ALUMINOUS CEMENT SLAG COMPOSITION

a.

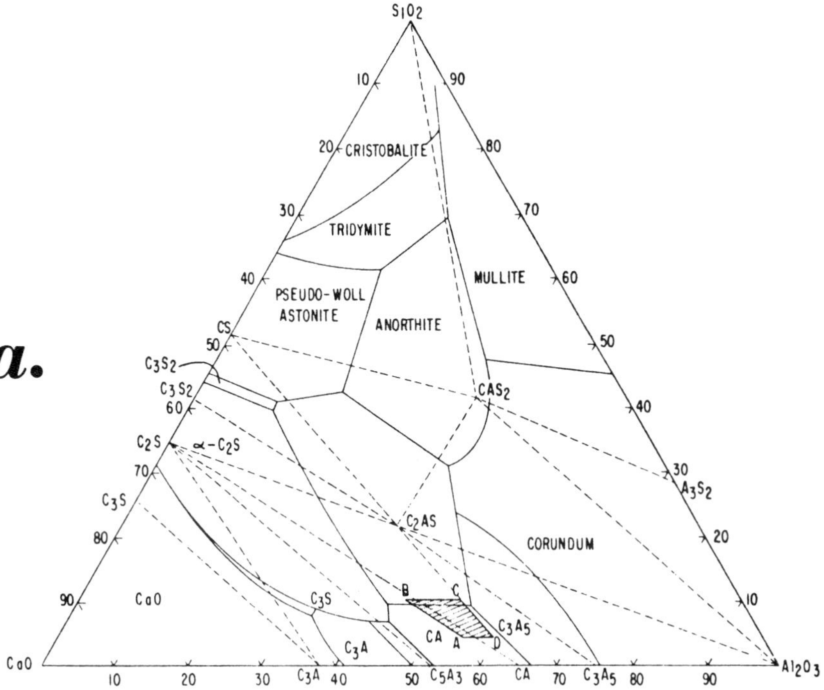

Liquidus Surface of the System CaO-Al_2O_3-SiO_2

b.

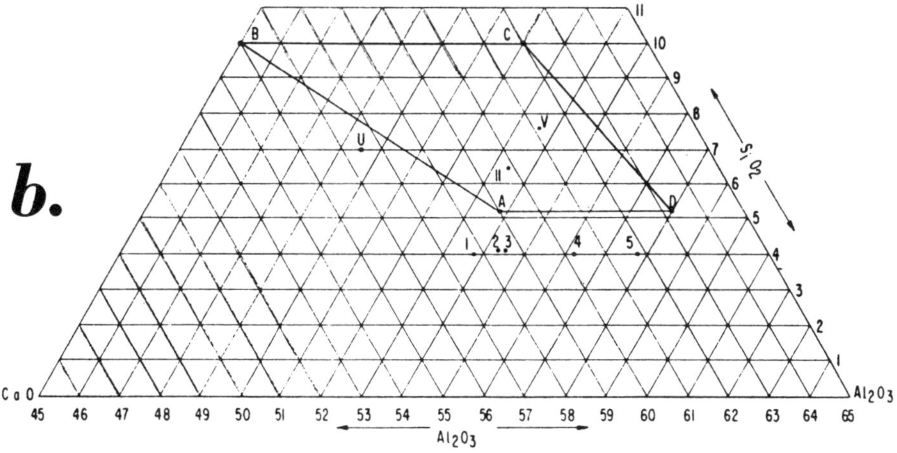

Enlarged View of Area **ABCD** from Figure 2.3a

Source: U.S. Patent 3,262,798

48.26% Al_2O_3; 5.80% SiO_2; 2.51% fluorine; 1.08% P_2O_5; and 0.28% Fe_2O_3. The slag was ground to a standard fineness, i.e., 3,000 cm^2/g as was determined by the Blaine air-permeability test. Cements containing no additive and 2% by weight of an additive mixture consisting of 7.7% sucrose, 15.4% $Ca(OH)_2$ and 76.9% silica were prepared and tested according to a standard ASTM procedure (ASTM C-109). The strengths of the cements with and without additive are summarized in the following table.

	Compressive Strength, psi			
	1 day	3 days	7 days	28 days
Without additive	1,845	2,847	2,232	-------
With additive	5,692	10,008	10,800	11,017

Example 2: A cement was prepared by preselecting a composition that would yield an alumina to lime ratio which, after compensating the gross composition for F, TiO_2, P_2O_5 and their CaO equivalents to CaF_2, $CaO\cdot TiO_2$ and $Ca_3(PO_4)_2$ would be within area **ABCD**. The burden consisted of 100.00 parts Florida hard rock phosphate (containing 34.85% P_2O_5, 49.57% CaO, 5.27% SiO_2, 3.66% F, 0.73% Al_2O_3 + TiO_2, 0.52% Fe_2O_3, 2.44% CO_2, and 0.39% moisture); 60.90 parts Surinam bauxite (containing 94.16% Al_2O_3 + TiO_2, 3.97% SiO_2, and 1.20% Fe_2O_3); and 18.81 parts coke (containing 86.00% fixed C, 5.32% SiO_2, 0.75% CaO, 3.37% Al_2O_3 + TiO_2, and 1.48% Fe_2O_3).

The slag formed after reduction of the phosphate to elemental phosphorus in an electric furnace contained by gross chemical analysis 42.10% CaO; 45.84% Al_2O_3 + TiO_2; 6.24% SiO_2; 1.73% P_2O_5; 2.76% F; and 1.55% Fe_2O_3. The slag was ground to a fineness of about 3,000 cm^2/g (Blaine air-permeability test) and tested with and without the same additive used in Example 1. Results of compressive strength tests (ASTM Test C-109) are summarized in the table below.

	Compressive Strength, psi			
	1 day	3 days	7 days	28 days
Without additive	1,385	1,924	1,924	-------
With additive	6,583	8,667	8,800	10,338

Example 3: Cements were prepared by combining and fusing Demarara bauxite (containing about 5.3% silica), technical grade lime and, where needed, calcium fluoride. A gross weight ratio of alumina to lime (inclusive of the CaO equivalent to CaF_2) of 1.22 was chosen. The fluorine content of the slag was varied from 0 to 4%. The corrected silica varied from 3.9 to 4.6%. Each fused charge was subsequently cooled from 1600° to 1000°C in 45 minutes and from 1000° to 400°C in 200 minutes.

The resulting slag was ground to a standard fineness and the cements therefrom were tested with and without an additive according to ASTM Procedure C-109. The additive consisted of ground sugar-lime-silica mixture containing the following: 7.7% sucrose; 15.4% $Ca(OH)_2$; and 76.9% silica. The amounts used were based on the weight of ground slag. The results are shown in the table below.

Compressive Strength of Various Cements

	Slag				Compressive Strength, psi		
	Al_2O_3/CaO ratio						
Number	Gross	Compensated	% F	Additive	1 day	7 days	28 days
1	1.23	1.27	0	None	4,800	8,150	8,150
2	1.22	1.30	1.1	None	550	175	150
3	1.17	1.31	2.3	None	688	288	250
4	1.22	1.41	3.1	None	3,125	3,113	2,250
5	1.24	1.51	4.1	None	1,000	2,925	2,425
1	1.23	1.27	0	2%	5,125	8,038	10,000
2	1.22	1.30	1.1	2%	3,700	5,450	6,050
3	1.17	1.31	2.3	2%	4,888	6,013	5,250
4	1.22	1.41	3.1	2%	8,000	13,000	13,600
5	1.24	1.51	4.1	2%	5,275	14,100	14,600

The dramatic improvement of the fluorine containing cements having the additives is readily evident from the above table. Note that the presence of the additive in the control cements (containing no fluorine) did not appreciably affect the final strengths of the cements.

Styrene-Butadiene Latexes for Reducing Slump

P.K. Mehta; U.S. Patent 3,895,953; July 22, 1975; assigned to The Regents of the University of California has found that in freshly mixed Portland cement concrete, slump loss, which normally occurs during transportation and handling, can be either prevented or considerably reduced by admixing small amounts of styrene-butadiene latexes, e.g., 0.01 to 0.15% of latex (on solid basis) by weight of concrete or 0.1 to 1.5% of latex by weight of cement. Thus, by maintaining plasticity of fresh (unhardened) concrete for prolonged periods of time, the small amount of latex mixture can permit transportation of premixed concrete over longer hauling distances, or permit longer handling time for placement, consolidation, and finishing of concrete in formwork.

A typical styrene-butadiene latex, stabilized with a nonionic surfactant has about 48% solids by weight, a specific gravity of 1.01 (at 25°C), a pH of 10.5, and an average particle size of polymer of about 2000 Angstrom units. The effect of styrene-butadiene latexes on the early slump loss of a wide variety of Portland cement concretes was tested. The following tables present results of examples showing reduction of early slump loss in various concrete compositions, when a styrene-butadiene latex was used. Unless otherwise stated, the ratio of styrene to butadiene of the latex used in the tests was 66:34.

Early Slump Loss of Type I Portland Cement Concrete

Latex, % by Wt. of Cement	Temperature of Mixing Water °F	Initial Concrete (mixed 5 minutes)		Final Concrete (mixed 30 minutes)		Slump Loss (−) or Gain (+) in.
		Temp., °F	Slump, in.	Temp., °F	Slump, in.	
0	122	84	6	77	3½	−2½
0.75	122	86	7½	78	8½	+1
0.50	122	86	7½	77	8	+½
0.50	75	75	8	78	8	0

Early Slump Loss of Type II Portland Cement Concrete

Latex, % by Wt. of Cement	Temperature of Mixing Water °F	Initial Concrete (mixed 5 minutes) Temp., °F	Slump, in.	Final Concrete (mixed 30 minutes) Temp., °F	Slump, in.	Slump Loss (−) in.
0	120	86	7	75	4¼	−2¾
0.25	120	84	7	75	7	0
0.25	75	75	8	77	8	0
0.10	120	85	7	76	6	−1
0.10	75	75	8	77	7	−1

Early Slump Loss of Type M Expansive Cement Concrete

Latex, % by Wt. of Cement	Temperature of Mixing Water °F	Initial Concrete (mixed 5 minutes) Temp., °F	Slump, in.	Final Concrete (mixed 30 minutes) Temp., °F	Slump, in.	Slump Loss (−) or Gain (+), in.
0	122	84	5½	77	2½	−3
0.75	122	84	7½	75	8½	+1
0.50	75	75	8	77	8	0

Early Slump Loss of Type S Expansive Cement Concrete

Latex, % by Wt. of Cement	Temperature of Mixing Water °F	Initial Concrete (mixed 5 minutes) Temp., °F	Slump, in.	Final Concrete (mixed 30 minutes) Temp., °F	Slump, in.	Slump Loss (−) or Gain (+), in.
0	122	85	6	76	3	−3
0.75	122	85	7¾	75	8½	+¾
0.50	75	75	8	78	8	0

Crystallized Magnesium Sulfate as Thixotrope

According to a process described by *R.G. Lange and W.D. Kobeski; U.S. Patent 3,847,635; November 12, 1974; assigned to United States Gypsum Company* cementitious mixtures of alpha calcium sulfate hemihydrate, Portland cement and a dispersing agent are rendered thixotropic by the inclusion of small amounts of crystallized magnesium sulfate.

Among the suitable dispersing agents or fluidizing agents, which are necessary to impart the initial dilatant characteristics, are Lomar D, which is a condensate of naphthalene sulfonic acids and formaldehyde, and other similar dispersing agents under the Lomar brand; Blancol which is a sulfonated condensate of formaldehyde and naphthalene; and guar gum. Any of the known fluidizing or dispersing agents for cementitious materials can be used in the practice of the process. The fluidizing agent is added in the customary amounts, generally about 0.5 to 0.8% by weight of the total dry mixture. Somewhat more or less may be used but without further substantial advantages. Although the magnesium sulfate ingredient may be in any of the different state of hydration forms, it is preferred to use the heptahydrate, commonly called crystallized magnesium sulfate and also commonly known as Epsom salts. The amount of crystallized magnesium sulfate to be used depends on the amount of fluidizing or dispersing agent also used in the cementitious mass, thus it is generally preferred to employ about 0.2 to 0.5% by weight of the total dry mixture; although somewhat more or less may be used but without further substantial advantage. The following examples illustrate the process.

Example 1: The following materials were dry blended to form a neat cementitious blend: 94% by weight alpha calcium sulfate hemihydrate; 5% by weight Type I Portland cement; and 0.5 to 0.75% by weight Lomar D dispersing agent. To aliquots of the above blends were added 0.25% (or about 8 pounds per ton) of crystallized magnesium sulfate. Then both the neat material and the material to which the magnesium sulfate had been added were mixed with water and expansion and temperature rise tests were conducted, with the following results:

	Neat Cementitious Mass	Cementitious Mass + $MgSO_4{\cdot}7H_2O$
Mix consistency, cc	24	24
300 gram Vicat set, minutes	26	16
Maximum expansion, percent	0.218	0.179
Maximum temperature rise, °C	61.7	66.5

From the above it is clearly evident that the magnesium sulfate addition produced a shorter set time and a higher temperature rise, and more importantly decreased the total expansion by 18.3%. To further show the effect of the magnesium sulfate in eliminating the dilatant property of the neat material, the following test on flowing characteristics and screedability of the material was conducted.

This test involved pouring material on a nonabsorbant surface at a 5% incline. Aliquots of the neat cement and cement plus magnesium sulfate prepared above were poured onto a felt covered piece of board that had three sides of the board provided with a ⅝-inch frame. The materials were poured onto the boards at the top and permitted to flow down the 5% incline. The frame around the board was left in place until the material set. While the material was setting attempts were made to screed the material. The neat material easily flowed all the way down the board and could not be screeded up the 5% incline because of its dilatant property at any time prior to setting.

The slurry containing magnesium sulfate had to be puddled using a spatula to get the slurry to flow. Once on the board, the slurry would not flow unless it was again puddled. After sufficient material was on the board and puddled slightly, the material was screeded to the full ⅝-inch thickness. When the screed bar was brought down the slope, the bottom frame member was knocked off. No material flowed out and all the material remained at its full ⅝-inch thickness, even though the board was at a 5% incline. This material screeded very smooth with no skip marks, pulling, or air pockets.

Once a final pass was made with a screed bar, no finishing had to be done because the material set to a smooth finish. After the final screeding pass on the board using the cement containing magnesium sulfate, all the frame members were removed; and even though the material had not set, the material remained at its full ⅝-inch thickness.

Example 2: To a neat cement slurry as set forth in Example 1 was added a quantity of crystallized magnesium sulfate. For this evaluation, the consistency used was intentionally dropped to 22 cc. Because this drop in consistency also affects the dilatancy, the amount of magnesium sulfate was reduced from 8 to 10 pounds per ton to 5 to 6 pounds per ton when the magnesium sulfate was added to the mixed slurry and 7 to 8 pounds per ton (for same wet distribution and flow characteristics) when the magnesium sulfate was dry blended into the neat cement.

This drop in consistency was to take maximum advantage of high green strength and dry strength for slush casting. Aliquots of the materials were placed into thin wall latex molds of the type used in slush casting shops. The molds had previously been treated with a slight amount of mold dressing to eliminate air bubbles on the surface of the cast.

The material performed well as a thixotropic material which would air-dry to a high compressive strength in a slush casting operation. It was observed that as long as energy was being applied, i.e., as the slush cast mold was being filled and agitated, the slurry flowed over the surface, filling all of the voids in the mold. Once the input of energy was ceased, i.e., as agitation was stopped, the material stopped flowing and started to body up thus giving a uniform covering of the mold surface inside the cast. After the cast had set, no artificial drying was required because of the low amount of excess water due to casting at almost theoretical minimum water level.

Epoxy-Containing Concrete

M.J. Hillyer; U.S. Patent 3,477,979; November 11, 1969; assigned to Shell Oil Company describes compositions which comprise cement, aggregate and an epoxy resin composition which contains a polyepoxide, an epoxy curing agent and a petroleum hydrocarbon fraction containing over 55% by weight aromatics. The presence of the aromatic hydrocarbon increases the strength of the concrete produced by this process over compositions containing a cured polyepoxide, but containing no hydrocarbon. Suitable amounts of hydrocarbon to be used are between 10 and 70% and preferably between 20 and 50% by weight of the epoxy resin composition. The total amount of epoxy resin composition used in preparing the cement mixture will vary depending on the strength of the concrete desired as well as economic considerations. Generally, an amount of epoxy resin composition between 0.5 and 20% and preferably between 1 and 10% is suitable where:

$$\text{Percent Epoxy Composition} = \frac{\text{Weight Epoxy Composition}}{\text{Weight Cement + Weight Aggregate}} \times 100$$

Example 1: A cement composition was prepared as follows. 1,504 grams of Portland cement Type III, 3,640 grams of top sand and 2,280 grams of pea gravel were placed in a Hobart mixing bowl which bowl was placed in a mixer. Water (650 grams) was added and thoroughly mixed in after which the bowl was removed from the mixer and the concrete towelled into forms. The concrete was allowed to cure into bricks measuring 2 x 2 x 12 inches used for flexural and compressive strength tests and dog-bones as described in ASTM C-190 used for tensile tests. The samples were covered with a Mylar sheet and cured for about 22 hours at 75°F, then released from the forms and cured additionally for 6 days at 75°F.

Example 2: (a) The same procedure described in Example 1 was repeated with the exception that 500 grams of water was used and 325 grams (4.4%) epoxy resin composition was added to the wet cement mixture and mixed in the mixing bowl immediately after the water had been added and then thoroughly mixed. The epoxy resin composition consisted of 33.33% of a polyepoxide (Epon 828) having an epoxide equivalent of 180 to 195 (grams of resin containing one gram equivalent of epoxide) and an average molecular weight of approximately 380, 20.0% of a polyamide curing agent (Versamid 140) prepared from dilinoleic acid and ethylene diamine and having the amine value of 375 and a specific gravity (25°/25°C) of 0.97, and a mixture of aromatic hydrocarbons, 16.67% Dutrex 298 and 30.0% Dutrex 739. Dutrex 298 and Dutrex 739 have the following properties:

	Dutrex 298	Dutrex 739
Viscosity:		
SSU/100°F	204	15,800
SSU/210°F	40	107
Gravity, °API	11.3	5.2
Specific gravity/60°F	0.9909	1.0351
Boiling range, °F	568-724	740-885
Aniline point, °F	22.5	13
Viscosity gravity constant	0.967	0.999
Molecular analysis, clay-gel, weight %:		
Polar compounds	9.0	23.0
Aromatics	80.9	70.3
Saturates	10.1	6.7

The epoxy resin compositions were prepared by mixing the polyepoxide, the curing agent and hydrocarbon. The epoxy resin compositions were then reacted for one hour at 75°F prior to their addition to the cement composition.

(b) The procedure of (a) was repeated except that the epoxy resin composition was not cured prior to addition to the cement compositions.

Example 3(a) and 3(b): The procedures of Example 2(a) and 2(b) were repeated respectively with the exception that only 400 grams of water was used to mix the cement composition.

Example 4(a) and 4(b): The procedure of Example 2(a) and 2(b) respectively was repeated with Dutrex 298 used solely as the hydrocarbon.

Example 5(a) and 5(b): The procedure of Example 4(a) and 4(b) was repeated respectively with the use of 400 grams of water in preparing the cement mixture.

The hardened concrete samples prepared in the Examples 1 through 4 were then tested for tensile strength on a Tinius-Olsen testing machine using a crosshead speed of 0.05"/min. The results of the test are set forth below.

Example	Tensile Strength (% of Example 1)
1 (control)	100
2(a)	174
2(b)	139
3(a)	179
3(b)	154
4(a)	165
4(b)	160
5(a)	154
5(b)	149

The concrete samples prepared in Examples 1 through 5 were then tested for flexural strength according to ASTM C293. The results are set forth below.

Example	Flexural Strength (% of Example 1)
1 (control)	100
2(a)	173
2(b)	118
3(a)	197
3(b)	145
4(a)	134
4(b)	115
5(a)	153
5(b)	122

The concrete samples prepared in Examples 1 through 5 were tested for compressive strength according to ASTM C116 with the use of bearing blocks 2" square. The results are given below.

Example	Compressive Strength (% of Example 1)
1 (control)	100
2(a)	124
3(a)	145
4(a)	112

Mixing Technique

R.W. Previte and D. Chin; U.S. Patent 3,812,076; May 21, 1974; assigned to W.R. Grace & Company describe a practical and efficient method for dispensing, transferring, and effectively dispersing low addition rate, solid, particulate, hygroscopic, difficultly-dissolvable mixtures for hydraulic cement compositions, such as high molecular weight polymers of ethylene oxide. The method provides for easy conveyance and optimum dispersion of the relatively small amount of solid particles throughout the large volume of the composition mix and is easily and conveniently adapted into the modern automated mixing operations.

The method ideally enables measuring and dispensing of the solid particulate at a point remote from the mixing vessel and simple automated transfer of the measured quantity in dispersed form to the mixing vessel for effective dispersion therein.

According to the method the exact desired amount of the powdered mixture is measured from a supply source by any convenient metering means. The measured amount of powder is then introduced to a moving, confined stream of gas, for example air under pressure, for instance by siphoning the solid particles into the moving stream. The moving gas-solid mixture or dispersion formed is then conducted to the mixing vessel into which it is released along with at least one of the remaining components of the cement composition, for example, the aggregate component.

WELL-CEMENTING COMPOSITIONS

SET MODIFIERS

Boric Acid and Salt of Lignosulfonic Acid

R.C. Martin; U.S. Patent 3,753,748; August 21, 1973; assigned to The Dow Chemical Company describes an aqueous hydraulic cement slurry and method of cementing wells employing it. The slurry contains as a required additive an improved retardant to the rate of setting thereof to a monolithic solid without adverse effect on ultimate compressive strength or other desirable qualities.

The additive is a previously unknown combination which is much more effective at the usually troublesome higher temperatures than would be expected from the behavior of each of the components of the combination when used separately in such slurry. The combination retardant is (a) an alkali metal, alkaline earth metal or alkali metal-alkaline earth metal salt of lignosulfonic acid and (b) boric acid or a borate, e.g., borax. The combination results in a truly synergistic effect.

Excellent results are obtained at between about 175° and 400°F. The process is commonly practiced at between 230° and 350°F. There is seldom a need for use of the retardant at lower temperatures unless an exceptionally long setting period is desired.

At temperatures above 400°F the time existing between mixing and emplacement may thereby be caused to be unduly short. However, if an extremely short setting period is desired, the process will fill that need. The amount of the additive retardant based on 100 parts by dry weight of the hydraulic cement present is between 0.5 and 5.0 parts of component (1), sodium calcium lingosulfonate, and between 0.5 and 5.0 parts of component (2), borax, to make a total of both (1) and (2) of between about 1.0 and 10.0 parts based on the dry weight of cement present. Between about 1 and 4.0 parts of each component are commonly and preferably employed. The following basic recipe was followed in the comparative tests which are not illustrative of the process.

100 parts by dry weight of Class H cement (as described in API RP 10B), 30 to 60 parts of water, and the amounts of either component (1) or (2), but not both (1) and (2), as shown in the following tables, were admixed. The borate used was a technical grade $Na_2B_4O_7 \cdot 10H_2O$. Class H cement is a high temperature cement and was used herein because it best represents practical operating or field usage under the conditions herein represented. Other cements can be used in the practice of the process but may not be the best choice because Class H is especially made for use at unusually high temperatures.

The examples of the process contain both components (1) and (2), water, and a Portland or aluminous cement.

Series One: Although silica flour was used in this series to impart additional strength to the set cement, it is not essential. Tests and examples of Series One contained 35 parts by weight, per 100 parts of cement, of silica flour having an average particle size of between about 100 and 200 mesh.

The composition employed both in the comparative tests and examples of the process contained 46 parts of water per 100 parts dry weight of Class H cement. Tests were conducted in accordance with API RP 10B as set out in Section 10 Schedule 11.

The results giving the thickening time of Class H cement containing 35% by weight of 100 to 200 mesh silica flour and of the indicated amounts of either (1) sodium calcium lignosulfonate or (2) borax or both (1) and (2) tested according to API casing cementing schedule 11 simulated 20,000 feet and 340°F circulating temperature are shown below.

Total Retardant, %*	Ratio (1):(2)	API Thickening Time** (hr:min)
0	-	0:52
1.0 of (2)	-	1:01
2.0 of both (1) and (2)	1:5	1:28
2.0 of both (1) and (2)	1:3	1:23
2.0 of both (1) and (2)	1:1	1:28
2.0 of both (1) and (2)	3:1	1:15
2.0 of both (1) and (2)	5.1	1:21
2.5 of both (1) and (2)	1:5	2:54
2.5 of both (1) and (2)	1:3	5:49
2.5 of both (1) and (2)	1:1	3:30
2.5 of both (1) and (2)	3:1	1:06
2.5 of both (1) and (2)	5:1	1:15
3 of (1)	-	1:36
3 of (2)	-	1:29
3 of chromium lignosulfonate	-	1:40
3.0 of both (1) and (2)	1:5	+6:00
3.0 of both (1) and (2)	1:3	+6:00
3.0 of both (1) and (2)	1:1	+6:00
3.0 of both (1) and (2)	3:1	+6:00
3.0 of both (1) and (2)	5:1	+6:00

*Based on 100 parts cement by dry weight.
**Tested by Pan-American Consistometer.

It is shown in the above table that the use of either sodium calcium lignosulfonate or borax alone does not adequately lessen the rate of setting. However, it does show that when both the sodium calcium lignosulfonate and borax were present in ratios of from 1:5 to 5:1, i.e., a ratio range of 1 part of component (1):1 part to 25 parts of component (2), the effect on the rate of setting was to slow such rate perceptibly as is desired. The chromium lignosulfonate test was run to show that other lignosulfonates alone were no better than the calcium species.

If either component (1) or component (2) had exerted the expected effect on the rate of retardation, as shown when used alone (i.e., if no unexpected synergistic effect occurred) then 1 part of each used together should have retarded the setting rate the cumulative sum of 52 and 36 minutes or 1 hour and 28 minutes. However, the table shows that the retardation caused by one part of each used together retarded the setting rate not less than 1 hour and 15 minutes and when a greater total amount than 1 part of each was employed the rate of retardation was in excess of 6 hours (actually setting at between 12 and 15 hours).

Series Two: This series of tests was conducted to ascertain the thickening time of Class H cement, again containing 35% by weight of finely divided silica flour as in Series One, tested according to various schedules in Section 10 of API RP 10B. The amount of retardant consisted of either one or both of the lignosulfonic acid salt and/or borax. The results giving the thickening time of Class H cement containing 35% of silica flour, by weight, retarded by the indicated amounts of (1) sodium calcium lignin and/or (2) borax as tested according to the API casing cement schedules indicated are shown in the following table.

	API Thickening Time** (hr:min)			
Total Retardant, %*	**7**	**8**	**9**	**10**
	172°F 12,000 ft	206°F 14,000 ft	248°F 16,000 ft	300°F 18,000 ft
0.5	3:08	-	-	-
0.6	3:21	-	-	-
0.7	4:48	-	-	-
0.8	-	2:18	-	-
0.9	-	2:06	-	-
1.0	-	2:14	-	-
1.4	-	2:08	-	-
1.6	-	+6:00	1:07	-
1.8	-	-	1:14	-
2.0	-	-	5:23	3:40
2.2	-	-	-	4:38

*Retardant consisted of 1 part (1) and 3 parts (2) based on 100 parts cement by weight.
**API schedules followed at indicated temperature and simulated depth.

It can be seen from the above table that a ratio of 1 part of sodium calcium lignosulfonate to 3 parts of borax, present in the total amounts set forth, and tested according to any one Schedules 7 to 10 resulted in greatly increased time of setting (i.e., a retarded setting rate). The greater the total amount of the two component retardant, the longer the time of setting, tested at any given schedule.

Pentaboric Acid Salts and Lignosulfonic Acid Salts

C.R. George; U.S. Patent 3,748,159; July 24, 1973; assigned to Halliburton Company describes additives for retarding the setting time of cement compositions at high temperatures which are basically comprised of mixtures of a lignosulfonic acid salt and a pentaboric acid salt.

The ratio of lignosulfonic acid salt to pentaboric acid salt can vary from about 1 to 4 parts by weight lignosulfonate per 1 part by weight of borate. Above a ratio of about 4 parts by weight lignosulfonate to 1 part by weight borate, the retarding effect on the cement composition is reduced, and below a weight ratio of about 1:1, gelation of the cement composition generally occurs.

A preferred additive is comprised of calcium lignosulfonate and potassium pentaborate with the ratio of lignosulfonate to borate being from about 1 to 2 parts by weight lignosulfonate per 1 part by weight of borate.

The retarder additives are employed in the cement compositions in an amount of from about 0.1% based on the weight of dry cement to about 10% based on the weight of dry cement. At concentrations below about 0.1% by weight little retardation is effected, and above about 10% by weight the retardation is generally too long.

However, the particular quantity of retarder additive utilized depends on a variety of factors such as the particular temperature and pressure to be encountered, the type of cement used, etc. As a general rule, the higher the temperature and pressure, the greater the amount of retarder additive required.

D-Gluco-D-Guloheptolactone

H.T. Harrison and K.J. Goodwin; U.S. Patent 3,686,008; August 22, 1972; assigned to The Dow Chemical Company describe an aqueous hydraulic cement slurry which may be retained in a pumpable fluid state for an indefinite time (and accordingly used as a well-drilling circulating fluid) by use of a heptolactone, particularly D-gluco-D-guloheptolactone, and just prior to ultimate emplacement of the slurry containing the additive, mixing in a polyvalent metal salt accelerator to the setting rate, e.g., $CaCl_2$.

D-gluco-D-guloheptolactone is considered to have the formula:

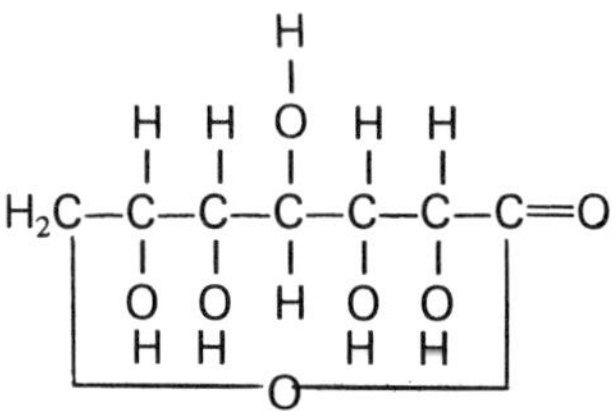

One commercial source is the Aldrich Chemical Company.

Field Example: To drill and then later cement a well in a field, the following procedure is illustrative of the process. The amounts set forth below prepare about 1,000 gallons of the aqueous drilling fluid convertible subsequently, to the cement composition of the process.

4,650 pounds of water are placed in a suitable mixing tank. 116.5 pounds of D-gluco-D-guloheptolactone (1.0% by weight of the dry cement to be used) are mixed with the water and 11,650 pounds of an API class A cement are mixed with the aqueous solution of the D-gluco-D-guloheptolactone and mixing continued until an aqueous composition of substantial homogeneity is made.

After thoroughly mixing the composition by means of the mud pumps, drilling of the well is commenced and the composition is circulated through the well by the mud pumps in the same manner as conventional drilling muds. As the well drilling operation progresses samples of the fluid composition are tested to determine the liquid and solids content. Any water lost to the formation, or lost by evaporation from the mud pits, is replaced to insure against undue increases in viscosity because of increased solids content. Drilling is continued without interruption until a depth of approximately 2,000 feet is reached at which point the well drilling is completed and is to be cemented.

582.5 pounds of $CaCl_2$, about 5% by weight of the cement solids, are added to 600 pounds of water at 85°F and mixing is continued until all solids are in solution. The resulting $CaCl_2$ brine solution is added to the drilling fluid composition, which is contained in the mud pit, and the mud pumps are operated for one hour to circulate all of the drilling fluid throughout the wellbore and mud pits, thereby insuring complete and intimate mixing of the cementitious fluid composition and the $CaCl_2$ accelerator solution.

The composition thus prepared is pumped or otherwise forced down the wellbore to the level where the cementing is desired to be done. Cementing equipment, including pumps, packers, or plugs is employed in the practice of the process in a similar manner to conventional cementing operations. The aqueous composition, after being emplaced in the well, is allowed to stand for a time sufficient for it to become a hard monolithic solid. The length of time necessary for the composition to remain undisturbed, according to practice, may be as little as 24 hours.

Modified Starch and Dextrin Mixtures

S. Maravilla and J.E. Kopanda; U.S. Patent 3,414,420; December 3, 1968; assigned to United States Steel Corporation describe a retarded oil well cement including an oil well Portland cement, and a retarder present in an amount equal to 0.08 to 0.13% by weight of the cement.

The retarder consists of 1 to 8 parts by weight of a modified starch to 1 part of dextrin. The starch has a cold water solubility range of 20 to 30% and the dextrin a cold water solubility range of 12 to 35%. The cement is made by grinding the modified starch at a maximum temperature of 100°F to such fineness that it will pass a 200 mesh sieve, and then mixing the three ingredients at a maximum temperature of 140°F.

Carboxymethylcellulose Derivatives

In a process described by *I.R. Dunlap and F.D. Patchen; U.S. Patent 3,245,814; April 12, 1966; assigned to Socony Mobil Oil Company, Incorporated* an aqueous slurry of a hydraulic cement containing a quantity of carboxymethylhydroxyethylcellulose is introduced into a well bore hole in the earth adjacent to a formation having a temperature of at least 180°F.

It has been found that an aqueous slurry of hydraulic cement containing carboxymethylhydroxyethylcellulose has an appreciably lengthened setting time. Thus, an aqueous slurry of hydraulic cement containing carboxymethylhydroxyethylcellulose may be pumped, as from the surface of the earth to a position within a well bore hole, for an appreciably longer time than in the absence of the carboxymethylhydroxyethylcellulose before setting of the cement will occur. Thus, pumping of hydraulic cement slurry may be carried out to greater depths and under more severe conditions of temperature and pressure without danger of setting prior to the cement arriving at its desired location.

R.P. Ordonez, A.A. Rosado, F.G. Petrirena and E. Rivera de Campos; U.S. Patent 3,905,826; September 16, 1975; assigned to Instituto Mexicano del Petroleo describe a composition for retarding the setting of cement which comprises a mixture of at least one of the group consisting of carboxymethylcellulose and alkali salts of carboxymethylcellulose, borax and dextrin, in the proportions by weight of 35 to 40%, 30 to 35% and 25 to 30%, respectively.

Water-Soluble Organic Polymer in an Organosolvent

In a process described by *L.H. Eilers and C.F. Parks; U.S. Patents 3,624,018; November 30, 1971; 3,845,004; October 29, 1974; 3,839,263; October 1, 1974; 3,847,858; November 12, 1974; all assigned to The Dow Chemical Company* the sealing of void spaces, e.g., as in geological formations and/or between metal shapes is accomplished with a liquid slurry of a particulate, water-soluble organic polymer in an organosolvent. The mixture of the polymer and solvent has a controlled set time to allow the emplacement of the slurry as a liquid. The slurry then sets to form a sealing cementitious material.

Organosolvents useful in the process fall into three general classes. Each class is characterized by unique gel forming characteristics and in the properties of the final hardened gel. Although the organosolvent used may consist of only organic materials it may also be mixed with water, or in fact, any desired liquid material compatible with the solvent to produce a variation within the process.

The first class of organosolvents, termed Group 1 solvents, is the liquid organic solvents which solvate or plasticize solid, water-soluble polymers. That is, the polymers, if they are not totally miscible with the solvent, absorb a significant proportion of the same. Solvents of this nature, suitable for use with a wide variety of water-soluble polymers, include for example, ethylene glycol, ethylenediamine, glycerol, dioxolane, formamide, pentaerythritol and acetic acid.

A second class of useful organosolvents, termed Group 2 solvents, includes a large number of organic materials which are not solvents for the polymers (nonsolvents), i.e., they do not solvate or plasticize the water-soluble polymers to form a continuous phase, but which, when mixed with a small amount of water, produce organosolvents useful for the purpose of the process.

Manifestly, such materials must be miscible with water. It has been found that as little as 1% by weight water dissolved in an organic liquid, which is not a solvent for the polymer, often renders the resulting mixture a suitable organosolvent for the purposes of the process. Sometimes the organic material may normally exist as a solid. In such instances, enough water, e.g., up to 50% by weight of the total solvent composition may be used to provide a liquid solution of the organic material. The resulting aqueous solution is an organosolvent suitable for the process. Since the set time of a formulation is decreased by increases in the water content of the organosolvent used, the amount of water is usually maintained at less than about 25% by weight of the total solvent composition.

Materials which are not normally solvents for the water-soluble polymers, but which on addition of water become such solvents at room temperature, e.g., 75°F, include diethylene glycol, dipropylene glycol, diethylene glycol monoethyl ether, dioxane, ethylene carbonate, ethanolamine, triethylene glycol, propylene glycol, sucrose, urea, dextrin, diethanol amine, triethanol amine, and diethylene triamine. A most useful class of organic materials for formulating Group 2 solvents includes the liquid alkylene oxide polymers, which are not solvents in themselves for the water-soluble polymers.

A third class of organosolvents, termed Group 3 solvents, is composed of mutual solutions of an organic material, which is not a solvent for the water-soluble polymers, and an organic material, which is a solvent for the polymer, i.e., a Group 1 solvent. In such mixtures, at least 0.05 part of the polymer solvent is employed for each part by weight of the nonsolvent. Such organic solvents allow convenient adjustment of the set times in nonaqueous cementitious formulations by increasing or decreasing the amount of polymer solvent relative to the nonsolvent.

Controlling the set times, or in other words gel times, of water-soluble polymer-solvent mixtures, where the organosolvent classifies with Group 1 solvents, is achieved through temperature control. The temperatures of the polymer-solvent mixture is decreased to a point at which gelation occurs after a given period of time as the temperature of the system equilibrates with its environment. The formulation temperature is most conveniently controlled by adjusting the temperature of the organosolvent.

The set times of formulations prepared with Group 2 solvents can be adjusted by the proportion of water incorporated into the organic materials which are nonsolvents for the polymers. As little as 0.5% by weight water, based on the total organosolvent, will appreciably shorten the set time of the formulation. Unless the polymer nonsolvent requires a higher proportion of water to exist as a liquid, it is generally not desirable in the interest of providing longer set times to utilize more than 25% by weight based on the weight of the total organosolvent.

The set times of formulations prepared with Group 3 solvents are controlled by the proportion of organic polymer solvent utilized conjunctively with the organic, polymer nonsolvent. With increases in the amount of solvent in relation to the nonsolvent, the set time is decreased. Organosolvents of this type are preferred for use in applications where control of set times cannot be achieved by control of temperature and a nonaqueous cement is desired.

Example 1: This example illustrates formulating cementitious gels using Group 1 and Group 2 organosolvents. To 100 ml of ethylene glycol at a temperature of 75°F was added 75 grams of a particulate carbamoyl polymer. The polymer was a polyacrylamide in which 7% of the initially available carboxamide groups had been hydrolyzed to sodium carboxylate groups. It was characterized by a molecular weight of at least one million. The polymer was stirred into the ethylene glycol to produce a uniform blend at 75°F. The composition set to a shape-retaining, rubbery gel in 45 minutes.

The set gel is useful as a cement in a variety of oil well treating applications. In a specific use, a slurry prepared as described above is placed in the annulus defined by two concentric oil well casings. When set, the composition provides a highly adherent and persistent annular seal.

To illustrate set time control with Group 3 organosolvents, a second formulation was prepared with a 50/50 mixture of ethylene glycol and diethylene glycol, which are a solvent and nonsolvent for the polymer respectively. To 100 ml of this mixture at 75°F was added 75 grams of the above polymer. After 4 hours, the composition achieved an initial set to produce a solid, rubbery cement. By way of comparison, 75 grams of the polymer in 100 ml of diethylene glycol remained fluid through 48 hours, after which the test was terminated. Through the conjoint use of the nonsolvent and solvent to prepare the organosolvent, the set time of the composition was extended by over 3 hours. The longer set time permits batchwise preparation of the formulation and otherwise more extensive handling prior to use.

Example 2: To exemplify the metal surface adhesion of cementitious compositions prepared in accordance with the process, a quantity of the formulation in the above example prepared with the Group 3 solvent was poured into a vertically aligned pipe, the lower end of which was plugged with a rubber stopper.

The composition was allowed to cure for 20 hours at 75°F. In a second run, the temperature was increased to 150°F. In this manner, plug seals about 12 inches long were produced in each pipe. The set time of the composition at these temperatures was determined as the length of time required for the organosolvent-polymer slurry to form a shape-retaining mass.

To measure seal quality, the upper end of the pipe was fitted with a high pressure cap and the rubber stopper removed. Nitrogen gas was applied gradually to the seal until it failed or withstood the maximum pressure for 10 minutes. The maximum pressure applied with this technique was 1,000 psi.

In still further runs similar to the above, a series of compositions were prepared in which portions of the polymer were replaced with corresponding amounts of magnesium hydroxide. The curing and testing conditions were as discussed previously. The compositions, set times for cures at 75° and 150°F, and adhesion strength properties for the foregoing operations are included in the table below.

	Formulation				Properties		
Run	Carbamoyl Polymer, g	Ethylene Glycol, ml	Diethylene Glycol, ml	$Mg(OH)_2$, g	Yield Strength, psi	Set Time, hr 75°F	Set Time, hr 150°F
1	75	100	-	-	650	0.75	0.05
2	75	50	50	-	650	4	0.2
3	74	50	50	1	400	4	0.2
4	73	50	50	2	500	4	0.2
5	72	50	50	3	1,000	4	0.2
6	70	50	50	5	1,000	4	0.2
7	68	50	50	7	1,000	4.5	0.25

FLUID LOSS CONTROL

Methacrylamidopropyltrimethylammonium Chloride Polymers

According to a process described by *L.J. Guilbault and F.A. Hoffstadt; U.S. Patent 3,931,096; January 6, 1976; assigned to Calgon Corporation* minor amounts of a polymer of methacrylamidopropyltrimethylammonium chloride are added to an aqueous hydraulic cement slurry to reduce the rate of fluid loss from the slurry to any fluid-absorbing medium with which the slurry may come into contact.

The polymers are high molecular weight, water-soluble polymers of methacrylamidopropyltrimethylammonium chloride and may be prepared in any convenient manner, as for example, in the manner taught by U.S. Patent 3,661,880 or by conventional solution or inverted emulsion polymerization techniques, as for example the procedures described in U.S. Patent 3,284,393.

The polymers should have high molecular weights, preferably of at least 100,000 and more preferably of at least 1,000,000 and should have a solubility in water of at least 0.25%. These polymers may be copolymers of methacrylamidopropyltrimethylammonium chloride and acrylamide containing from 10 to 90% by weight acrylamide.

While the chloride anion is the most preferred quaternary ammonium derivative, other anions such as fluoride, bromide, nitrate, acetate, hydrogen sulfate, and dihydrogen phosphate may be utilized with the cationic polymers of this process. The following example illustrates the process.

Example: Evaluation of Fluid Loss Reducing Properties – The evaluation procedure employed was that set out in API bulletin RP 10B, 19th Edition, January 1974, Section 8, pages 43-44. A cement slurry was made up containing 600 grams of Class H cement (API Class H cement has a fineness in the range of 1,400 to 1,600 square centimeters per gram and contains, in addition to free lime and alkali, the following compounds in the indicated proportions: tricalcium silicate, 52; dicalcium silicate, 25; tricalcium aluminate, 5; tetracalcium aluminoferrite, 12; calcium sulfate, 3.3), 210 grams silica flour and 6 grams fluid loss additive of the process (1.0% by weight of dry hydraulic cement) in 324 ml of 18% sodium chloride solution.

These ingredients were mixed in a Waring blender for 15 seconds at low speed, and then for 35 seconds at high speed. The resultant slurries were then mixed in a Halliburton Consistometer for 20 minutes at 190°F. The slurry samples were placed in a Baroid high pressure filter press cell maintained at 190°F.

In the filter press cell the slurry samples were forced against a No. 325 U.S. Standard Sieve Series screen with 1,000 psi pressure supplied by compressed nitrogen. The fluid removed from the slurry was collected and measured. Constant pressure was maintained and the filtrate collected over a 30 minute period. Results were reported as volume of filtrate (in milliliters) collected in a 30 minute period. The results obtained are illustrated in the following table.

Sample	Concentration of Additive (percent by weight of dry cement)	Fluid Loss (ml/30 min at 1,000 psi)
Neat cement	0.0	1000*
MAPTAC/AM 25/75		
(high molecular weight)	1.0	60.0

*For neat cement (hydraulic cement and water only), virtually all water present was removed in less than 1 minute, and the value indicated was obtained by extrapolatio· MAPTAC = methacrylamidopropyltrimethylammonium chloride AM = acrylamide.

Radiation-Induced Polymerization of Acrylamide and Sodium Acrylate

According to a process described by *B.L. Knight, J.S. Rhudy and W.B. Gogarty; U.S. Patent 3,937,633; February 10, 1976; assigned to Hercules, Incorporated* reduction in fluid loss from cement slurries as well as reduction of friction loss during the pumping of the cement slurry into the well bore is accomplished by incorporating within the aqueous phase of the cement slurry a water-soluble polymer obtained by radiation polymerization of acrylamide and/or methacrylamide and acrylic acid, methacrylic acid and/or alkali metal salts thereof.

The aqueous solution to be polymerized contains 10 to 60% by weight of monomer; the monomer preferably is 25 to 99% acrylamide and 75 to 1% sodium acrylate, by weight. Radiation intensity is preferably 250 to 1,000,000 rads/hr and the radiation dosage is preferably 500 to 300,000 rads.

The reaction product may be diluted with water and used directly, or the polymer may be extracted from the reaction product, dried and thereafter solubilized. Concentrations of about 0.001 to about 3.0 weight percent, all based on the aqueous phase of the slurry, are useful. The following examples illustrate the process. Unless otherwise specified, all percents are based on volume.

Preparation of the Copolymer: Polymers used for testing are prepared with cobalt 60 gamma radiation; radiation intensities and dosages are outlined in Table 1 on the following page. The process for preparing Polymer A is explained; preparation of the other polymers is similar except for variations indicated in Table 1.

To 24,000 grams of deionized water there are added 692 grams of sodium hydroxide. After cooling the solution to 30°C, 1,250 grams of acrylic acid are added. Thereafter, 5,000 grams of acrylamide are added while mixing and the pH is adjusted to 9.4. The resulting solution contains 75% by weight acrylamide (AAd) and 25% by weight sodium acrylate (NaAA) and has a total monomer concentration of 21.4% by weight.

The solution is purged with N_2 for 20 minutes and thereafter sealed. The sample is irradiated with cobalt 60 gamma radiation at an intensity of 18,000 rads/hr (R/hr) to a total dose of 8,800 rads (R). The resulting product is a gel-like mass.

A portion of the gel is weighed, and thereafter extracted with methanol to precipitate the polymer. The polymer is dried in a vacuum oven at 36°C and 0.02 psia for 24 hours and then to a constant weight at 110°C. Weight of the dried product divided by the theoretical weight gives a monomer conversion of 93%.

A portion of the gel is solubilized in water by first extruding it through a "meat" grinder; the "spaghetti" like extrusion is cut into "BB" size particles and then dissolved in water by agitating at a low rpm to prevent substantial shearing of the polymer.

The residue of the gel is recovered in dry powder form by first extruding the gel, then dissolving it in water and thereafter adding methanol to precipitate the polymer out of the solution. The polymer is then ground to less than 20-mesh size and finally dried at 60°C in a vacuum oven.

TABLE 1: INFORMATION ON POLYMER SAMPLES

Polymer	AAd/NaAA Wt Ratio	Monomer Concentration (%)	pH	Intensity (R/hr)	Total Dose (R)	Additive (%)	Monomer Conversion (%)	Intrinsic Viscosity Gel (dl/g)	Intrinsic Viscosity Powder (dl/g)	Huggins Constants Gel	Huggins Constants Powder
A	75:25	21	9.4	18,000	8,800	-	93	23.7	23.0	-	0.19
B	70:30	21	9.4	20,000	9,800	-	93	22	20	0.19	0.19
C	60:40	22	9.4	20,000	10,300	-	93	23.0	23.0	-	-
D	70:30	30	9.5	230,000	50,000	-	91	14	12.8	-	0.38
E	70:30	40	9.5	10,000	1,760	-	34	39.4	33	0.06	-
F	70:30	24	9.5	100,000	15,000	-	86	18.5	-	0.24	-
G	70:30	27	9.5	20,000	11,500	MeOH 9.1	91	12.4	11.7	0.31	0.38
H	70:30	13	9.5	220,000	44,000	MeOH 15	96.5	1.0	-	-	-
I	70:30	13	9.5	220,000	44,000	-	96.5	5.8	-	0.64	-
J	70:30	25	9.5	220,000	44,000	MeOH 15	84.0	6.9	-	0.52	-
K	70:30	24	9.5	20,000	7,660	-	86.7	28.2	-	0.13	-
L	70:30	30	9.5	20,000	2,667	-	54	31.0	-	0.04	-
M	90:10	40	9.6	10,000	1,350	-	24	53	-	>0.02	-

The intrinsic viscosity is measured at 25.5°C in a 2 N aqueous sodium chloride solution. The Huggins constant is measured by the method described in *Textbook of Polymer Chemistry,* Billmeyer, Interscience Publishers, New York, 1957, pages 128 to 139. The monomer used in Sample G is dissolved in water containing 9.1% by weight of methanol.

Example: This example is presented to show that the copolymers of this process impart unexpected results over polymers and copolymers of the prior art.

Fluid loss control properties are simulated in the laboratory by flooding water-saturated reservoir core samples with aqueous polymer solutions at a frontal velocity of 10 feet per day. The permeability reduction of the front section of the core is determined by first flooding the core with water (containing 500 ppm total dissolved solids) at the velocity of 10 feet per day and thereafter flooding the cores with the aqueous polymer solutions indicated in Table 2.

A high permeability reduction of cores is desirable to minimize fluid loss from a cement slurry. The aqueous polymer solutions are dissolved in water containing the amounts of total dissolved solids (TDS) indicated in Table 2.

TABLE 2: RESULTS OF POLYMER FLOODING IN 100-200 md SANDSTONE CORES

Run	Polymer	Brookfield Viscosity at 6 rpm (cp)	Initial Permeability (md)	Flushed Permeability (md)	Permeability Reduction
1	A	26.7	107	0.5	214
2	B	32.2	142	1.5	93
3	C	27.2	132	0.9	150
4	D	20.0	110	0.6	178
5	B	8.8	159	1.3	124
6	B	7.1	97	0.7	143
7	Partially hydrolyzed polyacrylamide	16.3	135	2.5	54
8	Copolymer No. 1	39.0	123	2.6	47
9	Copolymer No. 2	38.5	134	5.5	25

Runs 1 through 4 contain 700 parts per million polymer dissolved in water containing about 500 parts per million TDS. Run 5 contains 300 parts per million polymer dissolved in water containing about 500 parts per million TDS. Run 6 contains 700 parts per million polymer dissolved in water containing 18,000 to 20,000 parts per million TDS. Runs 7 through 9 contain 800 parts per million polymer dissolved in water containing about 500 parts per million TDS.

Copolymer No. 1, a commercially available, anionic acrylamide copolymer obtained by a chemically catalyzed polymerization reaction, has an intrinsic viscosity of 12.5 and a Huggins constant of 0.34. Copolymer No. 2, a commercially available very high molecular weight, strongly anionic copolymer of acrylamide obtained by a chemically catalyzed polymerization reaction, has an intrinsic viscosity of 22.0 and a Huggins constant of 0.18.

Partially hydrolyzed polyacrylamide, a commercially available, partially hydrolyzed, high molecular weight polyacrylamide obtained by a chemically catalyzed polymerization reaction, has an intrinsic viscosity of 12.7 and a Huggins constant of 0.56.

The data in Table 2 indicate that Runs 1 through 6, as compared to Runs 7 through 9, exhibit higher permeability reductions than do the commercially available polymers and copolymers. Such large permeability reduction is a direct indication of the fluid loss control properties of a cement slurry when such polymers are used.

Polymer E, which is listed in Table 1, at the same concentration and water conditions as copolymers No. 1 and No. 2, has a Brookfield viscosity at 6 revolutions per minute of 52 centipoises.

Lignosulfonate, Tartaric Acid and Sodium Naphthalene Sulfonate

W.J. Haldas and J.A. Faust; U.S. Patent 3,723,145; March 27, 1973; and U.S. Patent 3,772,045; November 13, 1973; both assigned to Lone Star Cement Corporation describe an additive composition for enhancing the properties of cement compositions, and a cement composition of enhanced properties, particularly useful in the cementing of wells, and methods of cementing geologic formations traversed by well bores.

The additive composition comprises a water-soluble lignosulfonate or lignosulfonic acid or mixtures, tartaric acid or a salt of tartaric acid or mixtures, and sodium naphthalene sulfonate, and preferably about three parts calcium lignosulfonate, one part tartaric acid and four parts sodium naphthalene sulfonate.

The cement composition includes a hydraulic cement and preferably about 2% of the additive composition based on the weight of the cement, although good results are obtained with from about 0.30 to about 5.00%.

The cement composition has good initial viscosity and strength, ideal thickening times, and a reduced amount of mix water may be used thereby increasing the weight of the cement slurry to about 18 pounds per gallon without using weight materials.

Ferrochrome Lignosulfonate

F. Mitchell; U.S. Patent 3,375,873; April 2, 1968; assigned to Mobil Oil Corporation describes a cement composition which comprises Portland cement containing ferrochrome lignosulfonate. The ferrochrome lignosulfonate enables the cement slurry to be pumped at temperatures in excess of 250°F and even up as high as 325°F without premature setting.

The ferrochrome lignosulfonate also provides fluid loss control without destroying the ability of the cement slurry to set at lower temperatures and to have adequate compressive strength.

Sulfonated Polymer with Glucoheptonic Acid and Amine Lignin

R.C. Martin; U.S. Patent 3,234,154; February 8, 1966; assigned to The Dow Chemical Company describes a hydraulic cement composition where a sulfonated polystyrene or a sulfonated polyvinyltoluene polymer, a glucoheptonic acid or its salt, and a stabilizer are used as effective additives to reduce the fluid loss and the setting time of the cement.

The fluid loss additive is a high molecular weight polymer of sulfonated polystyrene or sulfonated polyvinyltoluene, which is generally added in an amount of from 0.2 to 4.0 weight percent of the dry hydraulic cement, preferably from 0.7 to 1.6 weight percent of the cement. Polymers which have a molecular weight of 100,000 and above are most commonly used.

The glucoheptonic acid retardant is generally employed in an amount of 0.025 to 1 weight percent based upon the weight of the dry hydraulic cement. The retardant may be the glucoheptonic acid itself or a water-soluble salt thereof. These salts may be generally obtained by the reaction of the acid with an alkaline salt or a hydroxide.

Generally, when the acid is added to the cement a certain amount of salts are formed by the acid being neutralized by some of the constituents in the cement. Both the alpha and the beta form of the acids may be used. The sodium salt of the acid is generally preferred.

The stabilizer or lignin additive is an amine derivative of lignin. One of the most convenient and available sources of lignin is the paper and pulp industry. In pulping of woods, such as pine wood, the lignin is subjected to alkaline hydrolysis by sodium hydroxide and sodium sulfide.

As a result of the hydrolysis some thio groups are introduced into the lignin molecule in addition to the formation of a sodium salt of the lignin. The treated lignin thus obtained may be further treated to introduce an alkyl amine group into the molecule through a methylene linkage.

The amine groups are the lower alkyl amines having alkyl radicals of up to 4 carbon atoms and may be primary, secondary or tertiary amines. It is believed that the presence of an amine group in the lignin molecules is essential to obtain the desired result. The product may also contain sodium or other metal substituents.

The stabilizer or lignin additive is an amine derivative of lignin, preferably Indulin XW-1. The product contains 2 amine groups per 1,000 molecular weight and also contains 2 sodium atoms per 1,000 molecular weight which are reacted with phenolic groups located elsewhere in the structure.

The amount of the lignin amine derivative used will vary somewhat with the amount of retardant and the amount of fluid loss agent added. However, generally the most effective results are obtained with from 0.0125 to 0.5 weight percent of the lignin derivative, based upon the dry cement, although the amount used may vary from 0.003 to 5%.

To illustrate the stabilization effect of the lignin amine derivative on the sulfonated polymer fluid loss agent in the presence of the glucoheptonic acid type retardant, a series of runs was made where the fluid loss, setting time, and the ultimate strength of the cement were measured using various combinations and concentrations of the respective additives.

Standard procedures for testing oil well cement as given in the API publication RP 10-B, 9th Edition, January 1960, "Recommended Practice for Testing Oil Well Cements and Additives," were used. In making the tests 800 grams of the hydraulic cement were slurried with 320 grams of water to give a cement slurry. Various amounts of additaments were added to the cement, the concentrations of which are expressed in weight percent of the dry cement.

In determining the fluid loss, the cement slurry was run in a Halliburton thickening tester described in the above cited API publication, Section VIII, entitled, "Atmospheric Pressure Thickening Tests—A." As in Schedule 8, a 14,000 foot casing-cementing well-simulating test was partially followed. The Halliburton thickening time tester was operated at atmospheric pressure.

The slurry was stirred employing the temperature and time given in Schedule 8 until a temperature of 200°F had been reached after 80 minutes. This temperature was then maintained with continued stirring for an additional 30 minutes after which time the fluid loss of the cement slurry was determined. The procedure followed in determining the fluid loss was as specified under Section VI entitled, "Filter-Loss Test," given in the above cited publication using a high pressure filter press.

The filter medium was a 325 mesh U.S. Standard Sieve Series screen supported by a 14 mesh Sieve Series screen. The tests were made at 1,000 psi and 200°F. The thickening tests were made using the Pan American Petroleum pressure-temperature thickening-time tester and Schedule 19, the 14,000 foot squeeze-cementing well-simulation test described in the foregoing publication under Section VIII was used.

The compressive strength tests were determined by the procedure described in the foregoing publication under Section VI entitled, "Strength Test," Schedule 7S. A 12,000 foot depth well test was used employing a bottom hole circulating temperature of 260°F. The results obtained in the numerous runs made are given in the the table on the following page.

The lignin amine derivative used in the test was a high molecular weight amine derivative product containing 2 amine groups and 2 sodium atoms per 1,000 molecular weight (a compound such as Indulin XW-1).

The polyvinyl-toluene polymer had a molecular weight of 300,000 determined from the viscosity of a toluene solution of the polymer according to the well-known and accepted methods.

Test	Sodium Salt of Sulfonated Polyvinyl Toluene Polymer, percent wt of dry cement	Sodium Glucoheptonate, percent wt of dry cement	Lignin Amine Derivative, percent wt of dry cement	Cement, API Designation	Fluid Loss (ml in 30 min)	Thickening Time (hr:min)	24 Hour Compressive Strength, psi
Blank	-	-	-	Class E	>500	0:40	6,509
Blank	1.5	-	-	Class E	45	1:45	4,212
Blank	-	0.1	-	Class E	>500	2:28	8,900
Blank	-	-	0.5	Class E	>500	0:55	-
Blank	1.5	0.1	-	Class E	230	-	-
Blank	1.5	-	0.5	Class E	–	0:32	-
1	1.5	0.1	0.0125	Class E	99	-	-
2	1.5	0.1	0.05	Class E	40	-	-
3	1.5	0.1	0.1	Class E	36	2:39	9,034
4	1.5	0.1	0.25	Class E	31	-	-
5	1.5	0.1	0.5	Class E	27	2:16	-
6	1.5	0.2	0.05	Class E	30	-	-
7	1.5	0.2	0.1	Class E	42	2:50	-
8	1.5	0.2	0.2	Class E	28	>4:00	8,249
9	1.5	0.2	0.3	Class E	28	4:00	-
10	1.5	0.2	0.5	Class E	27	4:00	-
11	1.5	0.4	0.05	Class E	49	>4:00	-
Blank	-	-	-	Class A	>500	0:34	-
Blank	1.5	-	-	Class A	55	0:45	-
Blank	-	0.1	-	Class A	>500	>4:00	-
Blank	-	-	0.1	Class A	>500	0:40	-
Blank	1.5	0.1	-	Class A	250	>4:00	-
12	1.5	0.1	0.1	Class A	46	>4:00	-

Hydroxyalkylcellulose Ether and Sodium Chloride

F.E. Hook; U.S. Patent 3,483,007; December 9, 1969; assigned to The Dow Chemical Company describes a composition having low fluid-loss properties and relatively low viscosity, whereby loss of fluid to porous media is lessened by the mixing with water and cement, of a hydroxyalkylcellulose ether, whereby the usual increase in viscosity, due to the ether is lessened by using not less than 10% by weight of water present, of sodium chloride.

A method of preparing the water-soluble hydroxyalkylcellulose ethers for use in the process is as follows. An alkali cellulose may be first prepared by reacting a source of cellulose, usually cotton linters or wood pulp, with an aqueous solution of NaOH, e.g., one of 25 to 50% by weight, in a suitable reactor, from which air has been evacuated.

A preferred weight ratio of NaOH to cellulose is from 0.5 to 1.5, 0.9 to 1.1 commonly being used, and a water to cellulose ratio of from 1.0 to 2.0 of water to 1 of cellulose. Alkali cellulose so produced is reacted with a selected olefin oxide, e.g., ethylene oxide or propylene oxide or with 2-chloroethanol or 1-chloro-

2-propanol (it being theorized in the latter case that the chlorohydrin is converted in situ to the corresponding oxide). The range of the olefin oxide or chlorohydrin employed is that which results in the desired average DS (degree of substitution) which may be from 0.25 to 2.5. A substitution of at least 1.0 is provided where complete solubility of the ether in cold water is desired.

Example: A well which had been recently completed in the Postle Field in Texas County, Oklahoma to a depth of 6,610 feet was to have the casing cemented in place at a level between 5,790 and 6,586 feet. The bottom hole temperature was 126°F. The cementing was carried out according to the process as follows.

0.5 part of hydroxyethylcellulose, 18 parts NaCl and 46 parts of water were mixed with 100 parts of Portland cement API Class A, by weight, and emplaced in the well bore in the annulus between the casing and the face of the formation employing high pressure pumping equipment. After 24 hours, a "cemented bond log" was made and recorded by Go Jet Services, Inc. of Oklahoma City, Oklahoma.

The cement bond log measures the security of bonds formed at the cement-casing and cement formation interfaces. The log, which was measured in the interval between 5,650 and 6,586 (well bottom) feet showed excellent bonding between 5,790 feet to the well bottom without one defect therein, which would have indicated a weak bond, appearing on the log.

TURBULENCE INDUCERS

Polymaleic Anhydride

S.L. Adams, M.M. Cook and F.D. Martin; U.S. Patent 3,943,083; March 9, 1976; assigned to Calgon Corporation describe aqueous hydraulic cement compositions containing polymaleic anhydride as a flow-property-improving and turbulence-inducing additive.

The influence of the turbulence-inducing additives on the rheological properties of the total cement slurry composition is susceptible to mathematic characterization. Opinion varies as to the correct theoretical model, and thus the proper mathematical characterization of the rheological properties of cement slurries. They have been treated according to the principles of Bingham plastic fluids and the Power Law concept. The mathematic formulation is based on Fann Viscometer readings.

The behavior of cement slurries may be viewed as governed by the Power Law concept. For a mathematical characterization of this concept, two slurry parameters are determined: the flow behavior index (n') and the consistency index (K'). A Fann Viscometer may be used to make these determinations. The instrument readings at 600 and 300 rpm are recorded and the values for n' and K' are then calculated as follows:

$$n' = 3.32 \left(\log_{10} \frac{600 \text{ reading}}{300 \text{ reading}}\right) \qquad K' = \frac{N\ (300 \text{ reading})}{100\ (511)\ (n')}$$

where N is the extension factor of the torque spring of the particular Fann Viscometer instrument. The so-called Reynolds Number for a fluid moving through a conduit is the critical value at which the flowing fluid will begin to flow in turbulence. The Reynolds Number may be calculated according to the following equation:

$$N_{Re} = \frac{1.86\ V^{(2-n')}\rho}{K'(96/D)^{n'}}$$

where N_{Re} is the Reynolds Number (dimensionless), V is the velocity, feet per second, ρ is the slurry density, pounds per gallon, n' is the flow behavior index (dimensionless), K' is the consistency index, pound-seconds per square foot, D is the inside diameter of the pipe, inches. The velocity (V_c) at which turbulence may begin is readily calculated from the following equation, which is derived from the equation for the Reynolds Number, and assumes a Reynolds Number of 2,100:

$$V_c = \left[\frac{1129\ K'\ (96/D)^{n}}{\rho}\right]^{\frac{1}{2-n'}}$$

where the different elements of the equation have the same meaning as indicated above for the Reynolds Number Equation. Where the conduit through which the cement slurry passes is the annulus between the well casing and the well wall, $D = D_o = D_i$ where D_o is the outer inside diameter on hole size in inches, and D_i is the inner pipe outside diameter in inches. The Power Law concept equation will give higher flow rates or lower flow rates required to produce turbulence where the Fann Viscometer readings are reduced.

Example: A 38% Class H (API Class H cement has a fineness in the range of 1,400 to 1,600 sq cm/g and contains, in addition to free lime and alkali, the following compounds in the indicated proportions: tricalcium silicate—52, dicalcium silicate—25, tricalcium aluminate—5, tetracalcium aluminoferrite—12, and calcium sulfate—3.3) cement slurry was made up by adding tap water (228 ml) to a Waring blender, dissolving the indicated amount of polymaleic anhydride turbulence-inducing additive in the water, then mixing in the cement (600 grams) at low speed.

Molecular weight of the polymaleic anhydride was determined by gel permeation chromatography with a polyethylene glycol standard. The number average molecular weight was determined to be 404 and the weight average molecular weight was determined to be 515. The resultant slurry was then mixed at high speed on a laboratory gang-stirrer for 20 minutes.

The viscosity of the slurry was then immediately measured on a Fann Viscometer equipped with a No. 1 spring. The instrument readings for various speeds were recorded and are illustrated in the table of values shown on the following page.

The polymaleic anhydride flow-property-improving and turbulence-inducing additive of the process is employed to prepare cementing compositions which are readily pumped in turbulent flow during a cementing operation at a satisfactory low pump rate. The additive is also used to advantage simply to reduce the pumping pressure required for a given flow rate, or to obtain a higher flow rate at a given pumping pressure.

Turbulence-Inducing Additive	% Additive (based on weight of dry cement)	Fann Viscosities at Indicated rpm's 600	300	200	100
Control*	–	165	105.5	87.5	65
Polymaleic anhydride	0.5	103.5	45.5	31.5	16.5
Polymaleic anhydride	0.25	95.0	39.5	26.5	15.0

*Neat Class H cement.

The cementing composition which has improved flow properties and is easily flowed in turbulence is employed in the conventional cementing operation in the same manner as would a cementing composition which did not contain the flow-property-improving and turbulence-inducing additive, but which was otherwise the same composition.

Poly(2-Acrylamido-2-Methylpropane Sulfonic Acid)

S.L. Adams, M.M. Cook and F.D. Martin; U.S. Patent 3,936,408; February 3, 1976; assigned to Calgon Corporation describe a well cementing composition comprising hydraulic cement, water, and from about 0.01 to about 5.0% by weight of the cement of a flow-property-improving and turbulence-inducing additive comprising a polyamido-sulfonic compound.

Preferably, this composition contains from about 0.1 to about 2.0% by weight (based on dry weight of cement) of the polyamido-sulfonic compound additive. The polyamido sulfonic compounds employed as flow property improving and turbulence-inducing additives for well cementing compositions may be represented by the following formula

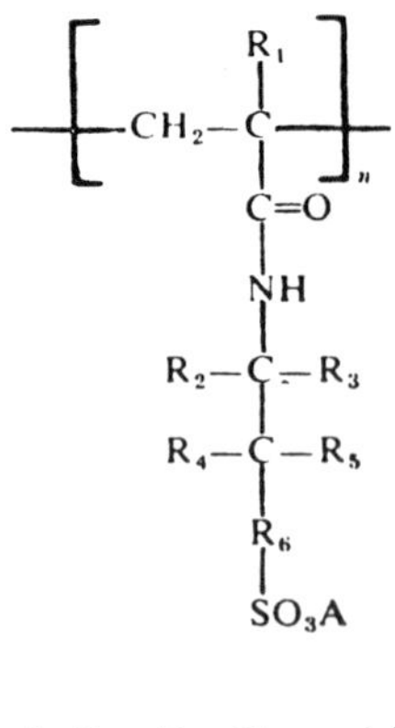

where R_1 is hydrogen or methyl; R_2, R_3, R_4 and R_5 are each independently selected from the group consisting of hydrogen, phenyl, straight or branched alkyl of from one to twelve carbon atoms, and cycloalkyl of up to six carbon atoms; R_6 is straight or branched alkyl of from one to twelve carbon atoms, cycloalkyl of up to six carbon atoms, phenyl, or is absent; A is hydrogen, alkali metal ion, or ammonium; and n is an integer of from 2 to about 100, such that the weight average molecular weight of the polyamido-sulfonic compound is from about 200 to about 10,000 and preferably from about 200 to about 1,500.

A particularly preferred polyamido-sulfonic compound for use in the process is polymeric 2-acrylamido-2-methylpropane sulfonic acid comprising recurring units of the following formula:

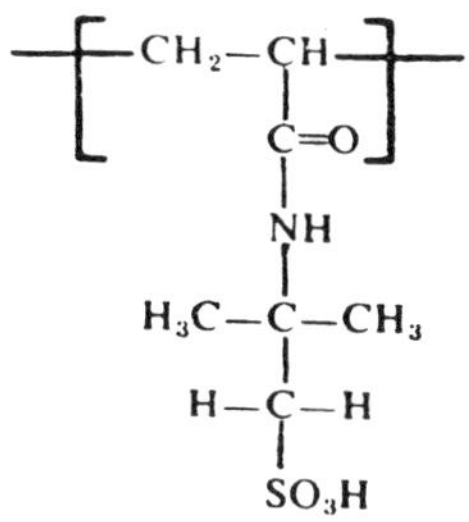

The polyamido-sulfonic compounds may be prepared by way of a number of conventional polymerization routes, and consequently any such method giving satisfactory yields of acceptably pure product will be suitable.

Example: A 38% Class H (API Class H cement has a fineness in the range of 1,400 to 1,600 sq cm/g, and contains, in addition to free lime and alkali, the following compounds in the indicated proportions: tricalcium silicate—52, dicalcium silicate—25, tricalcium aluminate—5, tetracalcium aluminoferrite—12, and calcium sulfate—3.3) cement slurry was made up by adding tap water (228 ml) to a Waring blender, dissolving the indicated amount of poly(2-acrylamido-2-methylpropane-1-sulfonic acid) turbulence-inducing additive in the water, then mixing in the cement (600 grams) at low speed.

Molecular weight of the poly(2-acrylamido-2-methylpropane-1-sulfonic acid) was estimated to be less than 1,000. The resultant slurry was then mixed at high speed for 35 seconds. The slurry was then stirred at high speed on a laboratory gang-stirrer for 20 minutes. The viscosity of the slurry was then immediately measured on a Fann Viscometer equipped with a No. 1 spring. The instrument readings for various speeds were recorded and are illustrated in the table of values below.

Turbulence-Inducing Additive	% Additive (based on weight of dry cement)	Fann Viscosities at Indicated rpm's			
		600	300	200	100
Control*	–	165	105.5	87.5	65
Poly(2-acrylamido-2-methyl-propane-1-sulfonic acid)	0.5	114.5	48.5	34.5	18.5
Poly(2-acrylamido-2-methyl-propane-1-sulfonic acid)	0.25	104	88.5	53.5	25.5

*Neat Class H cement.

The polyamido-sulfonic flow-property-improving and turbulence-inducing additive is employed to prepare cementing compositions which are readily pumped in turbulent flow during a cementing operation at a satisfactorily low pump rate. The additive is also used to advantage simply to reduce the pumping pressure required for a given flow rate, or to obtain a higher flow rate at a given pumping pressure.

The cementing composition which has improved flow properties and is easily flowed in turbulence is employed in the conventional cementing operation in the same manner as would a cementing composition which did not contain the flow-property-improving and turbulence-inducing additive, but which was otherwise the same composition.

Lithium-Sodium Mixed Salt of Naphthalene Sulfonic Acid and Formaldehyde

F.E. Hook, C.H. Kucera and L.J. Scott, Jr.; U.S. Patent 3,465,825; September 9, 1969; assigned to The Dow Chemical Company describe an aqueous cement slurry containing a turbulence inducing agent which imparts to the slurry the capacity of attaining a state of turbulence, while in motion in a conduit, at a slower rate of movement and with less expenditure of energy than is necessary for such attainment in the absence of the agent.

The slurry comprises, by weight, 100 parts of an hydraulic cement, between 35 and 65 parts of water, a turbulence inducer which is the lithium salt or the lithium-sodium mixed salt of the condensation product of mononaphthalene sulfonic acid and formaldehyde and optionally a fluid-loss control agent, of which water-soluble cellulose ethers are especially effective.

The lithium salt or lithium-sodium mixed salt of the condensation product required by the process is prepared by condensing formaldehyde and mononaphthalene sulfonic acid in a molar ratio to yield about 3 mols of water (which is conveniently removed as desired) and one equivalent of the desired intermediate, the product considered to have the formula

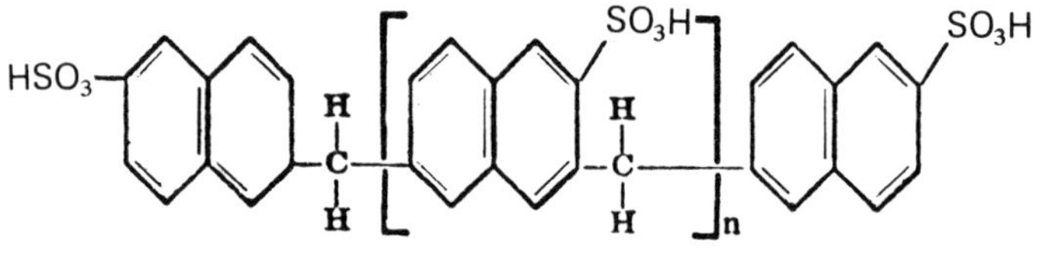

where n has a value sufficient to give a molecular weight of between 500 and 3,000, usually between 1,000 and 2,000. The product so made is subsequently neutralized by reacting it with LiOH, or a mixture of NaOH and LiOH where up to 85% of the salt of the condensate may be the sodium form.

In other words, the condensate salt may consist of all or substantially all the lithium form or it may be a mixture of the lithium form and the sodium form, such mixture consisting of as much as 85 parts of the sodium salt per 100 parts of the salt mixture, i.e., lithium:sodium proportions of 85:15, by weight.

Mixtures of the lithium and sodium salts of proportions between 75:25 and 25:75 of each are preferred.

The sodium salt-lithium salt mixture is conveniently prepared by first neutralizing the condensed acid product with only enough LiOH to convert between one-fourth and three-fourths, e.g., about one-half, of the acid product to the lithium salt. Thereafter, the balance of the condensed mononaphthalenesulfonic acid-HCHO product is reacted with NaOH to convert it to the sodium form. A synergistic effect is accomplished by employing the mixed salts of lithium and sodium of the condensation product.

Phosphate-Urea Pyrolysis Product

H.T. Harrison; U.S. Patent 3,409,080; November 5, 1968; assigned to The Dow Chemical Company describes a composition which comprises (1) an hydraulic cement, i.e., Portland, aluminous, pozzolan or high sulfate expansive cement, (2) an O,O-alkylene-O',O'-alkylenepyrophosphate-urea pyrolysis product (sometimes known as bisalkylenepyrophosphate-urea pyrolysis product) as a turbulence inducer, and (3) water in sufficient amount to make a pumpable settable slurry.

Bisulfite-Modified Phenol-Formaldehyde

C.H. Kucera; U.S. Patent 3,465,824; September 9, 1969; assigned to The Dow Chemical Company describes a well-cementing composition which comprises an aqueous hydraulic cement slurry containing a small but effective amount of a bisulfite-modified phenol-formaldehyde condensation product which serves as a turbulence inducer to the slurry while being moved in a confined passageway.

Polyalkylenepolyamine-Acid Reaction Products

L.H. Scott, Jr., D.L. Gibson, F.E. Hook and C.H. Kucera; U.S. Patent 3,511,314; May 12, 1970; assigned to The Dow Chemical Company describe an aqueous hydraulic cement slurry comprising hydraulic cement, water and a turbulence inducing, fluid-loss, control agent consisting of the reaction product of (1) an amino compound selected from the class consisting of polyalkylenepolyamines, polyalkylenimines, and their mixtures, and (2) an acidic compound selected from the class consisting of carboxylic acids, sulfonic acids, polymers having a carboxyl substituent, and polymers having a sulfonate substituent.

Silica-Filled

J.U. Messenger; U.S. Patent 3,558,335; January 26, 1971; assigned to Mobil Oil Corporation describes cement compositions comprising hydraulic cement in a mixture with a turbulence inducer and silica or diatomaceous earth particles having sizes within the range of 0.1 to 44 microns in diameter. The turbulence inducer and silica or diatomaceous earth particles are present in the composition in amounts within the ranges of 0.1 to 2.0% by weight and 0.5 to 10.0% by weight, respectively, based upon the amount of hydraulic cement present.

Any of the known turbulence inducers may be employed in the compositions. The turbulence inducer may be organic molecules which are ionic enough in character to adsorb strongly enough to effect a reduction in the yield point of the cement slurry. Suitable turbulence inducers are water-soluble alkylaryl sulfonates, polyphosphates, lignosulfates, lactones and gluconates, synthetic polymers, and organic acids.

More specific chemical compounds which are illustrative are sodium dodecylbenzenesulfonate, sodium hexametaphosphate, calcium lignosulfonate, γ-butyrolactone, gluconic acid, sodium polystyrenesulfonate, and oleic acid. The ammonium salts, as well as the other alkali metal salts, may be employed in the place of the named sodium salts as turbulence inducers also. Suitable turbulence inducers which are sold for that purpose by oil well servicing companies include the following Dowell's TIC series and Halliburton's CFR series.

TIC 1	condensed alkylnaphthalene sulfonate
TIC 2	anhydrous citric acid
CFR 1	d-glucono delta-lactone
CFR 2	condensed alkylnaphthalene sulfonate

In the condensed alkylnaphthalene sulfonates, it is preferred that the alkyl groups contain at least 6 carbon atoms and no more than 13 carbon atoms.

The following example illustrates the process. For convenient measurement and determination of plastic viscosity and yield point in this example, a two-speed Fann VG Meter was employed. The values of plastic viscosity and yield point are those determined by measurements made at 300 and 600 rpm on the Fann VG Meter. The plastic viscosity is the difference between the readings of the Vann VG Meter at the 300 and 600 rpm rate. The yield point is determined by subtracting the plastic viscosity from the reading of the Fann VG Meter at 300 rpm.

Example: In this example, Texcor cement was mixed with ground silica passing No. 100 screen, U.S. Sieve Series; Dowell's TIC-2 turbulence inducer; Dowell's D28R retarder, water-soluble lignosulfonates; Dowell's D-46 defoamer, a mixture of the well-known polyglycols; and a sodium chloride-saturated brine to form two cement slurries, a control slurry and slurry No. 1.

The proportion of ingredients in both the control slurry and slurry No. 1 was as follows. 600 grams of Texcor cement were admixed with 0.5% by weight of turbulence inducer, 0.2% by weight of retarder, 0.2% by weight of defoamer and 40% by weight of brine. The plastic viscosity and yield point were determined at a temperature of 150°F.

Therefore, the concentration of brine was based on weight of cement alone and not on the weight of cement and silica. Into the control slurry there was admixed 35% by weight of the 100 mesh silica and 5% of silica particles having a predominant size of 5 microns effective diameter, and having a distribution of particle sizes between 1 and 15 microns in diameter. The table on the following page summarizes the data obtained with the control slurry and with slurry No. 1.

As shown in the following table, slurry No. 1 has a lower filter loss than the control slurry, and a lower plastic viscosity at 150°F. Of most significance however, is the increased yield point. Slurry No. 1, having a yield point of one pound per 100 square feet, could be pumped into a well and emplaced with minimal risk of premature settling of solids, whereas the control slurry, with its yield point of -6 pounds per 100 square feet, could not, since the negative value indicates settling even during the test.

Slurry Properties	Control Slurry	Slurry No. 1
Density, lb/gal	16.0	16.1
API filter loss, cc/sec		
at 850 psi and 200°F	54/60	55/115
at 850 psi and 300°F	160/120	148/167
Plastic viscosity at 150°F, (cp)	37	31
Yield point at 150°F, (lb/100 sq ft)	-6	1

GENERAL COMPOSITIONS

Ground Anthracite Coal

F. Mitchell; U.S. Patent 3,376,146; April 2, 1968; assigned to Mobil Oil Corporation describes a cement composition containing Portland cement and, as the essential ingredient with respect to providing satisfactory properties at the desired low density, particulate carbon in an amount of from 25 to 233% by weight of the Portland cement in the cement composition. The particulate carbon is insoluble in oil. Thus, it does not deteriorate in the presence of hydrocarbons. Further, set cement in which the desired low density has been obtained using particulate carbon retains adequate compressive strength in an oil-containing subterranean formation.

The following example is illustrative of the process. The samples used were prepared employing a commercial cement of the Class A type. The particulate carbon employed was ground anthracite coal passed through a standard 20 mesh screen. The bentonite employed was Magcogel. The compressive strengths reported in the following data represent the strength of the sample after curing for the designated time interval at 140°F. The densities of the slurries were measured in accordance with Section IV of the "API Recommended Practice for Testing Oil-Well Cements and Cement Additives" of the American Petroleum Institute, API, RP 10B, 10th Edition, March 1961. The preparation of the samples is illustrated by Samples 1 and 2. The same procedures but different proportions were used in preparing the five other samples. The proportions and data on all seven samples are shown in the table on the following page.

Example: Sample 1 was prepared by mixing Portland cement and water in the ratio of 46 parts by weight of water to 100 parts by weight of cement. This afforded a slurry density of 15.6 lb/gal and a 24 hr compressive strength of 3,175 psi. Sample 2 was prepared by mixing a first cement composition and water in the ratio of 40 parts by weight of water to 100 parts by weight of cement composition. The cement composition was comprised of 50 lb of particulate carbon and 94 lb of Portland cement. This afforded a slurry density of 13.3 lb/gal and a 24 hr compressive strength of 1,380 psi.

It will be noted from the table that the ratios of water-to-solids, i.e., water/solids for the slurries using the particulate carbon and cement, were appreciably lower than the ratios of water-to-solids for the slurries using the bentonite and cement, even though the same low densities were not completely achieved with the latter. Compressive strengths of the set cement using the particulate carbon and cement were superior to the compressive strengths of the set cement using the bentonite and cement. Lowering the densities of the bentonite and cement slurries to achieve exactly the same densities as the carbon-cement slurries requires even more water and further lowers the compressive strength of the set cements.

Sample No.	Solids, or Cement Composition	Water/Solids (wt basis)	Slurry Density (lb/gal)	Compressive Strength 24 hr, (lb/sq in)
1	Cement	0.46	15.6	3,175
2	50 lb carbon 94 lb cement	0.40	13.3	1,380
3	8 lb bentonite 92 lb cement	0.79	13.3	1,168
4	100 lb carbon 94 lb cement	0.39	12.4	710
5	12 lb bentonite 88 lb cement	0.96	12.6	623
6	200 lb carbon 94 lb cement	0.37	11.3	225
7	20 lb bentonite 80 lb cement	1.29	11.3	152

Clay-Containing Compositions

A.F. Tragesser, Jr.; U.S. Patent 3,582,375; June 1, 1971; assigned to The Western Company of North America describes a cement composition which is preferably made up of Class H oil well cement, from 10 to 16% clay, and usually from 0.3 to 0.75% each of a dispersing agent and a retarding agent.

No fluid loss control additive is required or desirable in the slurry. The slurry contains substantially less water than slurries previously prepared, but yet is still of a comparatively low viscosity and remains pumpable for a substantially longer period of time. A preferred composition contains only about 80 to 90 pounds of water per 100 pounds of cement as opposed to the usual 46 pounds, plus 5.3 pounds per pound of clay.

The cements from which the cement compositions of this process are prepared are described in the publication entitled "API Specification for Oil Well Cements and Cement Additives" API STD 10-A, 6th Edition, January 1959, published by the American Petroleum Institute, and the preferred cement is a cement of Class H as described in the specification. Other Portland cements such as Classes A, B, E, F and G may also be used, however, under suitable circumstances.

The clay mixed with the cement in the preparation of the cement compositions is a colloidal hydratable swelling type clay. Such clays are of the bentonite group, particularly montmorillonite, and are widely used in the drilling of gas and oil wells to modify the characteristics of drilling muds.

The clays satisfactory for use in this process have been used commercially in the preparation of the low gel cements heretofore used in oil well cementing. The preferred clay is that known as high yield Wyoming bentonite, which is a relatively pure sodium montmorillonite. The results obtained from the compositions of the process are illustrated by the following tests.

Cement slurries were prepared from Class H cement, bentonite, the sodium salt of the condensation product of sulfonated naphthalene and formaldehyde ("D" in Table 1), calcium ligneous sulfonate ("R" in Table 1) and water. The slurries were prepared by the method set forth in API Bulletin RP 10B, and the fluid loss of each slurry was measured by the high pressure method described on pages 17 and 18 of that bulletin. The following results were obtained.

TABLE 1

Test	Gel	Percent D	R	Water	Fluid, loss, cc.
1	8	0.5	0.25	88	160
2	8	0.5	0.50	88	180
3	8	0.5	0.75	88	152
4	8	0.5	1.0	88	156
5	12	0.5	0.25	88	124
6	12	0.5	0.50	88	80
7	12	0.5	0.75	88	88
8	12	0.5	1.0	88	56
9	16	0.5	0.25	88	88
10	16	0.5	0.50	88	68
11	16	0.5	0.75	88	72
12	16	0.5	1.0	88	52
13	8	0	0.7	88	360
14	8	0.25	0.7	88	180
15	8	0.5	0.7	88	152
16	8	0.75	0.7	88	132
17	8	1.0	0.7	88	136
18	12	0	0.7	88	160
19	12	0.25	0.7	88	102
20	12	0.5	0.7	88	88
21	12	0.75	0.7	88	68
22	12	1.0	0.7	88	60
23	16	0	0.7	88	132
24	16	0.25	0.7	88	72
25	16	0.5	0.7	88	72
26	16	0.75	0.7	88	52
27	16	1.0	0.7	88	44
28	12	0.5	0.7	80	64
39	12	0.5	0.7	90	80
20	12	0.75	0.7	90	52

The slurry of Test 6 in Table 1, was allowed to cure for 72 hours at 170°F, and was found to have a compressive strength of 1,200 psi. Test 20 had a compressive strength of 1,125 psi after 72 hours at 170°F, and Test 28 had a strength of 1,535 psi.

It is apparent from the above that cements with only 8% gel have excessive fluid loss, in that it exceeds the desired maximum of 120 cc. It is also apparent that fluid loss is excessive from higher gel cements if no dispersing agent is used. No substantial improvement of fluid loss characteristics results from increasing either the dispersing agent or the retarder to above 0.75%. In order to determine limits of proportions of ingredients, the samples on the following page (Table 2) were prepared by the method of API Bulletin RP 10B. The API test showed fluid loss of each of these samples to be below 100 cc.

Figure 3.1 illustrates the period of time during which each slurry had a viscosity low enough for pumping, and whether or not the sample set up to a satisfactory compressive strength in 72 hours. The maximum satisfactory pumping viscosity is considered to be about 30 poises, and in multiple completion wells the cement slurry is required to be below this viscosity for at least 10 hours and preferably for 30 hours. Also, in multiple completion wells the cement is normally required to set to a compressive strength of at least 700 psi in 72 hours.

TABLE 2

No.	Gel	D	R	Water	Temperature, °F
	----------Percent----------				
A	16	0.25	0.7	90	200
B	16	0.5	0.7	90	200
C	14	0.35	0.8	90	200
D	14	0.35	0.8	100	200
E	14	0.5	0.7	90	200
F	12	0.5	0.7	80	170
G	12	0.5	0.5	90	200
H	12	0.5	0.7	90	170

FIGURE 3.1: PERFORMANCE OF CLAY-CONTAINING WELL-CEMENTING COMPOSITIONS

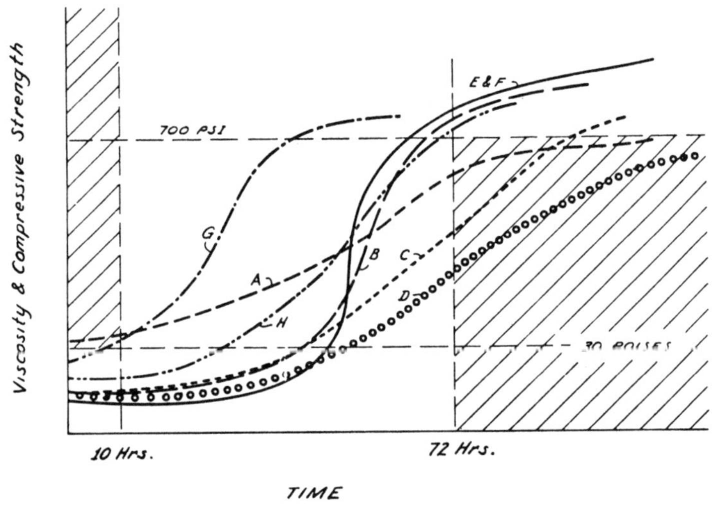

Source: U.S. Patent 3,582,375

The curves in Figure 3.1 are not intended to be accurate representations of the precise changes in the cements over the time periods shown, but only to indicate whether or not a particular cement meets the above requirements. Thus in Figure 3.1, the area above 30 poises and to the left of 10 hours is shaded, and the area below 700 psi and to the right of 72 hours is shaded. The curve of a cement must avoid these shaded areas to be considered satisfactory for multiple completion wells.

It is apparent from the curves that cements **C** and **D** are unsatisfactory in set strength. This is undoubtedly due to the high proportion of retarder used at the 200°F temperature, since cement **E**, with slightly less retarder and slightly more dispersing agent, set up to a satisfactory strength. Note that cement **D**, with 100% water, never did set up, although it might have set up at a higher temperature, say 230°F, which is about the highest that will usually be encountered in oil wells.

Figure 3.1 also shows that cement **G** thickened too rapidly to allow its use in cementing casing in multiple completion wells, although its set strength was satisfactory. This is apparently due to its low proportion of retarder. Cement **A** however, is unsatisfactory both as to the viscosity of the slurry and the strength in 72 hours. This cement apparently had too little dispersing agent, since cement **B**, which differed from **A** in only an increase in dispersing agent, is satisfactory in both pumpability and strength.

The other cements, all of which contain between 0.30 and 0.75% of dispersing agent, and between 0.30 and 0.75% of retarding agent, are all satisfactory both as to pumpability and 72 hour strength. Note, too, that these cements all contain 80 to 90% water and 12 to 16% clay.

It is apparent however, that some variations from this amount of water are possible. It is preferable to always use at least 75% water, and under some conditions as little as 65% may be used. Often as much as 100% water is used with success, but these amounts cause excessive fluid loss.

It is difficult to set precise limits on the proper proportions of the dispersing and retarding agents. The tests which have been run indicate that 0.25% of each is unlikely to be enough under almost any conditions, whereas 0.35% or more almost always results in a satisfactory cement. The tests also indicate that proportions up to about 0.75% appear to improve the cement for the purposes of this process, at the usual well temperatures of 140° to 200°F, whereas higher proportions are of little or no benefit at these temperatures and may be detrimental in many cases.

Well conditions, however, may necessitate some change in these proportions. For example, up to 1.0% retarder may be found to be advantageous at higher temperatures. For average conditions, however, and in most wells, it has been found that proportions of dispersing and retarding agents between 0.30 and 0.75% give good results.

Refractory Cement

C.H. Miller; U.S. Patent 3,595,642; July 27, 1971; assigned to Motus Chemicals, Incorporated describes a refractory cement in which an aluminum silicate, having from 15 to 27 weight percent of an aluminum oxide content, is combined with an hydraulic cement to impart resistance to high temperature changes. The refractory cement is particularly useful in cementing oil well casings.

Low-Temperature Formulation

A process described by *W.C. Cunningham, C.R. George and S.H. Shryock; U.S. Patent 3,891,454; assigned to Halliburton Company* relates to a composition and method for cementing pipe and casing in wells in low temperature earth formations and relates more particularly to the cementing of casing in oil and gas wells drilled through permafrost formations.

The composition comprises cement, calcium sulfate hemihydrate, a setting time retarding agent, a densifier, a freezing point depressant and a relatively small quantity of water. The composition provides a high strength cement having a short curing time which will not freeze before setting at low formation temperatures and which because of its low heat of hydration, will not melt permafrost formations.

The setting time retarder may be an alkali or alkaline earth salt of citric acid, citric acid or one of the sulfonates described in U.S. Patent 3,053,673. The most useful setting time retarding agents are sodium citrate and potassium citrate.

The dispersant useful in the process may be the dispersing agent described in U.S. Patent 3,359,225. This dispersant, commonly called a densifier, is available to the oil industry under the designation CFR-2. Other useful dispersing agents are available under the names Lomar D, Tamol SN and TIC. Such dispersing agents are generally sodium salts of naphthalene-sulfonic acid condensed with formaldehyde and may contain additives such as polyvinylpyrrolidone.

The dispersant CFR-2 allows the use of unusually small amounts of water in the cement slurry to which it is added. The densifier assists in dispersing particles in the water to make a useful slurry at water concentrations lower than those which would normally be acceptable. The reduced water concentrations have the effect of adding greatly to the strength of the cement after it has cured. The following are examples of the cement for wells in low temperature formations of the process.

To duplicate conditions under which mixing operations are carried out in low temperature environments, all materials were refrigerated at 30°F until just prior to mixing. Ice water was used for mixing and the molds for strength measurements were also refrigerated. Thickening times were determined at 32°F by circulating a refrigerated fluid through a Halliburton Company Consistometer. For compressive strength tests, samples of the compositions were cured by submerging molds containing the compositions in a refrigerated water bath.

The heats of hydration were calculated from plots of time versus temperature during the curing period. Weighted amounts of slurry were poured in replacement fillers for vacuum bottles, thermocouples embedded in the slurries and the bottles immersed in a 15°F water bath. The thermocouples were attached to a time temperature recording device and the resulting curves integrated with a planometer. This gives an average mean temperature for evaluating heat loss from the vacuum bottle replacement filler. The equation for calculation of heat of hydration used is as follows:

$$\Delta H\ (\text{Btu}) = \text{pounds mass} \times \text{specific heat} \times \Delta T + UA\Delta T\ \text{mean}\ T$$

An average value of heat loss for vacuum bottle fillers was determined. This value is combined with UA, and is 0.0477 Btu per hour per °F. Compressive strengths, thickening times, permeability, and heats of hydration tests were conducted in the laboratory using a cement slurry designated as Slurry A below. Three commonly used cement slurries were also tested which are designated as Slurries X, Y and Z below. A comparison of slurry properties is given in Table 1.

SLURRY A

Component	Weight, Lbs.	Composition, Weight %
Gypsum	41	42.95
API Class G Cement	26	27.24
Water	25	26.19
Sodium Chloride	3.0	3.14
CFR-2	0.33	0.35
Sodium Citrate	0.12	0.13
		100.00

SLURRY X

Component	Weight, Lbs.
API Class G Cement	47.00
Gypsum	50.00
Water	41.65

SLURRY Y

Component	Weight, Lbs.
Lumnite Cement	47.00
Pozzolan	37.00
Water	38.6
Sodium Chloride	3.35

SLURRY Z

Component	Weight, Lbs.
Ciment Fondu	43.75
Pozzolan	30.00
Water	33.2
Sodium Chloride	3.3

TABLE 1

COMPARISON OF SLURRY PROPERTIES

	Weight		Viscosity (Poises)		
Slurry	Lb/gal	Lb/ft³	0 Min.	10 Min.	20 Min.
A	15.40	115.5	12	11	11
X	15.37	115.0	5	5	8
Y	15.60	116.6	5	8	8
Z	15.21	113.7	3	9	10

TABLE 2

COMPARISON OF THICKENING TIMES AND COMPRESSIVE STRENGTHS

Slurry	Thickening Time (32°F) Hrs:Min.	Compressive Strength at 15°F. 16 Hr.	1 Day	3 Days	7 Days	14 Days
A	5:59	Not Set	905	905	1,535	2,695
X	1:05	843	450	508	494	531
Y	3:00+	Frozen	Frozen	200	415	760
Z	3:00+	Not Set	Frozen	120	315	360

From Table 2 above, it may be seen that the cement composition of the process (Slurry A) achieves relatively high compressive strength in as short a time as 24 hours. The other samples observed do not develop an adequate compressive strength within the required 24 hours or have too short a pumpable time to be useful.

TABLE 3

COMPARISON OF PERMEABILITY

Temperature at which Test Samples cured	15°F.
Time Test Samples Cured	7 Days
Slurry	**Permeability, Millidarcies**
A	0.18
Y	105.60
Z	56.60

From Table 3 it may be seen that the cement slurry of the process has a very low permeability after 7 days as compared to Slurries Y and Z.

TABLE 4

COMPARISON OF HEATS OF HYDRATION FROM TIME OF MIXING UNTIL HYDRATION COMPLETE

Slurry	Initial Temperature, °F	Final Temperature, °F	Mass of Test Sample, Lb.	Total BTU	BTU per Lb.
A	45°F	87°F	3.38	61	18
X	40°F	85°F	3.43	60.8	17.7
Y	43°F	134°F	3.46	197.3	57.0
Z	43°F	186°F	3.49	321.0	92.0

From Table 4 it may be seen that the cement slurry of the process evolves a low heat of hydration while setting.

Blast Furnace Slag, Quartz Sand and Magnesium Oxide

S.I. Danjushevsky, R.I. Liogonkaya and L.G. Sudakas; U.S. Patents 3,920,466; November 18, 1975 and 3,921,717; November 25, 1975 describe a binder for the cementing of wells at a temperature of 100° to 200°C and a pressure not over 1,000 atm includes blast-furnace slag, quartz sand and magnesium oxide with a refractive index between 1.722 and 1.734.

The presence of magnesium oxide in the binder as a component is conducive to an increase in the volume of the hardening mass of the binder in the course of hydration, mainly by virtue of the crystallization processes involved in the formation of magnesium hydroxide. Apart from that, the magnesium hydrosilicate forming under the conditions also adds to the volume of the binder.

To obtain the desired result, it is of importance to maintain the rate of chemical change during the process of hydration in agreement with the rate at which the formation of the structure of cement stone takes place. If the hydration of the magnesium oxide introduced is a rapid one occurring too early, the result will be an insignificant one since the hydration of magnesium oxide and all of the crystallization processes involved will take place in a mobile phase, which a suspension actually is, without affecting the volume. On the other hand, a retarded process of magnesium oxide hydration may bring about excessive straining of the cement stone which has formed, resulting in fracturing and eventual disintegration of the stone.

It has been found that the appropriate relationship between the rate of chemical change during the hydration of magnesium oxide and the rate at which the formation of the structure of cement stone takes place depends on the condition of the magnesium oxide introduced, the condition being controlled by the refractive index.

The use of a binder containing magnesium oxide with a refractive index between 1.722 and 1.734 in addition to other components makes the hermetic cementing of wells at a temperature of 100° to 200°C and a pressure not over 1,000 atm a practical possibility.

Example 1: A binder in an amount of 200 tons was obtained by grinding 180 tons of granulated blast-furnace slag with 20 tons of magnesium oxide, using a ball mill. The fineness of grinding was characterized by the fact that 15.4% of the particles were retained on a sieve with 4,900 meshes per square centimeter. The binder thus obtained was tested at a temperature of 150°C and a pressure of 400 atm. The results of the tests were as follows: flexural strength (after 48 hours), 26 kg/cm^2; setting time (with retarder), 1 hour 45 minutes; linear expansion, 0.4%.

Example 2: A binder in an amount of 1 ton was obtained by grinding 360 kg of granulated blast-furnace slag with 540 kg of quartz sand and 100 kg of magnesium oxide. The fineness of grinding was characterized by the fact that 1.2% of the particles were retained on a sieve with 4,900 meshes per square centimeter.

The binder thus obtained was tested at a temperature of 200°C and a pressure of 700 atm. The results of the tests were as follows: flexural strength (after 48 hours), 44 kg/cm^2; setting time (with retarder), 2 hours; linear expansion, 0.2%.

Zirconyl Chloride-Hydroxyethylcellulose Reaction Product as Thixotrope

J. Chatterji and G.W. Ostroot; U.S. Patent 3,804,174; April 16, 1974; assigned to Halliburton Company describe a thixotropic cementing composition for use in oil and gas wells which comprises a hydraulic cement slurry including as an additive a complex reaction product of a water-soluble carboxyalkyl, hydroxyalkyl or mixed carboxyalkyl hydroxyalkyl ether of cellulose, and a polyvalent metal salt, for example a reaction product of hydroxyethylcellulose and zirconyl chloride. The following examples illustrate the process.

Example 1: A dry blend is prepared by mixing 100 parts by weight of Portland cement (API Class H) with 0.25 part by weight of "Natrosol 250" hydroxyethylcellulose and 2.0 parts by weight of zirconyl chloride (zirconium oxychloride). The dry mixture is added to 46 parts of water with vigorous agitation, and the agitation is continued until a homogeneous slurry is obtained.

The initial viscosity of the slurry is 17 poises. The slurry is pumped into a simulated well for 10 minutes with substantially no change in viscosity. When the pumping is interrupted, the viscosity rises to 75 to 100 poises within a period of 20 minutes, but upon resumption of pumping the viscosity rapidly decreases again so that the slurry becomes pumpable once more. This sequence is repeated several times until the cement phase of the slurry hydrates and sets. Total setting time is about 2 hours.

Example 2: Proceeding as in Example 1, analogous tests were made using 2 parts by weight of each of chromium nitrate, lead chromate, and ferric chloride. The results were essentially similar.

Bitumen Particles for Lightweight Composition

B.B. Quist and J.J.M. Zuiderwijk; U.S. Patent 3,887,385; June 3, 1975; assigned to Shell Oil Company describe a dry mix for a lightweight cement. The mix comprises powdered cement mixed with bitumen particles of which more than 90% have sizes of less than 700 microns and aluminum silicate particles of which more than 90% have sizes of less than 30 microns.

A blown Qatar Marine propane bitumen, i.e., a bitumen derived from a crude oil from wells penetrating the sea bottom near Qatar and obtained by means of a precipitation treatment of a residual fraction of this crude oil, followed by blowing the bitumen obtained with air, was powdered by grinding it to a relatively fine powder.

In this way about 10 tons were obtained of a bitumen powder with a softening point (Ball and Ring) of 120°C and such a distribution of particle size that 100% of the particles was smaller than 500 microns, 80% by weight of the powder was formed by particles smaller than 300 microns, 60% by weight of the powder was formed by particles smaller than 200 microns, 30% by weight of the powder was formed by particles smaller than 100 microns and 10% by weight of the powder was formed by particles smaller than 70 microns. The average size of particles of this bitumen powder was 170 microns. 5% by weight of aluminum silicate powder was added to these 10 tons of bitumen powder. 95% by weight of the aluminum silicate powder had a particle size of less than 5 microns. The density was 2.15 and the specific surface was about 120 m^2/gram.

The 10 tons of bitumen powder were thoroughly mixed with the aluminum silicate powder and thereafter transported over a substantial distance in a cement bulk carrier. No detrimental caking of the particles took place during this transport since the aluminum silicate powder acted as an anticaking agent.

Subsequently, the mixture of bitumen powder and aluminum silicate powder was mixed by fluidization by means of air with 10 tons of oil-well cement of a type known under the commercial indication of API class G. The pneumatic transport both of the mixture of bitumen and aluminum silicate powder immediately before mixing with the cement and the final mixture of bitumen/aluminum silicate/cement did not present any difficulties regarding caking of particles or separation thereof, notwithstanding the fact that the density of the oil-well cement (3.2 g/cm^3) was considerably higher than that of the bitumen powder (1.0 g/cm^3).

The cement composition was subsequently mixed with the appropriate amount of water to form a pumpable slurry, which slurry was pumped through a cement string down into the hole to a level where the casing was to be cemented. The slurry entered the annular space around the casing and was retained therein to harden.

Based on the bitumen/aluminum silicate powder mixture prepared as described above, various mixtures were prepared with varying percentages of aluminum silicate with a weight ratio of cement/bitumen of 1.25 and a weight ratio of water/cement of 1.24. Provisions were made to ensure that each time the density of the fresh cement slurry was 1.3 g/cm^3.

The aluminum silicate had a double function, firstly to minimize the risk of caking of the particles of the bitumen powder, and secondly to increase the compressive strength of cement to which bitumen powder has been added for decreasing the density thereof.

The compressive strength of all mixtures was determined in special experimental samples of the cement slurry during 24 hours after hardening at a temperature of 20°C. However, for each mixture the temperature at which hardening took place was kept at three different values in different experimental samples, namely: 32°C, 43°C and 78°C.

The results of the measurements are incorporated in the graph shown in Figure 3.2. In this graph the compressive strength of the hardened cement is indicated in kilograms per square centimeters and plotted against the content of aluminum silicate in a weight percentage. Three areas have been indicated in the graph for the three hardening temperatures mentioned above. The graph clearly shows that the addition of aluminum silicate has a positive influence on the compressive strength of the oil-well cement for each hardening temperature.

FIGURE 3.2: COMPRESSIVE STRENGH vs % ALUMINUM SILICATE

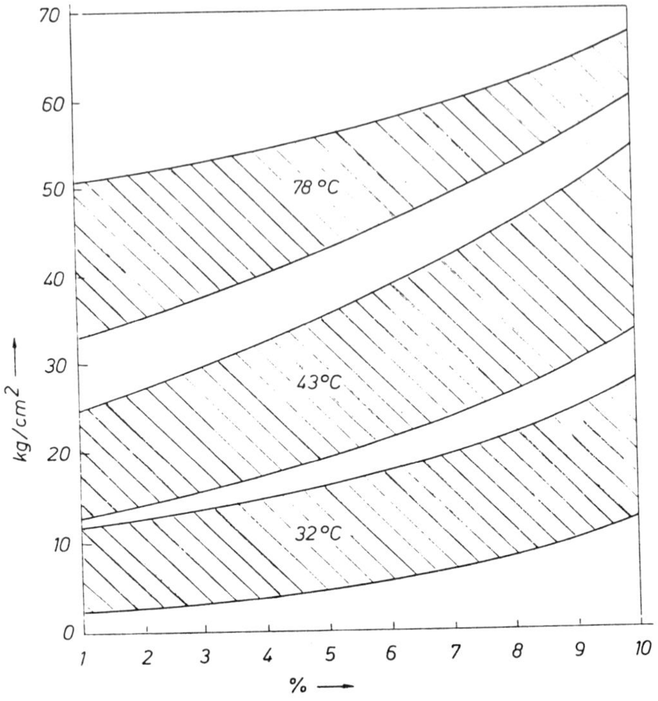

Source: U.S. Patent 3,887,385

It has also been found that a rise in the temperature at which hardening takes place will bring about an increase of the compressive strength.

Also, a batch of bitumen of the same type as described above, was ground less extensively, whereby 4 tons of bitumen powder were obtained with a particle size distribution such that 100% by weight of the particles was smaller than 2,000 microns, and 90% by weight of the powder was smaller than 1,000 microns.

The average size particle size was about 250 microns. The bitumen powder was subsequently mixed with 5% (by weight) of aluminum silicate (having a particle size less than 30 microns) and 5 tons of Pozmix cement. Once more no problems were encountered with mixing and pneumatic transport of the mixture with respect to caking and separation of bitumen/cement/aluminum silicate mixture.

Subsequently, various cement slurries with a water/cement ratio of 0.6 to 0.7 were tested. The compressive strength of these slurry samples was found to be below the requirement of 35 kg/cm^2 under standard conditions.

By replacing the 10% coarse bitumen particles, i.e., 10% particles greater than

1,000 microns, by the same weight of particles smaller than 1,000 microns, compressive strengths over 35 kg/cm^2 at the same conditions were found. The particle distribution of the bituminous component in these mixtures was the following:

100% by weight	$d < 1{,}000\ \mu$
95% by weight	$d < 700\ \mu$
75% by weight	$d < 500\ \mu$

The average diameter d_{50} was slightly less than 250 microns. If desired, the bitumen powder may be mixed together with the cement and the aluminum silicate instead of first mixing the bitumen powder with the aluminum silicate followed by mixing the mixture thus obtained with cement.

REINFORCED COMPOSITES

ASBESTOS-CONTAINING COMPOSITES

Thermal Shock Resistant Compositions

J.C.-S. Yang; U.S. Patent 3,933,515; January 20, 1976; assigned to Johns-Manville Corporation describes a thermal shock resisting asbestos-cement composition comprising in percent by weight of dry solids (a) 10 to 50% asbestos fiber; (b) 25 to 65% Portland cement; (c) 1 to 30% of a finely divided additive adapted to increase the porosity of the composition and (d) 0 to 40% of silica flour. This composition, when shaped to form as asbestos-cement article and autoclaved possesses thermal shock resistance superior to previous asbestos-cement articles, while maintaining the other desirable properties exhibited by prior autoclaved asbestos-cement articles. A major reason for the superior shock resistance is the increased porosity of the composition at high temperature which allows moisture to freely and gradually escape.

Conventional autoclaved asbestos-cement compositions contain up to 40% silica flour, but in accordance with this process, the silica content is either partially or entirely replaced by a finely divided organic or inorganic porosity increasing additive to provide increased thermal shock resistance. The compositions, after autoclaving, stay intact and retain appreciable strength under intense heating conditions, and when subjected to high temperatures while wet, or saturated with water.

The asbestos-cement compositions contain finely divided additives as stated above. One such group of additives may be inorganic mineral additives selected from the group consisting of asbestos tailings, asbestos floats, talc, serpentine rock, and mica to improve thermal shock resistance. Another group of additives may be organic additives such as, nylon fibers, orlon fibers, sisal fibers or the like. A third group of additives (either organic or inorganic) may be those which decompose or dehydrate at elevated temperatures, examples of such additives being wood fibers, diatomaceous earth, magnesium hydroxide, magnesium carbonate or the like. In any case, the additives are of the type which will increase porosity and, therefore, thermal shock resistance of the ultimate product when the latter is subjected to temperature approaching 700°C. The asbestos-cement formulations in which these additives are substituted as a replacement of the silica flour or quartz show exceptional thermal shock resisting properties while exhibiting good sulfate resistance and mechanical strength comparable or superior to regular asbestos-cement steam curable compositions. It has been observed, however, that when the additive/silica ratio exceeds 17.5/12.5, a marked decrease in ultimate strength occurs. In the following examples all percentages are by weight and all screen sizes are U.S. Standard unless otherwise stated.

Example 1: A number of astestos-cement compositions are prepared by forming dry furnishes containing 50% Portland cement, 20% asbestos, and 30% of a mixture of silica flour and a finely divided porous providing additive. A number of compositions containing no silica flour and 30% additive are also prepared along with a control composition containing no additive but 30% silica flour. Each of these compositions is mixed with water and shaped into 8 inch by 3 inch by ¼ inch samples.

These samples are cured at 100% relative humidity and room temperature for 24 hours, and are then autoclaved at 170°C and 100 psig for 16 hours. After autoclaving, the samples are saturated in water for 24 hours, and then placed directly in a furnace at 700°C for one hour to test their resistance to thermal shock. The results are summarized in the table on the following page which sets forth the various additives and the percent additive replacing the silica flour. As noted in this table, with no additive, the sample exploded in two to three minutes after being placed in the furnace. In most cases, with the addition of larger amounts of additive, the sample did not explode. These tests thus show that asbestos-cement compositions modified in accordance with the process exhibit greatly improved thermal shock resistance.

Example 2: A thermal shock resistance study is conducted using high asbestos fiber content Portland cement articles containing 44% asbestos fiber and 33% Portland cement on a dry solids basis. A number of autoclaved samples are prepared containing 10%, by weight of the total dry solids, of talc or asbestos floats as replacement for silica flour, as well as samples containing no replacement of silica flour. One face of each sample is exposed to a temperature of 100°F for 15 minutes on a hot plate after adjusting the moisture content to a normal level, about 12% by weight, or to the saturated level. The standard samples containing no talc or asbestos floats explode at both moisture content levels, whereas the samples modified with 10% asbestos floats or talc substituted for silica remain intact.

Asphalt Emulsion for Water Resistance

A process described by *R.C. Pomerhn and R.M. Johnson; U.S. Patent Application (Published) B 431,072; January 20, 1976; assigned to National Gypsum Company* involves the manufacture of low density asbestos-cement sheets incorporating a form of asphalt emulsion which permits the continuous production of these products without a troublesome amount of asphalt accumulating on the manufacturing equipment. The low density asbestos-cement sheets are produced from a formulation which includes an asphalt emulsion in which the internal asphalt phase consists of an asphalt having a softening point of at least 150°F (preferably from 180° to 190°F) and which is present in smaller than 10 microns.

Effect of Additives on the Thermal Shock Resistance of Steam-Cured Asbestos Products

Additive, %	0	1.0	2.0	5	7.5	10	10–12.5	15	17.5	18.8	20	22.5	30
Desert talc	X			*		X	X	*	*	*	*		
Silvery talc	X			X		*	*	*	*	*	*		*
C. S. Smith Talc	X					X		X	*		*		
Serpentine rock	X					*		*	*		*		
Calcined serpentine rock	X					*		*			*		
Asbestos floats (chrysotile)	X				X	X	X	*	*	*		*	
Asbestos tailings (chrysotile)	X			X	X	X	X	*	*	*	*	*	
Calcined asbestos tailings (chrysotile)	X			X		1 hr. cracked into 1		*					
Blue brand chrysotile	X					X		*					
Pulverized chrysotile	X			X	X	X	X	*	*	*	*	+	
Chrysotile tailings	X				X	X		*			*		
Pulverized chrysotile tailings	X			X	X	X	X	X	X	*	X	*	
Calcined chrysotile tailings	X			X		*		*					
Johns-Manville asbestos fiber(a)	X	X	X			*		*					
Hectorite	X					X		X					
Mica	X			X		X		*					*
Nylon	X	*	*										
Orlon	X	*	*										
Crimped orlon	X	*	*										
Silica (quartz)	X			X		X		X					

* = remains intact
X = exploded in 2 to 3 minutes
+ = small surface crack
(a) = Munro M-100 chrysotile asbestos fiber

The standard ASTM test for the softening point of asphalts, of a softening point between 85° and 347°F, involves a ring and ball apparatus in an ethylene glycol bath, and is designated D2398-71.

Example: A low density asbestos-cement sheet was made on equipment adapted for continuous production, using the following formulation:

	Parts by Weight
Portland cement	35
Asbestos fibers	15
Cellulose fibers	9
Asphalt emulsion (59.8% asphalt)	5
Inorganic siliceous fillers	36
Sufficient water as required in the cylinder method of wet forming.	

The ingredients other than the cellulose fiber, the asphalt emulsion and the water are thoroughly mixed in a dry powder ribbon mixer. The cellulose fiber is paper or wood pulp which has been thoroughly disintegrated in a pulper beater tank containing a water slurry of the paper or pulp of about 3% solids.

The cellulose fiber slurry is then transferred through a Reitz disintegrator to a measuring tank, and then to a wet-mixing tank. The dry powder is also transferred to the wet-mixing tank and mixed with the cellulose fiber slurry to form a blended slurry. The asphalt emulsion is then also added and thoroughly blended. The asphalt emulsion, must be made of an asphalt having a softening point of at least 150°F and the asphalt must be present in the emulsion in a form each portion of which is smaller than ten microns. A suitable asphalt emulsion is Bitusize BB7 (Chevron Asphalt). It is preferable to filter the asphalt emulsion through a 50 mesh screen as it is being fed to the wet-mixing tank.

The mixed total furnish is then transferred to a holding tank where it is held under continuous agitation and at no more than 110°F, as enough is continuously being withdrawn to keep the cylinders suitably supplied. The furnish is then continuously formed into a continuous web by being deposited on the cylinder screen and then transferred to a felt conveyor. A plurality of cylinders deposit a plurality of webs, laminated, onto the felt conveyor. The material is transferred from the felt conveyor to an accumulator roll where several plies are accumulated and formed into a sheet of desired thickness.

As the above-described mixed total furnish is withdrawn from the holding tank, to be fed to the cylinders, the furnish is directed to a trough, additional dilution water is immediately added along with a cationic retention aid such as a cationic, water soluble, high molecular weight polyacrylamide, or other suitable cationic retention aid, and the furnish, dilution water and cationic retention aid are thoroughly admixed as the materials progress along the trough toward the vats in which the cylinders operate. This addition of a cationic retention aid causes the asphalt emulsion to break, whereby the original individual liquid particles of asphalt are no longer completely and individually surrounded by water, but instead are free to attach themselves to the adjacent solids.

Just before the furnish, containing the broken asphalt emulsion, reaches the end of the trough leading to the cylinder vats, an anionic flocculating agent is added, such as an anionic, water soluble, high molecular weight polyacrylamide, or other suitable anionic flocculating agent. This anionic flocculating agent causes the asphalt to be attracted to the solids in the slurry, including the cement, the fibrous materials, and the inorganic siliceous fillers, and to become attached to these solids and also causes the solids to generally agglomerate. As the solids then proceed into the cylinder vats and then into the continuous web that is formed, the asphalt is carried by the solids into the web that is formed. The cationic retention aid and the anionic flocculating agent are each used in an amount of from about 0.05 lb/1,000 lb to about 0.3 lb/1,000 lb of mixed total furnish solids.

The sheets of the desired thickness removed from the accumulator roll are partially dried, smoothed by a pressure roller and then stored for a precure time of about 7 to 14 days. They are then autoclaved using about 120 psig steam, for about 20 hours. The finished sheet of this example has a thickness of ⅜ and may be used with or without coatings. If coating is desired, there is preferably applied a first clear sealer coat followed by a white primer coat, suitable for finish coatings of latex or oil paints, as described in U.S. Patent 3,413,140.

The product has a dry density of about 60 lb/ft^3. It will typically have a moisture content of something less than 15% when shipped. If submerged in water for 24 hours, this asphalt-containing product will absorb an amount of water equal to about 35 to 40% of its weight, as compared to an absorption of about 55% for a similar product having no asphalt.

Of particular significance is the fact that the cylinder screens, the felt conveyors, the accumulator roll and other apparatus pick up no more than a trace of asphalt during a typical production run which can easily be removed by making a short run afterwards of the same product having no asphalt content. Alternatively, a relatively easy cleansing, as with gasoline as a solvent, will remove all of the asphalt deposited on the equipment.

Alkanolamine Additives

S.M. Quint; U.S. Patent 3,553,077; January 5, 1971; assigned to Johns-Manville Corp. describes an improvement in the manufacture of asbestos-Portland cement pipe products formed by accumulating a thin wet sheet, or sheets, of asbestos-Portland cement stock ingredients through convolute winding of the sheet upon a rotating mandrel while compressing to consolidate the material.

The improvement is based upon the discovery that relatively minor proportions of a specific class of alkanolamines applied to the thin wet sheet of asbestos-Portland cement stock ingredients effectively and markedly quicken the initial setting and hardening of asbestos-Portland cement mixtures to the extent of reducing sag in resulting pipe products up to 85 or 90%, or more, over like produced pipe but without the treatment of this process.

Requisite alkanolamines for the process consist of the substantially water-soluble monoethanolamines, diethanolamines and triethanolamines, or mixtures thereof. Triethanolamine is preferred since it achieves optimum retardation of sag with a minimum of adverse effects. These alkanolamines were found to be highly effective in rapidly and substantially quickening the initial set of asbestos-Portland cement compositions in the manufacture of pipe when applied in such minimal proportions of about 0.01% by weight of the alkanolamine based upon the total weight of the dry asbestos-Portland cement stock composition, and typically for substantially any Portland cement manufacture, amounts of 0.02 up to 2.0% by weight of amine based upon the total dry stock. Preferred proportions for normally optimum results comprise about 0.02 or 0.3 up to 1.0%, since with most stock compositions the degree of sag reduction achieved with increased proportions is not proportionally or even discernibly extended, and because of the potential hazard of degrading the product or disrupting the manufacturing process.

Asbestos Fiber Treatment

J.P. Nicol; U.S. Patent 3,661,603; May 9, 1972; assigned to Asbestos Corporation Ltd., Canada describes a lightweight, rigid composition comprising mixing water, a cementitious material, asbestos fibers and an anionic surfactant capable of chemically opening the fiber and resistant to alkalis and saponification. The product may be cellular, in which case it may additionally include a different surfactant, a sequestering agent or a colloidal dispersion.

Commercial Canadian chrysotile asbestos as supplied by the mines and milled by normal milling methods has a specific surface area between 3,000 and 14,000 cm^2/g. Harsher physical opening of the fibers such as obtained by high speed pulverizing, e.g., Raymond mill, can increase the surface area to around 20,000 cm^2/g. Use of still harsher means such as ball-milling can increase the specific surface to around 25,000 cm^2/g as measured by conventional air permeability methods. However, severe physical opening of Canadian chrysotile asbestos results in some fiber destruction.

In comparison, chemically opened asbestos causes surface areas as high as 200,000 cm^2/g (calculated) with a percentage of the fibers having colloidal dimensions. The fibers in water exhibit hydrophilic colloidal properties including electrophoresis, high viscosity and film-forming characteristics.

The chemically opened asbestos fibers referred to are the subject of U.S. Patent 2,626,213. In accordance with the method described in that patent, chrysotile asbestos fiber bundles are opened and dispersed to separate, individualize and disseminate the fibers, and to produce an alkaline, gelatinous, stable, aqueous dispersion, composed predominantly of fibro-colloidal asbestos fibers or fibrils.

This is achieved by treating the fiber bundles with an aqueous solution or organic detergent surface-active material which is adsorbed on the surface of the asbestos and is effective in disrupting the molecular bond between fibers. An excess of surface active agent is employed to maintain the opened fibers in fluid dispersion. The asbestos is thus reduced in large measure to colloidal size particles and these are held in a state of colloidal dispersion.

The treatment of the asbestos fibers with a surface active agent in accordance with the process results in chemically opening the fibers. In preferred methods of the process the degree of chemical opening may be qualitatively assessed by stretching a drop of colloidal asbestos dispersion when a tenuous fibro-colloidal filament is obtained. A high degree of chemical opening is shown by a stable filament of about 1½ inches in length. A low degree of opening is shown by a filament of about ¼ inch in length. Another rapid method for assessment of the degree of chemical opening which is semiquantitative in nature, is measurement of the viscosity, of a colloidal asbestos dispersion. The higher the viscosity, the greater the degree of fiber opening. It has also been found that the greater the fiber length, the higher the viscosity. In semiquantitative terms:

$$\frac{\text{Viscosity}}{\text{Fiber concentration}} = \text{f (fiber length)/(fiber diameter)}$$

A suitable instrument for measurement of viscosity is a Brookfield Synchro-Lectric Viscometer, Model HBT. Preferred surfactants to chemically open chrysotile fiber are of the anionic type and include representative compounds from each group within the general class of anionic surface active agents. The classifications used are taken from McCutcheon's, Detergents and Emulsifiers, 1968 Annual.

The chemically open fibers are maintained in a state of colloidal dispersion. A sufficient amount of surface active agent must be present to maintain these fibers in a stable state of dispersion. The amount of surface active agent required depends on the concentration of active ingredient in the surfactant. 0.4 to 0.9% or more by weight (active ingredient) surfactant should be used based on the weight of cementitious material. However, as much as 2 to 4% should be used to obtain products of lower density. Since excess amounts of surface active agent will increase the required setting time at a given temperature, they should be avoided. The following examples illustrate the process. For ease of reference, chrysotile asbestos will simply be referred to as asbestos.

Example 1: (A) Water (6,000 cc) at 80°F was added to an open vessel of proper capacity equipped with a multilevel electric stirrer. Portland cement (1,500 g) was then added, followed by addition of the sequestering agent, sodium hexametaphosphate (30 grams). These ingredients were mixed together at low speed (450 rpm) for a short period (5 minutes). Addition of Duponol G (60 g) and asbestos 5R (300 g) followed. The mixture was then stirred again at low speed (450 rpm) for a short period, e.g., 5 minutes. The mixture was then subjected to vigorous agitation, e.g., 1,200 rpm until a volume of 12,000 cc was obtained. The time of stirring was 39 minutes and the product was fine celled and of a definite viscous consistency.

The viscosity as measured by a Brookfield viscometer, model HBT, using number H-1 spindle was 497 cp. The product was then poured into a wooden mold lined with polyethylene film. After 5 days it had set firm. At 7 days it was removed from the mold and transferred to a humidity cabinet for a further 14 days. It was then dried out in hot air. The time of initial set of 5 days at ambient temperature could be reduced to a matter of hours by heating the mold in a steam filled atmosphere or if the mold is equipped with a lid to retain the water, in hot air, e.g., 6 hours at 150°F. Shrinkage during curing is also reduced. The product obtained gave the following test data: shrinkage during curing, 19% (based on initial and final thickness measurements, lateral shrinkage small); density, 14.3 lb/ft^3; and flexural strength (modulus of rupture), 101 psi.

(B) To illustrate the use of other anionic surfactants, with sodium hexametaphosphate, Example 1 was repeated using 200 grams Asbestos 5R and 40 grams of one of the following active materials: Ninate 411, Tergitol 7, or Triton 770. A fine celled product of about 12,000 ml volume was obtained. The viscosity of material, the time of setting, the density and flexural strength are shown on the following page.

	Ninate 411	Tergitol 7	Triton 770
Viscosity (cp)	357	416	363
Setting time (days)	10	23	8
Density (lb/ft^3)	15.1	19.5	12.9
Flexural strength (psi)	88	142	48

Example 2: This illustrates the preparation of a high expansion material.

Ingredients A:	6,000 ml tap water 1,500 g Portland cement 30 g sodium hexametaphosphate
Ingredients B:	40 g Duponol G 300 g Asbestos 5R

Employing a similar procedure to Example 1 except that an extra propeller was employed, a product of volume 18,400 ml was obtained after 30 minutes stirring. The material took 9 days to set firm and when fully cured had a density of 9.5 lb/ft^3 and flexural strength of 39 psi.

Lightweight Thermal Insulation

R.F. Shannon; U.S. Patent 3,597,249; August 3, 1971; assigned to Owens-Corning Fiberglas Corporation describes a process for producing composites of inorganic fibers and inorganic crystalline binders. In the process a hardened binder is attached to the fiber, a slurry of these binder coated fibers is made, and a hardened product formed by uniting the hardened binder coating already on the fibers. The uniting of the hardened binder coating may in some instances be done by bringing the coated fibers into contact and then changing the crystalline structure of the binder as by autoclaving.

By this procedure, sufficient driving force is provided to not only modify the crystalline structure of the hardened binder coating but at the same time link up the coating with that on other fibers. In other instances, the hardened binder coating serves only as a nucleus for crystalline growth which bonds to the coating on adjacent fibers. Not only is less time required for the hardening of the finished product, but slurries made from the binder coated fibers retain water better and are creamier than are slurries of untreated fibers and binder. The slurries of the process are more uniform and creamy, and the products produced are more uniform and have sharper detail and greater strength. The following example illustrates the process.

Example 1: This example describes the production of a lightweight thermal insulation material having a density of approximtely 11 lb/ft^3, and a binder crystalline structure that is primarily tobermorite ($4CaO{\cdot}5SiO_2{\cdot}5H_2O$).

	Parts by Weight
Amosite asbestos	400
Chrysotile asbestos	100
Quicklime	760
Supersil (95% 325 mesh silica) Pennsylvania Glass Sand Co.	400
Clay	100
Celatom (diatomaceous earth) Eagle-Picher Co.	650
Limestone	735

In the above listed materials the quicklime is 0.94 active calcia, the celatom is 0.83 active silica, and the clay 0.45 active silica, to give a calcia to silica molar ratio of 0.794. A slurry of 6 to 1 water to solids ratio was prepared and kept agitated in a Hydropulper for approximately ¼ hour, and was then poured into molds to provide one inch block. The molds were placed in an autoclave whose pressure was raised to 175 lb/in^2 gauge over a period of 1 hour with saturated steam. The autoclave was held at this pressure for 2 hours, following which the temperature of the autoclave was raised to 550°F by circulating Dowtherm through heating coils within the autoclave. This temperature was held an additional 2 hours during which superheated steam was vented from the autoclave, and following which the autoclave was depressurized over a period of 1 hour. A product was produced whose binder was approximately 80% tobermorite.

The product was then placed in a Jeffery hammermill run at 900 rpm and fitted with a ½" screen. The action of the hammers cracks the crystalline binder without materially subdividing the fibers to leave a high percentage of the binder as a coating on the fibers. The following table indicates the fiber length as determined by the Bauer McNett fiber classifier testing procedure for the coated fibers as well as the asbestos used initially. It will be seen from the analysis that while some of the longer fibers are broken, the intermediate length fibers of the 14 and 35 mesh are retained.

	Percent Retained on:				
Material	**4 Mesh**	**14 Mesh**	**36 Mesh**	**100 Mesh**	**200 Mesh**
Amosite	62.7	14.1	8.2	11.7	3.3
Chrysotile	4	13	23	36	45
Hammermilled product	21	25.4	22.5	-	23.4

The fibers produced were then used to produce another slurry of the following composition.

Ingredients	Parts by Weight
Amosite	400
Chrysotile	100
Quicklime	740
Supersil	400
Celatom	650
Clay	100
Binder coated fibers	700

A slurry was made using a 6 to 1 water to solids ratio, and the slurry was then cast into one and a half inch thick blocks and autoclaved using the cycle described above, to produce a product having a modulus of rupture of 60+ and a crystalline structure which was more than 80% tobermorite. The second slurry made by using the binder coated fibers was much creamier in consistency, and had less tendency for solid separation than did the first formed slurry. The molded product had sharper corners, and approximately a 50% better strength than did the first formed material.

Example 2: A material having a xonotlite crystalline structure was made from the following ingredients.

Ingredients	Parts by Weight
Amosite	300
Chrysotile	200
Quicklime	750
Supersil	780
Red oxide	11
Limestone	165

This material had an active CaO to SiO ratio of 0.953 and 10,395 parts by weight of water was used. The molds were placed in an autoclave whose pressure was raised to 175 psi over a 1 hour period, held there for 4 hours following which the temperature was raised to 550°F. The temperature was held an additional 4 hours during which superheated steam was vented from the autoclave and following which it was depressurized over a 1 hour period. The product was then passed through the hammermill of Example 1. The water to solids ratio was 4.71. The crystalline structure produced was more than 80% xonotlite. The binder coated fibers were incorporated into a slurry of the following composition.

Ingredients	Parts by Weight
Amosite	300
Chrysotile	200
Quicklime	750
Supersil	780
Binder coated fibers	700

This material has an active calcia to silica ratio of 1.017 and when mixed with water to provide a water to solids ratio of 4.7 has much less tendency for solid separation, is creamier and easier to mold, and has greater strength than the first formed product. In the above examples, the binder coated fibers are used as a nucleus for growing additional binder of substantially the same crystalline structure. However, xonotlite can be grown over tobermorite and vice versa by using additional binder material of the appropriate calcia to silica ratio. It is also possible to form a product of the binder coated fibers without additional binder by autoclaving the binder coated fibers.

Example 3: The tobermorite coated fibers produced as in Example 1 were wetted with water without additional binder forming materials, and were autoclaved using the same cycle given above. A hard product was formed having a density of 20 lb/ft^3 and a modulus of rupture of 60 with over 80% of the binder being tobermorite.

Example 4: The process of Example 3 was repeated using the xonotlite coated fibers of Example 2 to produce a bonded product similar to that produced in Example 3 excepting that the structure was more than 80% xonotlite.

Example 5: 100 parts by weight of the binder coated fibers of Example 2 are mixed with 100 parts of an aluminous cement, purchased under the name of Lumnite. The fibers are dampened with approximately 200 parts of water and compacted into a mold and allowed to air dry for 2 days. A material having a density of approximately 30 lb/ft^3 is formed which has a greater strength than the material of Example 2 indicating that a good bond between the aluminous cement and the coated fibers is obtained.

ORGANIC FIBERS

Melt Spinning of Polymeric Resin-Cement Mixture

According to a process described by *S. Oya, H. Noro and K. Suzuki; U.S. Patent 3,865,779; February 11, 1975; assigned to Hiromichi Murata, Japan* the reinforcing additives to be applied to inorganic cements are prepared by mixing one or more polymeric resins in pellet or powder form selected from the group consisting of polyethylene, polypropylene, polyvinyl chloride, polyamide and a copolymer of styrene and acrylonitrile and 0.1 to 3.0% of a hydrophilic surfactant based on the weight of the resin and a small amount of stabilizer, such as a metallic salt of monovalent higher fatty acid, if necessary, and by further adding to the above mixture adequate amounts of a hydraulic cement such as Portland cement and white cement, water hardenable lime or water hardenable magnesia-lime, or inorganic powders which are stable under heating (to about 300°C) such as anhydrous gypsum silica sand or limestone and then thoroughly mixing the whole mixture, by

(1) spinning the molten mixture into fibers having solid, hollow circular or noncircular cross sections and drawing or crimping the fibers, or

(2) extruding the melt into a thin film, drawing and splitting it as fibers, and, by finally cutting the resulting fibers to adequate lengths.

A preferred range of cement to polymer is 10 to 50 parts by weight of cement to 100 parts by weight of polymer, and a preferred range of fiber to cement is 1 to 5%. An operable range of fiber length is 5 to 100 mm; preferably 10 to 15 mm for cement or mortar reinforcement 30 to 80 for concrete reinforcement. The preferred ranges of temperature of melt extrusion are 150° to 250°C for polyethylene; 220° to 290°C for polypropylene; 160° to 180°C for vinyl chloride; 220° to 260°C for nylon; and 230° to 250°C for acrylonitrile-styrene copolymer.

The range of proportions for each of the constituents is (per 100 parts of thermoplastic synthetic resin): hydrophilic surfactant, 0.1 to 3.0 weight parts; inorganic materials, 20 to 80 weight parts (preferred range for inorganic materials, 30 to 60 weight parts); the fiber thickness is 50 to 300 denier, preferably 50 to 150 denier for mortar and cement and 150 to 250 denier for concrete; the particle size of inorganic materials is between 5 and 50 microns.

Example 1: 69.5 parts by weight of polypropylene in pellet form, 0.5 part of triamine type surfactant and 30.0 parts of normal Portland cement were mixed. 69.5 parts by weight of nylon in pellet form, 0.5 part of a triamine type surfactant (AMS-313 available from Lion Yushi KK), and 30.0 parts of normal Portland cement were mixed. Each mixture was melted, spun, drawn and cut into a length of about 15 mm to obtain the respective reinforcing additives.

Each of the additives in the amount of 3 parts by weight was added to a cement mortar consisting of 100 parts of normal Portland cement, 300 parts of sand and 55 parts of water, and the resulting mixture was formed into shape. The formed matter was taken out of the mold after 24 hours and cured in a wet atmosphere for 6 days at 20° ± 2°C and at 93 to 95% relative humidity. The final product was tested with the Charpy impact testing machine, which gave the following results shown in Table 1 and Table 2.

TABLE 1: RELATION BETWEEN IMPACT ENERGY AND WIDTH OF CRACKS*

Impact energy (Kg-m)	0.60	1.50	2.25
No additive added	0.14	—	—
Additive containing polypropylene	0.01	0.25	0.93
Additive containing nylon	0.02	0.28	1.29

* (Width of cracks in mm)

TABLE 2: RELATION BETWEEN IMPACT ENERGY AND DEPTH OF CRACKS*

Impact energy (Kg-m)	0.50	1.00	1.50	2.25
No additive added	12.0	—	—	—
Additive containing polypropylene	0.0	8.0	12.0	14.4
Additive containing nylon	2.0	12.0	14.5	16.0

*(Depth of cracks in mm)

Example 2: 69.5 parts by weight of polypropylene, 30.0 parts of normal Portland cement and 0.5 part of a triamine type surfactant were mixed and melted. The melt was then spun into fibers that were drawn about 3 times as long and cut to 10 to 15 mm length. 3.0 parts by weight of the reinforcing additive thus obtained was mixed with a cement mortar consisting of 100 parts of normal Portland cement, 65 parts of water and 200 parts of standard sand from Toyoura and the whole mixture was formed into shape.

This was taken out of the mold after 2 days, enclosed in a polyethylene bag, and cured for 7 days in a thermostated room at 20° ± 2°C. Strength against stretching, bending and compressing of the final product mortar was tested according to the testing method of physical properties of mortar (JIS R-5201). The results are shown in Table 3 in the ratios to those obtained with a mortar which was prepared without any additive.

TABLE 3

Thickness of fibers (Denier)	Ratio of strength		
	Stretching	Bending	Compressing
No additive added	1.00	1.00	1.00
137 denier	1.21	1.17	1.08
215 denier	1.18	1.05	1.15

Example 3: The reinforcing additive prepared in Example 2, fibers of 100% polypropylene (spun from melt and drawn) and fibers prepared by spinning of a melt containing cement and polypropylene but not drawn, all of an approximately equal thickness, were used to make mortar formations. Strength of them was tested as in Example 2 and the results are shown in Table 4 in the ratios to those of the mortar to which no additive was applied.

TABLE 4

Fibers admixed	Thickness (Denier)	Ratio of strength Stretching	Bending	Compres-sing
No additive	—	1.00	1.00	1.00
100% Polypropylene	133	0.87	0.88	0.84
Not drawn	130	0.89	0.86	0.85
Additive of this invention	137	1.19	1.18	1.08

It is concluded from the above experiments that the reinforcing additives to cement prepared by the process not only markedly improve the impact resistance but also increase the resistance to stretching, bending and compressing of mortars of cements, and thus exhibit an effect entirely different from that expected from addition of synthetic fibers themselves. The thickness and the length of the reinforcing additives depend, of course, on the material of fibers and the object to which they are added. But fibers over 50 denier approximately are considered suitable in view of the process of manufacturing and the effect obtained by the additive.

Example 4: To a thorough mixture consisting of 69.5 parts by weight of pellets of nylon 6 and 0.5 part of a triamine type surfactant, was added 30.0 parts of white Portland cement and the whole mixture was mixed thoroughly. This was melted at 220° to 230°C and was spun into fibers of a circular cross section of about 0.4 mm diameter. After being cooled, the fibers were drawn 4 times as long and the resulting fibers of about 0.2 mm diameter were cut to 15 mm length to make a reinforcing additive to be applied to inorganic cements.

Example 5: To a mixture of equal weights of polyethylene and polypropylene in pellets in the total amount of 69.5 parts by weight, 0.5 part of a triamine type surfactant was added and mixed, to which 30.0 parts of normal Portland cement was added and mixed thoroughly. The whole mixture was melted at 280°C and extruded into a film. After being cooled the film was drawn about 6 times as long and split off as fibers and cut to 10 mm lengths which are to be used as a reinforcing additive for inorganic cements.

Example 6: To a thorough mixture consisting of 69.5 parts by weight of polyvinyl chloride in powder, 4.0 parts of zinc stearate and 1.5 parts of sorbitan mono-stearate, 25.0 parts of fine powders of silica sand was added and mixed thoroughly. The whole mixture was melted at 150°C and spun into fibers of hollow section and cooled. The fibers were drawn 2 times as long then cut to 15 mm lengths, which are used as a reinforcing additive to be applied to inorganic cements.

Polyester Fiber

According to a process described by *G.R. Jakel; U.S. Patent 3,899,344; August 12, 1975; assigned to California Cement Shake Co., Inc.* a reinforced, relatively flexible concrete product is formed by: (a) combining Portland cement, polyester fibers, selected aggregate and water to form an aqueous mixture, and (b) forming the product from the mixture and curing the product. The process provides a reinforced, relatively flexible, concrete material for such products as slabs, roofing, siding and shingles.

Best results are achieved when the constituents of the mixture are present in the following relative amounts:

180 to 195 lb of Portland cement
72 to 90 lb of Perlite
4 to 8 ounces of Darex
38 to 46 gallons of an aqueous mixture of polyester fiber and cellulose
7 to 9 gallons of slaked lime
0.6 to 1 gallon of an accelerator

Portland cement is used to provide fireproofing, and Perlite makes the product lightweight and permits sawing and nailing.

Fibrillated Polypropylene Film

In a process described by *J.J. Zonsveld and R.F. Salmons; U.S. Patent 3,591,395; July 6, 1971; assigned to Shell Oil Company* concrete, mortar, cement or plaster of Paris are reinforced to provide products of improved bending strength by addition of up to 2% by weight of fibrillated polypropylene film to the mass prior to or during mixing. The following example illustrates the process.

Example 1: A cement/sand/water mixture in a ratio by weight of 1:3:0.5 was mixed in a conventional concrete mixer and to this mixture 1% by weight of reinforcing elements formed from a stretched and fibrillated polypropylene fiber was added. The total mixing time was no longer than is usual for this type of cement/sand/water mixture without the fibrous material. The polypropylene fiber was fibrillated twine of approximately 25,000 denier chopped in pieces 7.5 cm in length. Four test beams each 10 x 15 cm in cross section and 120 cm in length were cast in wooden molds from the resulting mixture, the filled molds being vibrated to allow air to escape prior to setting. The density of the resulting cast beams was between 2,270 and 2,340 g/l.

For comparison, four similar beams were made from the same cement/sand/water mixture but without the addition of the reinforcing elements. The flexural strength of each of the eight beams was measured by leading each beam in the center while it was supported at its ends. The average bending stress at failure for the four control beams was 55.5 kg/cm^2; the average for the four beams with the reinforcing elements was 73.8 kg/cm^2, an improvement of over 30%. The reinforced beams did not show a clean break at failure, the fibers still keeping the cracked beam together when subjected to a deflection of 30° and more. The results indicate that for many applications adequate strength of a cast or molded article can be obtained with additions of less than 1% by weight of the fibrous reinforcing elements of the process.

Example 2: A number of slabs were cast from a 3:1:0.5 mixture of Portland cement/sand/water containing as reinforcing elements chopped polypropylene baler twine. The slabs, which were 150 cm long and 70 cm wide, were cast in two thicknesses, namely 2 cm and 4 cm. The slabs were mechanically tested for impact strength after aging for 28 days to ensure thorough hardening. The baler twine had a weight in the range 23,000 to 30,000 denier and a staple length of approximately 7 cm; it was made from unpigmented polypropylene by twisting stretched polypropylene tape. Other slabs of similar dimensions were made under otherwise identical conditions except that no baler twine was added to the mix. In all cases, the sand and Portland cement components were dry mixed for 2 minutes after which the water was added and mixing continued for 1½ minutes.

When baler twine was added the mixing was continued for a further 2½ minutes during the first minute of which the twine (0.8% by weight of the total sand, cement and water) was added. Mixing was carried out in a 50-liter Eirich counter-current mixer. The slabs were cast in wooden molds, the mix being poured in, compacted in accordance with a standardized procedure for reinforced concrete and then shaken for 1 minute on a vibrating table at 4,000 cycles per minute to remove air. The slabs were hardened in the mold for 1 day under polyethylene sheeting and then aged for 7 days under water and then for 21 days at 20°C and 50% relative humidity.

The sand used was a dried river sand of which 80% weight was of a size in the range 0.3 to 2.8 mm (diameter of sieve mesh). The 2 cm thick slabs were impact tested by a falling weight test in which a steel ball 1 kg in weight was dropped onto the intersection of the diagonals of a slab from various heights, the slab being supported over its full width by two 70 cm wide trestles situated 125 cm apart.

The 4 cm thick slabs were tested by a pendulum test in which a 25 kg leather sandbag 30 cm in diameter and suspended by a rope 3 meters long was dropped against the intersection of the diagonals of a slab from various initial positions of the bag, the slab being held vertically against two rigid girders by clamps 125 cm apart, each line of clamping extending over 70 cm, i.e., the width of the slab. The results obtained were as follows:

The 2 cm Thick Slabs –The test slab made without baler twine broke completely at the impact load corresponding to a drop height of 50 cm, whereas the slab made in accordance with the process withstood without damage this impact load and showed only a crack at a drop height of 75 cm. A deflection of 8 mm was observed on the center of the slab when the crack appeared and this deflection increased to 20 mm as the drop height was increased, by 25 cm increments, up to 250 cm. However, the slab did not break.

The 4 cm Thick Slabs – The test slab made without baler twine broke completely at the impact load corresponding to a drop height of 27 cm. The slab in accordance with the process withstood without damage impact loads corresponding to a drop height of 104 cm at which point in the test three cracks formed over the whole width of the slab. However, the slab did not break up to the maximum drop height of the test which was 300 cm.

It will be seen from these results that the impact strength of slabs made from sand, cement, and water mixtures can be improved considerably by the addition of reinforcing elements formed from a stretched and fibrillated polypropylene film and that such slabs show only cracking at the failure point as compared with the complete breakage encountered with conventional slabs.

Moreover, the pieces of the slab remain connected together after the appearance of the crack(s) which allows the possibility of repair by grouting, for example, with an epoxy resin based adhesive in instances in which, with conventional slabs the pieces would become separated and therefore not repairable by grouting; and the slabs could withstand impact loads higher than the impact load at first failure without complete breakage occurring.

S. Goldfein; U.S. Patent 3,645,961; February 29, 1972 describes an impact resistant concrete product comprising a mixture of inorganic hydraulic cement and selected aggregate, together with a plurality of selected fibers substantially uniformly distributed throughout the mixture. The mixture of this process may constitute a mixture of cement, sand and gravel in a conventional ratio of 1:2:4, or otherwise, with fibrous material taken from the group consisting of nylon, polypropylene, polyvinylidene chloride and polyethylene. Less than 3% by weight of fibrous material taken from the abovementioned grouping will not only greatly increase the impact strength but will substantially improve the flexural strength as well.

Nonround Filaments

A process described by *C.G. Ball, A.C. Grimm and T. Melville; U.S. Patent 3,650,785; March 21, 1972; assigned to United States Steel Corporation* involves the discovery that certain filament configurations result in an increase in strength disproportionately greater than previously used filament forms. A part of the process is the finding that the strength improvement is related to the total surface area of the reinforcing filaments. Round filaments do not have as much surface area as nonround, i.e., rectangular or flat, filaments of the same cross-sectional areas. The geometric surface relationship between round and rectangular cross sections, for example, of equal cross-sectional area, is such that a minimum of 12.8% increase in surface area exists with the nonround filaments. This 12.8% increase is with respect to a square cross section for a width-to-thickness ratio of one.

Thus, increasing the total surface area for a given quantity, i.e., weight of filaments/unit volume of matrix, can be accomplished by decreasing the width-to-thickness ratio of the filaments used for reinforcing. A decrease in the width-to-thickness ratio will result in higher strength of the reinforced composite since the reason for the strength increase is that the total surface area is increased while the quantity of filaments, i.e., weight of filaments, unit volume of matrix, remains the same. In accordance with the process, discontinuous nonround reinforcing filaments with a width-to-thickness ratio (R) of not greater than about five are used to reinforce a castable matrix.

The terms nonround and round filaments or fibers as used refer to the cross section of the filament without regard to filament length. Thus, round filaments are those with circular cross section, usually made of cut wire lengths, and nonround filaments are those having definite width and thickness dimensions, albeit that the thickness for usual flat filaments may be measured in thousandths of an inch. Thus, it has been found that the addition of equivalent weights of round and nonround filaments of equivalent cross-sectional area/ft^3 of concrete mortar results in an increase in the tensile strength of the concrete with nonround filaments of about twice that achieved by the round filaments. Thus, for the same weight of filaments, nonround, i.e., flat filaments, are disproportionately more effective in strengthening a matrix than are round filaments. As a corollary, equivalent strength increase can be achieved with less (lb/ft^3) of nonround filaments than round filaments.

Thus, there is described a method of making a reinforced composite, such as of concrete, which comprises preparing a substantially dry admixture of cement, aggregate and a plurality of discontinuous reinforcing filaments having a nonround configuration, width and thickness dimensions, and a width-to-thickness ratio of not greater than about five. The substantially dry mixture is adequately mixed to randomly distribute the reinforcing filaments and is then ready for the addition of water to hydrate the cement after which it is remixed. Several specific examples are discussed in considerable detail in the patent.

GLASS FIBERS

Furane Coated Glass Fibers

It is known that the tensile strength of cement and concrete may be considerably improved by the incorporation of the precursors therein of an appropriate amount of glass fiber material. However, the normal glass fiber used (E glass) is very prone to corrosion by the high alkalinity generated during the hydration and setting reactions of Portland cement, the most common cement used in concrete manufacture, and the effect of such reinforcement is often short lived. This corrosion problem is normally overcome by adopting one or more of four basic techniques. These are:

(1) The use of a high alumina or similar cement which does not produce high alkaline conditions during the setting process.

(2) The addition to Portland cement of an agent to reduce the alkalinity generated during setting to an acceptable level.

(3) The reformulation of the glass used to produce an alkali-resistant glass fiber.

(4) The coating of the glass fibers with an impervious and alkali-resistant substance which bonds strongly to both the glass and the cement phases.

R.W. Wilkinson, J. Newton and D.M. Stitt; U.S. Patent 3,839,270; October 1, 1974; assigned to A.C.I. Technical Centre Pty. Limited, Australia describe a method of increasing the alkali-resistance of a glass fiber material suitable for use in reinforcing concrete which comprises coating the glass fiber material with a furane resin and subsequently curing the coating. In order to minimize cracking of the resin when it is used to protect reinforcing glass fiber in concrete, it is preferable that the resin should have an extension to failure equal to or greater than either the glass fiber or cement.

As an example, a suitable furane resin may be prepared by reacting furfuryl alcohol with furfuraldehyde in the presence of a catalytic amount of phosphoric acid. The resin should be of low viscosity to facilitate the coating and impregnation operation, and a satisfactory resin can be obtained by carrying out the condensation reaction at a temperature of 10° to 15°C for 24 hours. The material so produced may be kept at ambient temperature prior to use.

Furane coated and impregnated glass fiber rovings may be cured under conditions which may readily be determined by experiment. Thus, for example, satisfactorily cured impreganted rovings were obtained by heating at 110°C for

2 hours or at 80°C for 18 hours. The cured resin was a hard dark-brown to black material, highly resistant to alkali attack and strongly adherent to the glass fibers of the rovings. These glass fibers, a standard commercial product, had received the normal treatment with methacrylatochromic chloride or PVA to promote bonding to synthetic thermosetting resins.

Furane resins of the classes described as unmodified furfuryl alcohol polymers and aldehyde-modified furfuryl alcohol polymers are examples of suitable resins, particularly those consisting predominantly of chains of furane nuclei linked by methylene bridges. Such condensation polymers may be produced by alkaline or acidic catalysts.

In particular the acid catalyzed condensation products of furfuryl alcohol with furfuryl alcohol, furfuraldehyde and formaldehyde are preferred. For simultaneous layup of cement and glass, the impregnated rovings are used in short chopped lengths, preferably 1 to 1½ inches long. For sheet-cement products below 1 inch in thickness rovings having up to 60 ends (12,000 filaments) are preferred. The heat-curing conditions vary with number of ends as indicated in the examples. The reinforcement may also be laid up as rods, which may be up to 1 inch in diameter.

Furane resin coated fiber glass reinforcements are especially suitable for lightweight and aerated concrete products in that the density of such reinforcements is comparable to that of the cement matrix and considerably less than that of steel reinforcements, thus gaining an advantage in weight over a conventionally reinforced product. In addition, these furane-resin-coated glass rovings may be advantageously used in some situations where the use of steel reinforcement is precluded, e.g., in thin sheets (where the minimum protective covering for steel or 1½ inches of cement cannot be maintained) and in reinforced aerated concrete (where the foamed structure leads to rapid corrosion of steel). The performance of these furane-resin-coated rovings is illustrated in Figure 4.1.

Figure 4.1a shows the improved resistance of a 204 filament E glass strand **(A)**, coated with furane resin, as a function of immersion time in an alkaline test bath held at 75°C compared with a polyurethane coated strand **(B)** and an uncoated strand **(C)**. This simulated cement environment increases the rate of alkaline attack at least 50 times. Figure 4.1b shows the increase in strength of cement products reinforced with the impregnated rovings and the retention of that strength with time. Samples were Portland cement containing 2% by weight of impregnated aligned rovings (1% by weight of glass fiber). The table shows the resistance of the coating to cement environment when autoclaved in 120 psi steam for 7 hours, yielding a fully-reacted cement product which cannot produce further degradation of the reinforcement by alkali attack. The basic matrix was 30% Portland cement, 50% silica sand and 20% fine siliceous material.

Furan Rovings, wt %	Ultimate Modulus of Rupture, psi
0	1,100
1.2	1,500
1.9	1,950
2.3	2,700
3.6	3,150
4.0	3,500
5.0	3,600
7.0	3,750

FIGURE 4.1: GLASS FIBER REINFORCED CONCRETE

a.

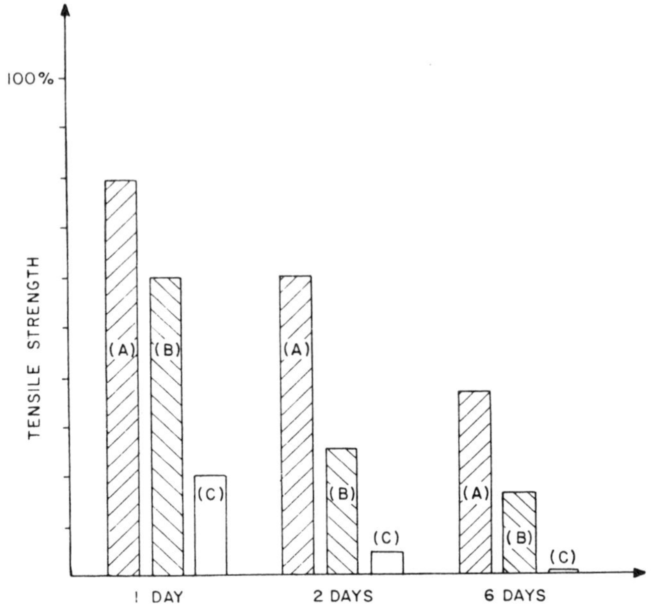

Tensile Strength in Alkaline Environment

b.

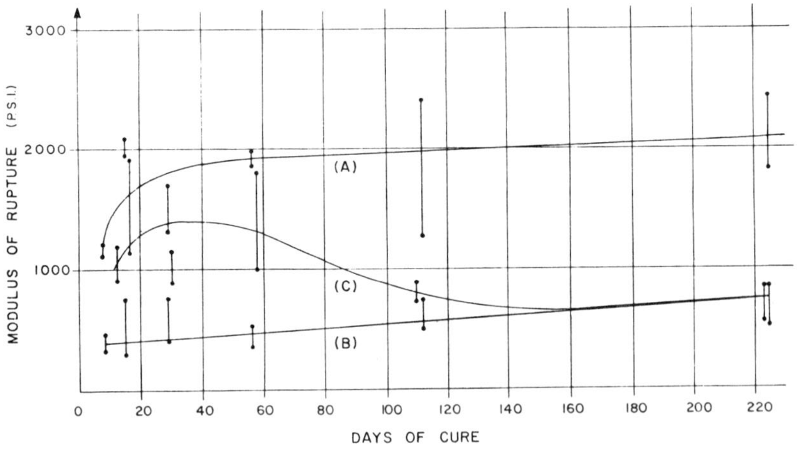

Strength vs Time for Reinforced Concrete

Source: U.S. Patent 3,839,270

Example 1: The furane resin is prepared by reacting 50 parts by volume of furfural with 2 parts of 85% phosphoric acid in a cooled vessel at 10° to 15°C. After 15 minutes, 100 parts of furfuryl alcohol is added. The mixture is allowed to react for 25 hours and is then ready for use. A 10-end E-glass fiber roving is drawn through the resin bath, then through a constriction orifice to remove any excess polymer and into a long curing oven. Hot air is passed down this oven countercurrent to the resin-coated roving, entering at 140°C and exiting at 80°C. The roving leaves the oven continuously in either a fully-cured or surface-cured condition, depending upon the residence time of the roving in the oven. Residence times of approximately 4 minutes yield a surface-cured product, and a time of 15 minutes yields a fully-cured product.

Fully-cured roving can be wound directly onto a drum or taken through a chopper to produce a desired length chopped roving. Surface-cured rovings are chopped directly and the short lengths fed into a tumble-heater to complete the curing process. This latter technique has the advantage of allowing some resin to flow from the inside of the chopped roving onto the exposed ends to produce a more fully-covered roving, and inhibit penetration of corrosive fluids into the coated fiber bundle.

Example 2: To 100 parts by volume of furfuryl alcohol contained in a chilled reaction flask (0°C) is slowly added 5 parts of 1:1 mixture of 85% phosphoric acid and water, the whole being stirred. The mixture is allowed to react at ambient temperature until of dark red-brown color, when it is ready for use. A 60-end roving is drawn through a coating apparatus by means of guides and rollers, then through a constriction to remove excess resin. The coated roving is wound onto a rectangular frame of rods, which when filled is placed in an oven at 110° to 120°C for 8 hours, resulting in a fully-cured product. The cured roving is unwound from the frame and may be cut into the desired size of chopped roving.

Example 3: An epoxy-modified furane resin is prepared by mixing 70 parts of the furfural-furfuryl alcohol resin in Example 1 with 30 parts of an epoxy resin, such as that manufactured by Shell Chemicals P/L and supplied under the name Epikote 828, which may be thinned with 5 parts of acetone prior to addition to the furane resin. A 60-end roving, having been drawn through a coated and constriction, is continually wound around two spaced parallel rods until a bundle containing the required number of loops is obtained. One of the winding rods is then rotated twisting the loops into a single length of twisted rovings.

The rod is refixed and tension applied to the twisted rovings. After excess resin has drained off, the rovings and rods are placed in an oven at 100° to 120°C for 8 hours, thus producing a reinforcing rod with spatulate ends and diameter dependent on the number of rovings originally wound onto the frame. The spatulate ends are of advantage in providing a mechanical bond in the concrete or cement structure which augments the frictional bond to the rod surface, and allows maximum stress-transfer from the concrete to reinforcement. This epoxy-modified furane resin showed improved flexibility over the resin of Example 1.

Example 4: A bundle of uncoated looped rovings as described in Example 3 is secured at one end and drawn into a cylindrical mold. One end of the mold is connected to the furane resin bath and vacuum applied at the other causing the resin to flow into the mold. When completely filled, the vacuum is removed and the mold sealed. The mold containing the resin and the glass fiber is then placed in an oven to cure at 100° to 120°C for 12 hours after which the finished rods are removable from the mold. This method is suitable for making reinforcing rods.

The resin-coated rods or mesh described in the examples may be incorporated into cement mortars in any of the ways commonly used for the reinforcement of mortars by steel rods, wire, or mesh. An outstanding advantage of the short chopped strands over similar strands of uncoated alkali-resistant glass is their rigidity, allowing placement of the reinforcement by random scattering, followed by pouring of cement mortar. This process of easy infiltration of a random matrix of reinforcing strands is not possible with nonrigid fibers, such flexible chopped fibers acting as a filtermat and preventing incorporation.

Alkali Resistant Glass Fibers

A.J. Majumdar; U.S. Patent 3,887,386; June 3, 1975; assigned to National Research and Development Corporation, England describes fiber reinforced cementitious products comprising glass fibrous material distributed throughout a cement matrix, in which the glass is one having per se a degree of alkali resistance such that when tested in the form of an abraded fiber of length 2½ inches and diameter of from 0.4–1.0 x 10^{-3} inches the fiber has a tensile strength of at least 100,000 psi after treatment with saturated aqueous $Ca(OH)_2$ solution at 100°C for 4 hours followed by successive washings at ambient temperature with water, then with aqueous hydrochloric acid (1%) for 1 minute, then water, acetone, followed by drying. The fiber experiences not more than 10% reduction in diameter during the test. Alkali resistant glasses useful for production of these products comprise the CaO-MgO-Al_2O_3-SiO_2 glasses including those in the anorthite field, silica zirconia glasses, and silica stannic oxide glasses. The products have outstanding durability and impact resistance.

For reinforcing cementitious matrixes at temperatures higher than ambient it is deisrable, moreover, that the fibers exhibit resistance to attack by alkali metal hydroxides, in addition to attack by $Ca(OH)_2$. A test for this resistance, similar to the previously mentioned test for attack by $Ca(OH)_2$, is as follows.

An individual glass fiber having a diameter which is in the range of 0.4–1.0 x 10^{-3} inches and a length of 2½ inches is exposed to attack by 1 N NaOH solution for 1½ hours at 100°C. After exposure, the fiber is taken out of the solution at room temperature, washed three times with distilled water, then with dilute HCl (1%) for ½ minute, followed by several washes with distilled water.

Finally, the fiber is washed with acetone twice and dried, after which its tensile strength is determined by measuring the breaking load with an Instrom testing machine and estimating the fiber diameter by an optical microscope.

A glass fiber exhibiting a tensile strength of at least 100,000 and preferably at least 200,000 pounds per square inch after the test, and experiencing during the test a reduction in diameter no greater than 10%, has the required properties for reinforcing the cementitious matrixes at higher than ambient temperatures.

In assessing the suitability of glass fibers for reinforcing cementitious matrixes, it is also instructive to study the behavior of the fibers in a cement effluent solution at different temperatures and different ages. The exact composition of the solution phase produced when Portland cement is mixed with water varies greatly with cements from different sources. A synthetic analogue of such a solution can, however, be prepared in the laboratory which will have ionic concentrations of Na^+ and K^+ similar to those found in the solution phase of average Portland cements. If this solution is then made saturated with respect to $Ca(OH)_2$, it represents approximately the compositions of the solution phase of a Portland cement slurry regarding the concentrations of hydroxides. A typical cement effluent composition is:

Alkali	Concentration, in g/l
NaOH	0.88
KOH	3.45
$Ca(OH)_2$	0.48

The desired alkali resistance properties are exhibited by fibers of certain glasses of CaO-Al_2O_3-MgO-SiO_2 type, wherein the oxide ratios are selected so that the composition of the glass lies within the anorthite primary phase volume of the CaO-Al_2O_3-SiO_2-MgO phase diagram or in its immediate vicinity, and the glasses are substantially boron free, i.e., have virtually no B_2O_3. Such phase diagram is described in the article "Quaternary slags CaO-MgO-Al_2O_3-SiO_2: initial crystallization and fields of crystallization at constant magnesia planes" by G. Cavalier and M. Sandrea-Deudon, *Revue de Metallurgie,* 57, 1143-1157 (December 1960). Within this group of glasses one preferred class of glasses is represented by those in which the main oxide percentages are selected to be in the following ranges:

Oxide	Weight %
SiO_2	50–65
Al_2O_3	12–21
CaO	10–21
MgO	3–10

Minor amounts of other metallic oxides such as also impart alkali-resistance may be present also, for example, TiO_2, Cr_2O_3, ZrO_2, Fe_2O_3, ZnO, BaO and MnO. The total amount of these additives will normally be less than 10% by weight of the overall oxide mix of the glass. For locating the compositions of the more complex glasses in the CaO-Al_2O_3-SiO_2-MgO phase diagram, it may be assumed that optional substituent ZrO_2 substitutes for SiO_2; optional substituents TiO_2, Fe_2O_3, Cr_2O_3 for Al_2O_3; and optional substituents BaO, MnO, ZnO for CaO.

The optional substituents, if their percentages are high, may sometimes change the composition of the glass in such a way that it falls outside the anorthite primary phase volume of the CaO-Al_2O_3-MgO-SiO_2 system but still exhibit adequate alkali-resistance. An example of such a glass is one containing a relatively high total percentage of MgO + MnO, say up to 20%. The following examples illustrate the process.

Example 1: Glasses of composition as set out in Table 1 were pulled into single filaments with diameters in the range 0.4–1.0×10^{-3} inches, and initial abraded strengths in the region of from 150,000 to 400,000 lb/in^2. They were then tested for alkali resistance according to the method previously described, with the results given in the table.

TABLE 1

Glass No.	CaO	SiO_2	Al_2O_3	MgO	TiO_2	MnO	CaF_2	% reduction in diameter	Tensile strength after test (lbs/sq. in.)	Tensile strength before test (lbs/sq. in)
	Analysis of the raw mix							Alkali-resistance		
1	19	55	21	5	—	—	—	Nil	130,000	155,000
2	15	56	21	5	—	—	3	4	120,000	200,000
3	18	55	17	5	—	5	—	7.5	180,000	255,000
4	10	60	15	5	5	5	—	Nil	190,000	170,000
5	15	55	15	3	6	6	—	Nil	250,000	430,000
E								9	70,000	250,000

TABLE 2

Glass No.	CaO	SiO_2	Al_2O_3	MgO	MnO	% reduction in diameter	Tensile strength after test (lbs/sq. in.)	Tensile strength before test (lbs/sq. in.)
	Analysis of the glass					Alkali-resistance		
6	17.0	54.5	11.5	7.2	9.8	3	175,000	315,000

TABLE 3

Glass No.	After 7 days in water	After 28 days in water	After 90 days in water	After 7 days in water and 21 days in air	After 7 days in water and 83 days in air
1	6,000	4,900	5,000	6,550	4,600
	5,890	5,260	5,030	5,200	4,810
E-Glass	4,470	4,350	3,930	4,950	3,300

The table also provides by way of comparison corresponding results obtained for a filament of a standard low alkali borosilicate glass known in the trade under the name E-glass, and containing 8% B_2O_2. The results in this table indicate that the glass fibers are particularly suitable for reinforcing Portland cement structures, and are substantially superior in this respect to fibers of E-glass.

Example 2: Glass similar to those in Table 1 but containing a much higher proportion of MnO is also suitable as reinforcing fibers. The performance of a glass fiber made from typical compositions of this family is shown in Table 2. In a preferred way of making fiber-reinforced cement-based composites, a high water to solid ratio is used initially for uniform dispersion of fibers and the excess water is removed by suction followed by pressing. Complicated shapes can be made by spray mixing of cement slurry and glass fiber on the surface of perforated molds connected to a vacuum pump.

Composites measuring 4 x 1 x ¼ inches have been made by this method using single-filament uncoated fibers derived from some of the glasses referred to in the above examples and Portland cement. The proportion of glass fiber in the composites was 0.5 gram of glass to approximately 30 grams of cement. The initial water to cement ratio in the slurry was 0.8, which fell to 0.3 after suction. Glass fibers measuring 4 inches in length were placed by hand in the tension zone of these composites during casting.

A perforated mold was used and the excess water removed by suction. After demolding, the composite test specimens were stored in a constant temperature (64°F), constant humidity (90% RH) room. Different curing conditions were used and the flexural strength of the test specimens was determined at different ages. Results are given in Table 3 for a typical alkali-resistant glass, the glass No. 1 in Table 1, in comparison with the previously mentioned E-glass. The values given are for flexural strengths in terms of lb/in^2.

In further comparison, asbestos-cement boards of ¼ inch thickness and containing 10 to 15% asbestos gave an ultimate flexural strength of 4,000 lb/in^2. It should be noted that no attempt was made to optimize the fiber content of composites in Table 3 to yield the highest possible flexural strength; also, the strengths given in Table 3 are average values only and do not reflect the wide scatter observed in the tests. In composites made with single filament glass fiber it is difficult to avoid uneven dispersion of the fibers in the mix; hence the wide variation in results.

Example 3: A glass of composition as set out in the accompanying Table 4 Glass No. 1 was pulled into single filaments with a diameter ranging from $0.4-1.0 \times 10^{-3}$ inch. A filament was tested for alkali-resistance according to the method previously described when using $Ca(OH)_2$ as alkali, in comparison with a similar filament of a standard low alkali borosilicate glass known under the name E-glass, with the results given in the table.

Results are given in Table 5 for the resistance exhibited at 80°C by filaments of Glass 1 and E-Glass using as alkali in the test previously described with an alkali metal hydroxide, a cement effluent solution of the composition specified hereinbefore. The strength values of the two glass fibers before testing as listed in this table are higher than those in Table 4.

TABLE 4

Glass No.	Analysis of the Raw Mix SiO_2	Al_2O_3	ZrO_2	Na_2O	Li_2O	Alkali Resistance % Reduction in Diameter	Tensile Strength After Test, psi	Tensile Strength Before Test, psi
1	71.0	1.0	16.0	11.0	1.0	Nil	190,000	210,000
E						9	70,000	255,000

TABLE 5

Glass No.	Diameter of Fiber Before Test (in)	Tensile Strength, psi Before Test	After 24 Hr	After 48 Hr	After 72 Hr
1	0.45×10^{-3}	415,000	340,000	265,000	180,000
E	0.48×10^{-3}	425,000	110,000	40,0000	40,000

The higher strength values are because the fibers in Table 5 were tested soon after they were pulled and did not have the chance to reach their abraded strength, which is always substantially lower than the virgin strength. The diameter of the fiber glass No. 1 remained unchanged after the tests. E-glass fibers were so badly attacked in the test that their diameters could not be measured very accurately afterwards. However, in each case they were found to be definitely smaller than those of the fibers before the test.

In calculating tensile strength of E-glass fibers at all ages after the test, it has been assumed that the diameter remained the same as before the test. Table 6 provides comparable alkali-resistance test results in the instance when the alkali employed is 1 N NaOH as described above.

TABLE 6

Glass No.	Alkali Resistance % Reduction in Diameter	Tensile Strength After Test, psi	Tensile Strength Before Test, psi
1	5	185,000	210,000
E	59	260,000	255,000

It will be noted that the apparent increase in strength due to NaOH attack in the case of E-glass is due to the very large reduction of the diameter of the fibers. If the attack progresses as vigorously as is indicated by this test, there would shortly be no glass fiber left at all to reinforce the cementitious matrix. The results in Tables 4, 5 and 6 indicate that fibers of the glass are particularly advantageous for reinforcing Portland cement structures.

Further tests have shown that at this temperature (80°C) fibers produced from Glass No. 1 when kept immersed in the cement effluent solution for a period of 2 weeks still retain some measurable tensile strength. After 96 hours' exposure, the E-glass fibers could not be tested.

METAL REINFORCEMENT

Protective Coating for Metal Reinforcement

K.O. Rönnmark and B.J.M. Schmidt; U.S. Patent 3,806,571; April 23, 1974; assigned to Internationella Siporex Aktiebolaget, Sweden describe a method for the manufacture of metal reinforced, steam-cured, light-weight concrete where the metal reinforcement is provided with a protective coating prior to being embedded in the concrete. The protective coating comprises an aqueous mixture of 60 to 90% by weight of Portland cement, 10 to 50% of water, and 2 to 20% by weight of a copolymer selected from the group consisting of copolymers of acrylic acid ethyl ester, methacrylic acid methyl ester, methacrylic acid butyl ester, and methacrylic acid isobutyl ester. The following examples illustrate the process.

Example 1: A coating composition was made up of the following: 100 parts Portland cement, 30 parts water and 12 parts of a copolymer of acrylic acid ethyl ester and methylacrylic acid methyl ester. The components were mixed thoroughly and worked. The coating composition was applied to reinforcing rods for light-weight concrete by immersion, whereupon it was found that two immersion operations were sufficient to obtain a very good coating. The applied coating was then allowed to dry in air for some hours.

In this manner a dense, well adhered elastic film was obtained on the reinforcing rods, which were then embedded in the light-weight concrete compound. The resulting product was subjected to a steam curing process at a pressure of 10 kg/cm^2 for approximately 20 minutes, whereby the coating was converted to a hard, very dense layer, which was effectively bonded to the surrounding light concrete.

Example 2: A coating composition was made up of the following: 100 parts of Portland cement, 35 parts of water, 9 parts of a copolymer of methacrylic acid butyl ester and methacrylic acid isobutyl ester, and 0.15 part of ethylhydroxyethylcellulose. In a similar manner to that described in Example 1 a hard and dense layer was obtained as well as an effective bond with the surrounding light-weight concrete subsequent to applying the composition to reinforcing rods intended for light-weight concrete and embedding the rods in the concrete compound and steam hardening the resulting product.

Example 3: The following coating composition was prepared of 100 parts Portland cement, 20 parts of water, 18 parts of a copolymer of acrylic acid ethyl ester and methacrylic acid butyl ester and 4.2 parts methylcellulose. In a similar manner to that described in Example 1 a hard and dense layer was obtained, effectively bonded to the surrounding light-weight concrete, subsequent to applying the composition to reinforcing rods of the light-weight concrete and embedding the rods in the concrete compound and steam hardening the resulting product.

Sodium Nitrite, Calcium Formate and Triethanolamine

According to a process described by *G. Whitaker; U.S. Patent 3,801,338; April 2, 1974; assigned to Fosroc AG, Switzerland* an additive for hydraulic cement comprises a major amount of sodium nitrite in admixture with a minor amount of calcium formate, optionally with triethanolamine or sodium benzoate. The additive for the hydraulic cement is usable with metal reinforcement and provides a cement which can be used at low temperatures. The additive gives improved concrete compressive strength. Examples of two specific additives (parts are by weight) are as follows:

Additive A: Calcium formate, 48%; and sodium nitrite, 52% (application rate, 1.8 lb/cwt dry cement).

Additive B: Calcium formate, 19.5%; sodium nitrite, 77.9%; and triethanolamine, 2.6% (application rate, 2 lb/cwt dry cement).

LIGHTWEIGHT AND FOAMED PRODUCTS

LIGHTWEIGHT PRODUCTS

Wet Surface Treatment of Polystyrene Beads

According to a process described by *A.D. Bowles and S.J. Parsons; U.S. Patent 3,869,295; March 4, 1975* lightweight concrete and plaster are prepared by a method which assures that the aggregate is uniformly mixed with the cementitious material and other relatively heavy ingredients of the concrete and plaster mixes. This is accomplished by wetting the surfaces of the lightweight aggregate particles with an aqueous medium, mixing the wet aggregate particles with dry finely divided cementitious material to form a coating thereon, and thereafter adding additional aqueous medium in an amount to produce a coherent formable uncured concrete or plaster matrix. The uncured concrete or plaster matrix may be formed into a desired configuration, and then is allowed to set in the usual manner. The addition of hydrated lime improves the cohesive properties of an uncured concrete matrix.

Increased strength in cured lightweight concrete may be obtained by mixing pozzolan, hydrated lime and/or finely divided inert inorganic fillers such as sand with the uncured concrete matrix. A lightweight aggregate including expanded polystyrene beads is preferred, and further increased strength may be obtained by using polystyrene beads expanded in hot water.

The aqueous medium that is used to wet the particles of aggregate prior to mixing with the dry inorganic cementitious material, and/or the additional aqueous medium that is mixed with the cement coated aggregate particles may be water, an aqueous emulsion of an organic binder for the aggregate particles, or mixtures. When employing an aqueous emulsion of an organic polymeric binder, it is usually preferred that the particles of aggregate be wetted, followed by coating the wetted aggregate particles with the inorganic cementitious material, and thereafter mixing water with the cement coated aggregate particles to provide the additional water for hydration, rehydration or crystallization required for hydration of the cementitious material, and to produce the desired formable cohesive concrete matrix or plaster matrix. The organic polymeric binder may be, for example, an aqueous emulsion or latex of polyvinyl chloride, copolymers of polyvinyl chloride with other ethylenically unsaturated monomers such as vinyl acetate or vinyl alcohol, acrylic resins, polyamides such as nylon, epoxy resins and the like. Aqueous emulsions of organic polymers used in prior art water-based paints, such as, for example, Super Kem-Tone are suitable.

The use of aqueous medium which contains an organic polymeric binder in emulsion form to wet the aggregate aids in stabilizing the water content thereof, and allows a uniform, controlled thickness of cementitious material to be coated thereon. It is also possible to use organic polymeric binders of this type in subzero weather. The heat of hydration has kept the matrix from freezing at +5°F. The following examples illustrate the process.

Example 1: This example illustrates the preparation of lightweight concrete when using first stage polystyrene beads expanded in dry steam as the lightweight aggregate. The expanded polystyrene beads weighed 2 pounds per cubic foot, and had diameters of approximately one-eighth inch to one-fourth inch.

The aqueous medium was water, and the hydraulic cement was an early strength Portland cement known in the art as SS-C-192, type 3. The pozzolan, hydrated lime, sand and perlite were commercial grades conventionally used in the preparation of concrete. The air entraining agent was Derex.

Concrete mixes were prepared from the ingredients set out in Table 1 below employing a modified pug mill as the mixing apparatus. The blade of the pug mill was modified to slice into the matrix and to fold it toward the center of the mixer with the least possible drag and subsequent entrapped air. The order of addition and amounts of ingredients was as they appear from left to right in Table 1. In general, the expanded polystyrene beads were placed in the mixer, and then the wetting water was added. The beads were mixed with the setting water for approximately 1 minute.

At the end of this mixing period, the polystyrene beads were uniformly wetted with the water, and then the dry Portland cement was added and the mixing was continued for approximately 1½ minutes. The matrix turned a gray color upon mixing with the Portland cement, and the beads had a uniform film of Portland cement thereon at the end of the mixing period. The air entraining agent was added to the water of hydration, and the mixture of water and air entraining agent was added and admixed.

The pozzolan was mixed with the Portland cement, and the Portland cement-pozzolan mixture was used in coating the wetted beads. The hydrated lime, sand and perlite were added in that order after addition of the water of hydration and admixed.

The concrete matrix for each run was placed in a test mold, cured for seven days, and tested for compressive strength which was recorded in pounds per square inch. The weight of the concrete was determined and recorded in pounds per cubic foot. The data thus obtained appear in Table 1.

TABLE 1

Test	Polystyrene Beads		Water (lbs)			Cement (lbs)	Pozzolan (lbs)	Derex (Grams)	Hydrated Lime (lbs)	Sand (lbs)	Perlite (lbs)	Weight (lbs/ft³)	Compressive Strength (p.s.i.)
	Lbs.	Cu. Ft.	Wetting	Hydration	Total								
1	0.4	0.2	1.6	0.09	1.69	3.92	—	—	—	—	—	24.7	95.6
2	0.4	0.2	1.6	0.38	1.98	3.14	—	—	0.47	—	—	23.8	108.5
3	0.4	0.2	1.6	0.43	2.03	3.53	—	—	3.90	—	—	23.8	84.3
4	0.4	0.2	1.6	1.10	2.70	2.74	—	—	1.18	—	—	30.8	108.5
5	0.4	0.2	1.6	1.76	3.36	1.96	—	—	1.96	—	—	32.3	72.9
6	0.4	0.2	1.6	0.25	1.84	3.12	—	—	0.31	3.0	—	38.0	113.7
7	0.4	0.2	1.6	0.58	2.18	3.81	—	—	0.38	3.0	—	45.6	270.5
8	0.1	0.05	0.025	0.348	0.373	0.870	—	—	—	1.740	—	57.1	448
9	0.1	0.05	0.025	0.408	0.433	1.044	—	2.5	—	1.044	—	45.5	435
10	0.1	0.05	0.025	0.408	0.433	1.044	—	10.0	—	1.044	—	47.8	442
11	0.1	0.05	0.025	0.408	0.433	1.044	—	10.0	—	—	—	29.0	172
12	0.1	0.05	0.025	0.557	0.582	1.392	—	10.0	—	—	—	40.0	316
13	0.1	0.05	0.025	0.696	0.721	1.740	—	10.0	—	—	—	49.2	353
14	0.1	0.05	0.025	0.408	0.433	1.044	—	10.0	—	0.522	—	38.2	229
15	0.1	0.05	0.025	0.557	0.582	1.392	—	10.0	—	0.696	—	53.2	459
16	0.1	0.05	0.025	0.696	0.721	1.740	—	10.0	—	0.870	—	68.5	453
17	0.1	0.05	0.025	0.702	0.727	1.392	—	—	—	—	0.150	44.4	450
18	0.1	0.05	0.025	0.654	0.679	1.392	—	—	—	—	0.100	33.8	327
19	0.1	0.05	0.025	0.702	0.727	1.392	0.08	—	—	—	0.150	46.0	468
20	0.1	0.05	0.025	0.654	0.679	1.392	0.08	—	—	—	0.100	45.3	309

TABLE 2

Test	Polystyrene Beads		Water (lbs)			Cement (lbs)	Pozzolan (lbs)	Hydrated Lime (lbs)	Weight (lbs/ft³)	Compressive Strength (p.s.i.)		
	Lbs.	Cu.Ft.	Wetting	Hydration	Total					7 Day	14 Day	28 Day
1	0.05	0.025	0.013	0.209	0.222	0.522	—	—	33.7	169	—	—
2	0.05	0.025	0.013	0.278	0.291	0.696	—	—	45.0	585	—	—
3	0.05	0.025	0.013	0.348	0.361	0.870	—	—	52.6	856	—	—
4	0.05	0.025	0.013	0.319	0.332	0.522	0.250	—	48.8	723	—	—
5	0.05	0.025	0.013	0.388	0.401	0.696	0.250	—	57.5	923	—	—
6	0.09	0.025	0.013	0.209	0.222	0.522	—	—	34.4	199	298	248
7	0.09	0.025	0.013	0.278	0.291	0.696	—	—	41.6	423	485	549
8	0.09	0.025	0.013	0.348	0.361	0.870	—	—	49.0	1050	1090	1095
9	0.09	0.025	0.013	0.242	0.255	0.522	0.078	—	36.5	261	262	201
10	0.09	0.025	0.013	0.323	0.336	0.696	0.105	—	47.6	700	768	1065
11	0.09	01025	0.013	0.278	0.291	0.522	0.157	—	41.4	554	646	537
12	0.09	0.025	0.013	0.369	0.382	0.696	0.209	—	48.5	1142	1183	886
13	0.09	0.025	0.013	0.316	0.329	0.522	0.104	0.052	43.5	840	672	924
14	0.09	0.025	0.013	0.422	0.435	0.696	0.139	0.070	51.1	942	989	1265

TABLE 3

Test	Polystyrene Beads		Water (lbs)			Calcined Gypsum (lbs)	Hydrated Lime (lbs)	Weight (lbs/ft³)	Compressive Strength (p.s.i.)
	Lbs.	Cu.Ft	Wetting	Rehydration	Total				7 Day
1	1.5	1.0	0.5	11.85	12.35	19.58	4.64	28.2	65
2	1.5	1.0	0.5	11.60	12.10	27.50	—	31.2	138
3	1.5	1.0	0.5	13.75	14.25	31.30	—	35.1	165
4	1.5	1.0	0.5	18.40	18.90	34.35	—	39.65	215

Example 2: The general procedure of Example 1 was repeated with the exception of using polystyrene beads which had been expanded in hot water having a temperature of 200° to 212°F. The hot water expanded polystyrene beads weighed 3.6 pounds per cubic foot, and apparently had some water entrapped within the beads. In general, the order of addition of ingredients was as in Example 1, i.e., as they appear when reading from left to right in Table 2. The data thus obtained appear in Table 2.

Example 3: This example illustrates the preparation of lightweight plaster when using expanded polystyrene beads as the lightweight aggregate. The expanded polystyrene beads weighed 1.5 pounds per cubic foot, and had diameters of approximately one-eighth to one-fourth inch. The aqueous medium was water, and the cementitious material consisted of calcined gypsum in three runs. In an additional run, the cementitious material was a mixture of hydrated lime and calcined gypsum.

Plaster mixes were prepared from the ingredients listed in Table 3 employing a modified pug mill as a mixing apparatus. The pug mill was modified as set out in Example 1. The order of addition and the amounts of ingredients were as they appear from left to right in Table 3.

In general, the expanded polystyrene beads were placed in the mixer, and wetting water was added. The beads were mixed with the wetting water for approximately 1 minute. At the end of this mixing period, the polystyrene beads were uniformly wetted with the water, and then the dry finely divided calcined gypsum was added and the mixing was continued for approximately 1½ minutes. At the end of the mixing period, the beads had a uniform coating of the gypsum thereon. The water of rehydration was added, and the mixing was continued until a uniform formable coherent plaster matrix was produced. In the one run where a mixture of calcined gypsum and hydrated lime was used as the cementitious material, the hydrated lime was added after the water of rehydration and admixed in the matrix.

The plaster matrix for each run was placed in a test mold, cured for 7 days, and tested for compressive strength which was recorded in pounds per square inch. The weight of the plaster was determined and recorded in pounds per cubic foot. The data thus obtained appear in Table 3.

Coated Polystyrene Beads

According to a process described by *L. Unterstenhoefer and W. Krieger; U.S. Patent 3,899,455; August 12, 1975; assigned to BASF Wyandotte Corp.* adhesion of foamed particles, e.g., foamed polystyrene beads, in a lightweight concrete composition is improved by surface coating the foamed particles prior to incorporation into the lightweight concrete mixture with a phenol containing (1) coal tar or (2) anthracene oil and crosslinkable phenolic resin blend.

The following examples are included to illustrate the preparation of the closed cell foamed particles of this process and their use in lightweight concrete. Unless otherwise specified all parts are by weight and all temperatures are expressed as degrees centigrade. In the examples foamed polystyrene of 15 grams/liter density with a particle diameter of the beads of from 1 to 3.0 millimeters was used. The hydraulic binder was type 350 Portland cement.

For testing the compositions prepared in each example, three cubes of 100 x 100 x 100 millimeters and three beams of 700 x 150 x 100 millimeters dimensions were produced according to the standard conditions for concrete preparation. As the above molded specimens were cast, a densification of from 5 to 8% occurred. The finished sections were tested according to DIN 1048. The test specimens were removed from the mold 24 hours after filling, put in water and after 7 days they were removed from water. The cubes were dried at room temperature, 20°C, and the beams were kept wet to the testing day with wet cloths.

On the 28th day after preparation of the test specimens, the cubes were used for density determination and compression strength determinations, and the beams were used for determining the bending tensile strength. The tests were carried out at a pressure increase of 1,000 kilograms/minute.

Example 1: In this example a lightweight concrete of density 0.2 kilogram/liter was prepared. A coating for the closed cell foam particles was prepared by emulsifying together with an agitator the following ingredients: 84.0 grams anthracene oil, 42.0 grams phenol resin (alkaline condensation product consisting of 1 mol of phenol and 2 mols of formaldehyde) and 4.2 grams of a mixture consisting of 76 parts of formamide, 19 parts of butyrolactone and 4.8 parts of boric acid (phenolic resin crosslinker).

The above emulsified coating was then applied to 38 liters of foamed polystyrene (15 g/l) in a free fall blender by spraying and mixing until all particles were covered. Then 5.15 kg of Portland cement PZ 350 and 2.68 kg of water were added, and by blending a homogenous mixture was produced. The mixture was charged into the molds to obtain the above described test specimens. After curing the molded particles were tested. The results are tabulated in the table.

Example 2: For preparing a lightweight concrete of density 0.5 kg/l, the same procedure as described in Example 1 was followed, however, for each 38 liters of foamed polystyrene of 15 g/l density, 13.5 kg of Portland cement PZ 350 and 6.16 kg of water were used. The test results obtained on the cast molded articles are listed in the table.

Example 3: For preparation of a lightweight concrete of density 0.8 kg/l, Example 1 was repeated, however: 35.9 liters of coated foamed polystyrene (15 g/l), 16.2 kg of Portland cement PZ 350, 6.97 kg of regular sand, and 7.30 kg of water were used. The results obtained again are found in the table.

Example 4: For preparing a lightweight concrete of density 0.2 kg/l, 90 grams cable tar was sprayed onto 38 liters of foamed polystyrene of 15 g/l density in a a free fall blender until all particles were completely covered. After this 5.15 kg of Portland cement PZ 350 and 2.68 kg of water were added and blended until a homogenous mixture was produced, which was charged into the molds. The specimens were cured and tested as described above. The result is reported in the table.

Example 5: Following the procedure as described in Example 4 a lightweight concrete of density 0.5 kg/l was prepared with 38 liters of coated foamed polystyrene of 15 g/l density, 13.5 kg of Portland cement PZ 350 and 6.16 kg of water. The composition was charged to the molds and test specimens prepared. The test results are listed in the table.

Example 6: Following the procedure described in Example 4 a lightweight concrete of density 0.8 kilogram/liter was prepared with 35.9 liters of foamed polystyrene of density 15 grams/liter, 16.2 kilograms of Portland cement PZ 350, 6.97 kilograms of regular sand, and 7.3 kilograms of water. The composition was charged to the molds and test specimens were prepared and tested. The test results are stated in the table below.

Test Results of Examples 1 Through 6

Example	Pressure	Tensile Strength kg/cm^2 Bending Tensile
1	2.00	1.80
2	15.00	7.80
3	31.40	12.50
4	2.00	1.80
5	13.00	7.50
6	27.00	12.50

Vermiculite and Sodium Silicate

A. Race; U.S. Patent 3,847,633; November 12, 1974 describes a lightweight building material having a density in the range of 26 pounds per cubic foot which is moldable so as to be useful in forming prefabricated interior wall panels and the like, to be assembled into a room at a construction site. The material is inorganic and includes lightweight aggregate. Portland cement and a property-enhancing mixture of vermiculite and sodium silicate.

In one example of the process, the following materials were dry mixed:

	Pounds
Type 3 cement	4.2
Fly ash	1.0
Fine ground vermiculite	0.2
Vermiculite, 7 pounds per cubic foot	0.4
Perlite, 3 pounds per cubic foot	1.4
Perlite, 7½ pounds per cubic foot	1.0
Fine fiber glass	0.2
Fiber glass fibers (1-2 inches)	0.3
Sodium silicate	0.3
Plaster of Paris	1.0

To this mixture 3 pounds of water, or 30 percent by weight of the dry mix, was blended with the dry mix. This blend was then cast into a body for curing. With the body still in the mold it was subjected to a two-stage cure in which forced air less than 100°F was directed over the mold for 3 days. After three days the body in the mold was oven-cured at 360°F until a density of 26 pounds per cubic foot was attained. The final density is equivalent to the density of the dry mix products. This body was then subjected to heating and mechanical tests and found to exhibit characteristics and properties suitable for use as a building material.

Resin Treated Crushed Coke

M.L. Davis; U.S. Patent 3,900,332; August 19, 1975 describes compositions which consist essentially of crushed coke and Portland cement, to which a small amount of lime may be added as a coloring agent or whitener. A preferred form of the composition involves treatment of the crushed coke with a waterproofing solution containing a resin. The coke size used ranged from a powder to particles of one-half inch in diameter.

In order to provide a composition in which the coke particles are waterproofed, so that concrete made from the composition is waterproof, a mixture is utilized consisting of the following ingredients in the proportions by weight indicated:

	Range, %	Preferred, %
Coke	44.14-67.63	56.36
Portland cement	29.78-44.68	37.23
Lime	0-6.38	5.32
Thermosetting Resin	0.87-1.31	1.09
Total	100%	100%

In preparing the mixture, the coke is first immersed in a bath of a thermosetting resin, which has been diluted with a small amount of benzene to render it more flowable. The resin is preferably a polyester resin or a phenol formaldehyde resin, although other types of thermosetting resins may be employed.

After the coke has been thus treated, the excess resin is permitted to drain off, and the coke dried, so that the coke particles are rendered waterproof by the resin which coats the particles. In practice, the weight of the coke particles, following such treatment, is increased only slightly by the addition of the resin, that is to say, the weight is increased about 2%. Following such treatment, the resin-coated coke is mixed with the Portland cement and lime to form a dry composition, the formula of which is within the above described ranges.

In making a concrete, utilizing the above preferred or optimum formula, approximately two gallons of water are added to 100 parts by weight of the mixture consisting of 56.36% coke, 37.23% Portland cement, 5.32% lime, and 1.09% resin.

As an example of such a dry mixture, 53 pounds of coke are treated with a solution consisting of two gallons of benzene and one pound of resin, thereby producing coke, the weight of which has been increased about 1 pound by the addition of the resin coating, that is to say, the weight of the coated coke is about 54 pounds, all of the benzene having evaporated as a result of the drying action. The 54 pounds of coated coke is then thoroughly mixed with 35 pounds of Portland cement, and 5 pounds of lime, forming a dry mixture weighing a total of 94 pounds, which is bagged.

In order to form a concrete mix, there is added to the 94 pounds of dry mix, approximately two gallons of water which is thoroughly mixed with the dry mixture and then poured.

In some cases, as where concrete, in ready-mixed form, is to be delivered from a concrete plant to the site at which the concrete is to be poured, and in which the delivery vehicle or truck is equipped with a rotating mixer into which the concrete-making ingredients are placed, the following ingredients, in the proportion by weight indicated, were placed in the rotating mixer:

	Range	Preferred
Coke	38.90 – 58.34%	48.62%
Portland Cement	25.68 – 38.52%	32.10%
Lime	0 – 5.52%	4.60%
Thermosetting Resin	5.87 – 8.81%	7.34%
Water	5.87 – 8.81%	7.34%
	100%	100%

In the course of thus preparing the ready-mixed concrete, coke, which has not been coated with the resin, is used, but in the course of the mixing, the coke particles become coated with the resin. Compression tests were performed on cylinders prepared from the ready-mixed concrete made in accordance with the preferred formula, consisting of 48.62% coke, 32.10% Portland cement, 4.60% lime, 7.34% thermosetting resin, and 7.34% water.

These tests were made in accordance with ASTM-C 39 (CSA A 23, 2-13) conducted by Gulick-Henderson Testing Laboratories. For this purpose, the cylinders were 5¾ inches in diameter and 12 inches in height or length. These cylinders were tested on a Forney 12 inch Cylinder Press, Serial LT-500-5546. The results of the tests were as follows:

Cylinder Mark	Weight of Cylinder	Cylinder Press Load (lbs) At Breaking Point of Cylinder	Unit Load (lbs/psi) At Breaking Point of Cylinder	Age at Test Date
CB-1	18.10	41,000	1450	7 days
CB-2	19.00	44,000	1555	7 days
CB-3	18.75	79,000	2795	28 days
CB-4	18.85	89,000	3150	28 days
CB-5	18.92	82,000	2900	28 days

It is thus seen that the process provides a concrete which is relatively light in weight, but has high strength in relation to its weight.

FOAMED PRODUCTS

Styrenated Chlorinated Polyester Resin and Aluminum Flakes

J.L. Head; U.S. Patent 3,925,090; December 9, 1975 describes a cellular cement composition which is obtained from reactants consisting essentially of Portland cement, water, a styrenated chlorinated polyester resin, aluminum flakes, and an alkali. The following example is typical of the manner in which the cellular cement is made.

Example: Thirty-two gallons of water (266.6 pounds, 24.3% by weight) and eight 94-pound bags of Portland cement, type 1 (752 pounds, 68.6% by weight) were mixed vigorously for two minutes at ambient temperature. Then 75 pounds (6.9% by weight) of a styrenated chlorinated polyester precursor mix consisting of 65 parts of chlorinated polyester resin (Hetron 197, Hooker Chemical Co.), 35 parts of styrene, 0.5 part of cobalt naphthenate, and 1.0 part of methyl ethyl ketone peroxide was mixed vigorously with the foregoing water-cement mixture for five minutes. Then, 0.7 pound of fine aluminum flakes (300-400 mesh; 0.06% by weight) was added to the foregoing and vigorous mixing was continued for about 30 seconds. Finally, 1.4 pounds of sodium hydroxide (0.12% by weight) was added. Vigorous stirring was continued until incipient gelation was achieved, that is, until the mixture had plastic strength such that discrete bubbles remain separate and continue to grow in size. This stage was usually reached within about 30 seconds of vigorous stirring after the addition of the sodium hydroxide. The mixture was then poured into an appropriate construction form and allowed to set overnight. The forms were then removed.

It will be understood that the critical time in the mixing operation is that during which the generation of the foaming agent takes place, that is, between the addition of the aluminum flakes and the pouring step. Under temperature conditions between about 60°F and about 90°F, this time usually varies from about one minute to about five minutes. The mixing times given above for the prior steps are normally unaffected by ambient conditions as long as the temperature is above about 50°F. Other formulations which can be used in the process include:

A	B
35 gal of water	30 gal of water
8 94-lb bags of Portland cement, type 3	8 94-lb bags of Portland cement, type 3
70 lbs of polyester precursor mix	77 lb of polyester precursor mix
30 parts styrene	40 parts styrene
70 parts chlorinated polyester	60 parts chlorinated polyester
0.7 parts manganese naphthenate	0.3 parts cobalt naphthenate
0.7 parts cumene hydroperoxide	1.2 parts dicumyl peroxide
0.5 lb aluminum flakes	0.9 lb aluminum flakes
1.8 lb potassium hydroxide	1.2 lb sodium hydroxide

Water-Dispersible Aluminum Paste

In a process described by *M. Hauska and F. Kontra; U.S. Patent 3,551,174; December 29, 1970; assigned to Chemolimpex Magyar Vegyiaru Kulkereskedelmi Vallalat, Hungary* aluminum powder and aluminum paste are provided, which are characterized by improved dispersibility in water, extended shelf life with no deterioration and without danger of explosion or fire during manufacture, storage or use. These materials are used generally where, as a result of reaction of the aluminum with alkaline or acidic media, the resulting gas formation is utilized as a blowing agent, e.g., in the manufacture of foamed concrete.

In the process an acidic anionic wetting agent reactive with ground aluminum, is used during milling in an organic solvent, then the excess wetting agent is washed out and the desired solids concentration of the resulting mass is adjusted. If aluminum powder is to be made, the adjusting of the solids concentration includes evaporation of the organic solvent. When an aluminum paste is desired, a substantially neutral pH wetting agent suitable for creating an aqueous emulsion of organic solvents and having no substantial anionic or cationic activity is added to the adjusted mass to obtain thereby the paste product.

Nitrogen Based Foaming Agents

K. Diggelmann and R. Serena; U.S. Patent 3,591,394; July 6, 1971; assigned to Kaspar Winkler & Co., Switzerland describe a method of producing injection mortar or porous concrete which comprises adding to a cement-containing mixture at least one nitrogen delivering compound which produces an expansion of the injection material or a formation of pores in the concrete material. It is known that nitrogen delivering compounds are used in the plastics industry, e.g., as expanding agents for rubber products, but there the operating temperatures are above 100°C in order to obtain the nitrogen delivery.

The nitrogen delivery of the agents used in the method is possible in the cement-alkaline medium at all temperatures at which the cement can normally set. The kind of cement used is not of primary importance. Activators such as aluminates and copper salts are used, which have a decomposing action on the agents used according to the process and containing at least one nitrogen-nitrogen bond. Organic as well as inorganic compounds are suitable, preferably azo- and hydrazine compounds. The nitrogen delivering compounds can be used advantageously as a mixture with further additions such as lignin sulfonates, albumen-decomposition products, hydroxy- or polyhydroxy-carboxylic acids and their derivatives, with activators for gas delivery and with fillers, such as stone powder, quartz powder, bentonite, infusorial earth or chalk.

The efficiency of the method is improved with respect to the methods in which treated metals, for example, passivated aluminum are used, since a time consuming, expensive and hardly controllable working process for the production of the addition agent will become superfluous. Also, according to the process, there is no formation of hydrogen. A mixture suitable as a cement addition contains for example sodium aluminate, soda, calcium hydroxide, calcium carbonate as filler, copper sulfate, an alginate concentrating agent, lignin sulfonate and a hydrazide, such as diphenylsulfone-3,3'-disulfohydrazide and/or benzene-sulfohydrazide.

The process is explained with reference to the following examples. The classification of the various mortar and concrete mixes is effected according to the usual rules of the building trade. All examples had been based on a cement mortar of the following composition:

Portland cement	2.000 kg
Water	0.780 kg
Additive	0.004 kg

By means of a flow according to Rilem-standard the flow times were measured after a mixing time of four minutes. The values obtained correspond to the mean values of three flow times. Subsequently cylindric plastic boxes having a diameter of 10 cm were filled to a height of 10 cm with the cement mortar, and by means of a depth gauge the level of the mortar was measured at six points. The mean value of these measurements was ascertained immediately after the filling into the boxes, and after three, six and twenty-four hours. Simultaneously with the measurements for determination of expansion and/or shrinkage of the mortar, the amount of water settled on the surface (bleeding) was determined. This was effected with a graduated measuring pipette.

For ascertaining the compression strengths of the injection mortar, cubic test specimens of a size 10 x 10 x 10 cm were produced and stored at 100% relative moisture and at 20°C until the test date, so that the expansion could be considered as impeded, as this is the case for sheath tubes with prestressing cables. In order to illustrate the efficiency of the additives, test specimens were produced for each example at a mixing temperature of 20°C. The zero test contains water and cement at a weight ratio 39:100 without further additives. The indicated percentages by weight refer always to the weight of the cement, when no other indications are given.

Example 1: To the mixture of the zero test was added 2% by weight of an additive which contains 2% by weight of aluminum powder as a gas delivering component and 4% by weight of lignin sulfonate as a plasticizing means. The measured values obtained serve to characterize one of the usual methods for the production of injection mortar.

Example 2: 2% by weight of an additive was added to the mixture of the zero test, the additive containing 3% by weight of benzene-sulfohydrazide and 1% by weight of methylcellulose as a water retaining component. Besides, 5% by weight of a basic activator, 20% of which consists of potassium copper tellurate, had been added to the additive.

Example 3: Instead of potassium copper tellurate, 9% by weight of copper sulfate and 5% by weight of sodium aluminate were used. For the remainder the composition corresponds to that described in Example 2.

Example 4: The benzene sulfohydrazide and potassium copper tellurate used in Example 2 were replaced by 2% by weight of dihydrazine-sulfate and 20% by weight of copper sulfate. For the remainder the composition corresponds to the mixture as described in Example 2.

Example 5: In this example, benzene sulfohydrazine was added as nitrogen delivering compound at a dosage of 0.05% by weight. Lead peroxide at a dosage of 0.068% by weight was used as activator. Besides, 0.05% by weight of an alginic acid ester was added to the mixture as a water retaining component.

Example 6: Instead of lead peroxide used in the Example 5, potassium persulfate was used at a concentration of 0.08% by weight. For the remainder the composition corresponds to the mixture as described in Example 2.

Example 7: To a concrete mix of 10 kg Portland cement, 30 kg aggregate (greatest grain size 3 mm), 5 kg water, 200 grams of an additive containing 3% by weight of benzene sulfohydrazine, 3% by weight of sodium perborate and 3% by weight of lignin sulfonate as active component were added.

Example 8: In this example, the additive contained 30% by weight of p-nitrobenzene diazonium salt of the naphthalene disulfonic acid-1,5 and 7.5% by weight of iron-II gluconate. Besides, the composition corresponds to the mixture described in Example 6.

Mixtures	Water/cement factor	Temperature of mixture in degree Celsius	Flow time through standard funnel in sec.	Separation of water in percent by volume after--			
				1 hr.	3 hrs.	7 hrs.	24 hrs.
Zero test	0.39	20	Standstill	1	3	2.5	2
Example:							
1	0.39	20	17.5	1.5	2	4.5	4
2	0.39	20	18.0	0	0	0	0
3	0.40	20	15.4	0	0	0	0
4	0.40	20	16.6	0	0	0	0
5	0.40	20	15.6	0	0	0	0
6	0.40	20	16.6	0	0	0	0

Compression strength of cube after 7 and 28 days, stored at 100% humidity, 20° C. (kg./cm.²)			Modification of volume of mortar in percent, increase = + and decrease = − after--			
Mixtures	7 days	28 days	1 hr.	3 hrs.	7 hrs.	24 hrs.
Zero test		414	−2.3	−3.8	−3.6	−3.5
Example:						
1	211	294	+0.5	+0.8	+1.8	+2.6
2		354	0.2	+1.0	+2.6	+4.3
3	247	352	+0.6	+1.3	+3.4	+4.7
4	253	364	+2.0	+3.3	+3.5	+3.5
5	258	341	+2.6	+4.5	+5.2	+5.5
6	413	302	+5.3	+5.5	+5.8	+6.0

Concrete mixtures	Properties of fresh concrete			Compression strength in kg./cm.² after—	
	Water/cement factor	Unit weight, kg./l.	Air in percent	7 days	28 days
Zero test	0.5	2.23	2.3	390	435
Example:					
7	0.5	2.21	5.8	322	361
8	0.5	2.20	6.0	311	354

Saponin and Cellulosic Foam Stabilizer

A process described by *S. Uogaeshi; U.S. Patent 3,834,918; September 10, 1974; assigned to Teijin Limited, Japan* provides a raw batch containing homogeneously dispersed fine bubbles, which is suitably used for the building of porous architectural structures, the raw batch comprising a homogeneous mixture of 100 parts by weight of a hydraulic substance, 0.001 to 0.05 part by weight of a blowing agent, 0.01 to 0.8 part by weight of a foam stabilizer and 70 to 200 parts by weight of water. Saponin is the most preferred blowing agent for use in the process.

The preferred foam stabilizers include hydroxypropylmethyl cellulose, polyvinyl alcohol and the salts of polyacrylic acid. The following examples illustrate the process.

Example 1: A mixing tank equipped with a stirrer was charged with 100 parts of Portland cement, 30 parts of sand, 0.01 parts of saponin, 0.5 parts of hydroxypropylmethyl cellulose and 100 parts of water, following which the mixture was stirred and mixed at a stirring speed of 250 rpm to prepare a raw batch. This raw batch had stably incorporated therein fine bubbles which were homogeneously dispersed throughout the batch. In addition, this raw batch possessed good fluidity.

When this raw batch was poured into a form and allowed to set, a porous structure was obtained without sinkage of its placed height. On microscopic examination of this structure, fine voids were found to be homogeneously distributed throughout the structure. Hence, there was also uniformity in the mechanical strength of this structure.

Control 1 – When, by way of comparison, the experiment was carried out exactly as in Example 1 but without using the hydroxypropylmethyl cellulose, a raw batch having fine bubbles homogeneously dispersed throughout the batch could not be obtained. The reason for this was that the fine bubbles only formed in the upper portion of the mass that was being stirred and mixed, and moreover these bubbles tended to become disintegrated and disappear, with the consequence that the lower portion of the mass was maintained in a pasty or slurry state not containing any bubbles. When this raw batch was placed to a height of 300 millimeters, a separated layer of a height of about 65 millimeters was formed.

Control 2 – Again, by way of comparison, the experiment was carried out as in Example 1, except that saponin was not used. In this case, there was a drop in the air entrainment capacity of hydroxypropylmethyl cellulose, and fine bubbles could not be formed adequately. Hence, a raw batch such as obtained in Example 1 could not be obtained.

Example 2: A slurry was prepared by mixing 100 parts of Portland cement, 50 parts of sand and 75 parts of water with stirring. Separately, a homogeneously foamed mixture was prepared by adding 0.012 part of saponin and 0.6 part of hydroxypropylmethyl cellulose to 75 parts of water followed by vigorous stirring. Next, the slurry was introduced all at once to this foamed mixture, and the resulting mixture was kneaded together with stirring to prepare a raw batch containing fine bubbles homogeneously dispersed therein.

When this raw batch was placed to a height of 300 millimeters and allowed to set, only a sinkage of 1 millimeter was noted. The resulting structure had a compressive strength of 40 kg/cm^2. On microscopic examination of this structure, fine voids were found to be homogeneously distributed throughout the structure.

Example 3: A slurry was prepared by mixing with stirring 100 parts of white Portland cement, 25 parts of sand and 70 parts of water. Separately, a homogeneously foamed mixture was prepared by adding 0.005 part of calcium stearate and 0.5 part of sodium polyacrylate to 80 parts of water and vigorous stirring of the mixture. This was followed by introducing this foamed mixture all at once to the foregoing slurry and then kneading the mixture with stirring to prepare a raw batch containing fine bubbles homogeneously dispersed therein. When this raw batch was placed to a height of 300 millimeters and allowed to set, a sinkage of only 3 millimeters was noted. The resulting porous structure had a compressive strength of 35 kg/cm^2.

Magnesia Cement with Silicone Stabilizer

J.L. Edwards and J. Macoustra; U.S. Patent 3,573,941; April 6, 1971; assigned to BP Chemicals Limited, England describe an expanded magnesia cement containing a volume of gas at least equal to the volume of the cement in its unexpanded state. The cement is formed by mixing magnesium oxide, magnesium chloride and water in the presence of a foam-forming surface active agent and in the presence of a foam stabilizing water-soluble silicone, entraining a gas into the resulting mix to form a flowable foam and allowing the foam to set.

Particularly useful cement from an economic and structural point of view have three to twelve times the volume of gas present. The process is illustrated by reference to the following examples.

Example 1: 100 parts of heavy magnesium oxide (first setting time of 5 hours according to British Standard 776:1963) and 100 parts of magnesium chloride solution obtained by dissolving 60 parts of magnesium chloride hexahydrate in 40 parts water were added to a Hobart planetary mixer together with 1 part of Nansa HS80 (dodecyl benzene sodium sulfonate) and 0.5 part of CGA1 (silicone fluid comprising ethylene glycol/siloxane block polymers). Air was supplied to the bowl of the planetary mixer at a pressure of about 5 psi. Mixing was carried out for 10 to 20 minutes by which time the mixture had entrained therein a large volume of air and was in a flowable plastic state.

The plastic mass was poured into two molds, one the shape of a cube 6" x 6" and the other in the form of a sheet 1" thick. They were allowed to set and the set cement had excellent structural strength. Both forms were expanded to five or six times the volume of the unexpanded cement and had as a consequence a density six or seven times less than unexpanded magnesia cement.

Example 2: Four samples of foamed magnesia cement prepared according to the directions given in Example 1 were set into 4" cubes and their compressive strength measured. The density of the cubes and their respective compressive strengths are given in the following table:

Sample	Density [1]	Compressive strength [2]
1	20	350
2	20	350
3	21	350
4	20	300

[1] Pounds per cubic foot.
[2] Pounds per square inch.

Gypsum Castings

E.G. Foster and M.S. Bloom; U.S. Patent 3,454,688; July 8, 1969; assigned to Imperial Chemical Industries Limited, England describe a continuous process for casting gypsum building blocks or panels directly from product gypsum.

The process for producing quick-setting lightweight foamed gypsum castings requiring little or no drying, comprises the steps of mixing calcium sulfate hemihydrate with water to form a pourable or pumpable slurry, preferably of approximately plastering consistency, introducing into the slurry, or into the slurrying water, calcium carbonate and sulfuric acid in quantities sufficient to generate sufficient foam-forming carbon dioxide gas in situ within the slurry to reduce the density of the final product to within desired limits, the sulfuric acid being added to sufficient excess to accelerate setting of the slurry to within desired limits, and casting the foaming slurry, either continuously or batchwise, in a mold.

Cellulosic Waste and Magnesia Cement

A.N. Vassilevsky; U.S. Patent 3,816,147; June 11, 1974; assigned to V.R.B. Associates, Inc. describes a hydraulic cement binder consisting of magnesium oxide, magnesium sulfate and calcium chloride in proportion such that a magnesium oxychloride/magnesium oxysulfate/calcium sulfate hardenable mass is produced upon addition of water and setting. The binder is combined with cellulosic waste (municipality trash) which is treated with HCl or H_2SO_4 to partially hydrolyze the cellulosic material and thereafter with magnesium oxide or carbonate to neutralize it. For decreased brittleness and increased waterproofing characteristics, the binder contains sodium silicate and silicofluoride.

The system is particularly advantageous in the conversion of municipal and area wastes, especially cellulose-containing garbage, into useful materials. More particularly, it has been found that cellulose-containing garbage, i.e., garbage containing paper, cardboard, vegetable wastes and, in general, municipal garbage of all forms from which indigestible materials such as metal and synthetic resin have been removed, can be digested or hydrolyzed using hydrochloric acid and/or sulfuric acid and/or silicofluoride acid to yield a product which can be combined with cement, or ingredients thereof, to provide a hardenable composition of high compressive strength, light weight and low cost. One of the most surprising facts is that the waste-containing concrete of the process need contain no mineral aggregate. The hydrolyzed filler may be present in an amount up to 70% by weight of the overall composition (hardened concrete) and may be used in lesser quantities as desired.

Cement No. 1: 5 parts by weight of aluminum silicate or silica, 40 parts by weight of previously calcined magnesium oxide, or an equivalent amount of previously calcined magnesite, 20 parts by weight of barium chloride and 20 parts by weight of calcined magnesium sulfate are mixed, ground to a fine powder, and stored in polyethylene bags or other suitable airtight containers. The calcination is carried out at 750° to 780°C.

Cement No. 2: 2 parts by weight of aluminum silicate or silica, previously calcined at 250° to 300°C, are mixed with 5 parts by weight of calcined magnesia or magnesite and the mixture is ground (Powder A). Separately, 2 parts by weight of previously calcined barium chloride and 1 part by weight of strontium chloride are mixed and ground together (Powder B). Thereupon the Powders A and B are mixed in a weight ratio of 2:1, and 3 parts by weight of magnesium sulfate are added to the mixture which is then stored as in the case of Cement No. 1.

Cement No. 3: The ingredients enumerated below are separately subjected to a preliminary calcination at 350°C and then, after cooling, are mixed in the following proportions:

5 parts by weight of silica
48 parts by weight of calcined magnesite
29 parts by weight of calcined magnesium sulfate
10 parts by weight of calcined calcium chloride
8 parts by weight of calcined strontium chloride

The mixture is ground to a fine powder and, if necessary, subjected to additional drying. Thereupon it is stored as in the case of Cement No. 1. The following examples are given to illustrate the manner in which these cements are used in the preparation of hardened compositions.

Example 1: 80 parts by weight of Cement No. 1, 2 or 3 is mixed with 50 parts by weight of water to a consistency of heavy cream. Then there is added 30 parts by weight of sawdust, and the slurry is stirred to uniform consistency. It is then poured into molds and allowed to set in the open air without heating.

Example 2: 20 parts by weight of Cement No. 1, 2 or 3 is mixed with 12 parts by weight of water to a consistency of a heavy cream. Then there is added 7 parts by weight of sawdust and 5 parts by weight of sand. The resulting slurry is placed in a mold with application of pressure as the mixture is fairly dry and does not flow freely.

Example 3: 20 parts by weight of Cement No. 1, 2 or 3 is mixed with 12 parts by weight of water, and to this mixture 10 parts by weight of stone powder is added. The mixture is stirred until a thick slurry is obtained, and then treated as in Example 1. In addition to casting or molding, the cement in slurry may be used to impregnate cardboard, plywood or wood panels, thereby producing noncombustible, waterproof, stony panels of superior strength.

The materials prepared in accordance with this process are well suited for a wide variety of applications. In the field of building construction, for example, hollow bricks made from these materials provide good thermal and acoustic insulation. Sheets and panels can be used for flooring, in the construction of driveways, walks, stairs, and terraces. Building materials prepared with these cements lend themselves ideally to prefabricated-house construction. Further, since most of the components used in the preparation of these materials are relatively impervious to gamma radiation, they offer an added protection against such rays when employed in the construction of radiation shelters.

When mixed with cellulosic and plastic materials, or both, these cements produce, in many instances, a product of greater strength than the cellulosic or plastic substances themselves, and much less expensive. Since the cements of the process bind any mineral, including stones or earth, along with any wood or cellulosic scraps into a homogeneous mass, crushers set up near the site of a demolished building can prepare new material right on the spot. Finally, and by way of further examples illustrating the possible applications of the cements, textile, burlap or nylon materials added to the cement mixture produce panels similar to sheetrock but of higher strength than the latter, possessing the added advantage that they are cleaner to work than sheetrock. The equipment required for the manufacture of these cements is very simple, consisting merely of an oven for the calcination, a grinding mill, mixers, and drying and packaging apparatus.

Example 4: A hydraulic binder is prepared by individually calcining calcium chloride, magnesium oxide and magnesium sulfate at a temperature of 750° to 780°C. Each of the calcined components is ground and combined with the others in the following proportions: 3 parts by weight of magnesia MgO, 3 parts by weight of anhydrous magnesium sulfate and 1 to 2 parts by weight of anhydrous calcium chloride $CaCl_2$. The powder is placed in a polyethylene bag and stored in a humid environment for several weeks. No deterioration of the cement was observed.

A comparison was made between a Portland cement composition prepared in accordance with ASTM procedure with 1 part Portland cement and 4 parts of clean washed and sieved sand. This composition was compared with a soil cement prepared by mixing 1 part of the powder stored as above with 4 parts by weight of ordinary garden earth containing substantial proportions of organic material in addition to the usual mineral matter. The results and characteristics of the tests are set forth below:

(A) Flexural Strength — Tests were conducted in accordance with ASTM C348. Test specimens for each type of mortar were aged for 9 days at room temperature.

	Flexural Strength, psi	
Test Specimen	**Soil Mortar, this process**	**Portland Cement Mortar**
1A	309	382
1B	281	418
2A	281	400
2B	315	437
3A	304	406
3B	270	448
Average	293	415

(B) Resistance-To-Abrasion Wear — The abrasion-wear tests were conducted after test specimens had cured for 9 days at room temperature. The Tinius Olsen Wearometer, as specified in Military Specification, Mil-D-3134F, was used. The wear after 1,000 cycles for each type mortar is shown below. Two test specimens for each mortar were used.

Soil Mortar—The Loss in Thickness After 1,000 Cycles of Wear

Specimen No. 1	0.101 inch
Specimen No. 2	0.096 inch
Average	0.099 inch

Portland Cement Mortar—The Loss in Thickness After 1,000 Cycles of Wear

Specimen No. 1	0.094 inch
Specimen No. 2	0.089 inch
Average	0.091 inch

The foregoing results demonstrate that an expensive Portland cement composition using high quality sand can effectively be replaced by an inexpensive composition such as the soil cement of the process using ordinary earth available at all necessary locations. Best results are obtained when 20 to 40 parts by weight of earth or clay are mixed with 10 parts by weight of the cement composition of the process. Moreover, it was found that the magnesium oxide (anhydrous) content should be confined between the critical limits of 4.5 to 5.5 parts by weight, the anhydrous magnesium sulfate should be confined to an amount between 2.8 and 3.3 parts by weight while the calcium chloride (anhydrous) should be present in an amount of 1 to 3 parts by weight. Approximately 300 to 400 cm^3 of water is required when 1 lb of the cement composition is admixed with 3 lbs of earth. The cement may be admixed with water first and then combined with the earth, an exothermic reaction occurring immediately upon addition of the water to the cement.

Tests performed with samples having soil as a component have shown a compression resistance between about 3,000 and 3,500 lbs/in^2, or better than 200 kg/cm^2.

Example 5: The hydraulic cement binder produced as described in connection with cement No. 1, is combined with 4 parts by weight of sodium silicofluoride (Na_2SiF_6) and about 6% by weight of sodium silicate (Na_2SiO_3), previously dried to an anhydrous state prior to combining with the composition and after calcination thereof. When this cement binder is used in the systems of Examples 1, 2, 3 or 4, the composition is found to have waterproof characteristics and to be less sensitive to crumbling.

Magnesium Oxychloride Cement and Polyurethane Foam

A process described by *H.C. Thompson; U.S. Patent 3,951,885; April 20, 1976; assigned to Thompson Chemicals, Inc.* provides a method for making a foamed fireproof product of magnesium oxychloride cement. A porous substrate is impregnated with a foaming mixture of magnesium chloride, magnesium oxide, and frothing agent in water. The mixture hardens with small voids throughout the porous substrate, thus providing a fireproof product of low density. The cement mixture may be cofoamed with polyurethane foam. The fireproof products of relatively low density are particularly valuable for building and construction purposes. Magnesium powder is the preferred frothing agent which, in combination with a surfactant, induces a large volume of small bubbles that remain in the composition as it is set.

Example 1: In the preparation of the froth, 19¼ pounds magnesium chloride were added to 4¼ gallons of water. The specific gravity was measured and additional magnesium chloride was added to adjust the specific gravity to 22° Baume. The combined total made about 5 gallons. 200 grams of lactic acid were added to the aqueous mixture. Magnesium oxide was then measured in an amount equal to 5.5 pounds magnesium oxide to each gallon of magnesium chloride in water. Then, 32 grams of powdered magnesium, which is about 0.1% of the dry weight of the magnesium oxychloride cement mixture and 100 grams of surfactant (Rohm and Haas X-100 defined by the manufacturer as octylphenol ethylene oxylate with an average of nine to ten mols of ethylene oxide per molecule) were added to the magnesium oxide component.

The surfactant constituted about 0.5% of the dry weight of the total mixture. The dry magnesium oxide component was added to the aqueous magnesium chloride component and the mixture immediately began to froth.

Example 2: In this example the cement was cofoamed with an NCO type polyurethane foam. 22° Baumé magnesium chloride was intimately reacted with magnesium oxide in a ratio of six parts oxide to ten parts chloride by weight. The combined oxide and chloride weight is referred to as cement weight and used as the basis for other ingredients. Polyurethane foam ingredients in the form of equal parts by weight of Decor foam No. 227 components A and B were used. The total weight of polyurethane foam components was 12.5% of the cement weight. The magnesium chloride was heated to a temperature of 98°F. A filler of vermiculite in the amount of 75% of the weight of cement was added to the oxide and the two components were mixed with the chloride.

The B component of the polyurethane was added along with ½ of 1%, based on the weight of the cement, of surfactant (G.E. Silicone SF-1079). Upon mixing there was a slight rise in the liquid level because of the foaming agent in the B component. Cement expander (magnesium powder) was added in the amount of 1% of the cement and blended for two minutes. The A component of the polyurethane foam was then added and an immediate and vigorous reaction took place. Upon blending, the foaming mixture was immediately poured into a mold where it set in approximately 20 minutes into a low density, nonburning product. Complete cure of the cement does not take place for several hours even though the molded product may be handled.

The product of Example 2 is lightweight and has excellent fire resistance. When used as a 2 inch core for a door or wall, it has a rating of approximately two hours. Even the chlorinated polyurethanes commercially available as fire resistant will melt at 300°F. In contrast, the cofoam of this process can withstand 1800°F at which temperature it gradually spalls. There is essentially no smoke.

Example 3: In this example a much heavier product is made, yet it is still half the weight of wood and completely fireproof. Cement ingredients were mixed in the ratio of six parts magnesium oxide to 10 parts by weight magnesium chloride. A polyurethane foam used to simulate wood and sold under the designation Decor Urewood was prepared in a separate container in an amount equal to 10% of the weight of the cement. Chopped glass fiber was mixed with the cement ingredients in an amount equal to 5% of the weight of the cement. Immediately after the A and B components of the polyurethane were mixed, it was blended with the remaining ingredients for 1 to 2 minutes and poured into a mold having graining resembling wood. The mixture gels within 30 minutes and is hard enough to be removed from the mold after one hour. Curing continues for at least 4 hours and the strength improves each day for several days, indicating that reaction continues long after removal from the mold. The product resembles wood, but is completely nonburning and will pass any fire test for building materials.

Foamed Structure

According to a process described by *M.T. Ergene; U.S. Patent 3,867,159; February 18, 1975; assigned to The Stanley Works* cellular concrete structures are made by mixing water and cement under conditions sufficient to produce a high degree of hydration of the cement particles, followed by the introduction of a foam formed under pressure from a mixture of water, air, foaming agent and chloride accelerator. The foam mixture and cement mixture are blended to a substantially homogeneous, foamed cement slurry, which is cast into a mold and cured to form a lightweight cellular concrete structure.

In the structure, the voids are relatively spherical in shape and have an average diameter of about 0.005 to 0.050 inch. The foaming agent and cure accelerator are concentrated at the surface of the concrete matrix about the voids, and the structure has a uniform density of about 25 to 75 pounds per cubic foot. All parts are on a weight basis, unless otherwise indicated. The following examples illustrate the process.

Example 1: Into a paddle mixer having a 10 cubic foot capacity are charged 188 pounds of Type 1 Portland cement (sold by Penn Dixie Company), 83 pounds of tap water and about 6 pounds of a water-soluble partially hydrolyzed polyvinyl acetate resin (Polyco 2119) and the charges are mixed at 35 rpm for a period of about 15 minutes to produce a hydrated cement slurry. A foam having a density of about 4.37 pounds per cubic foot is produced in a Mearl No. AT-10-5A foam generator fitted with a 36 inch long nozzle, from 73.4% water, 25.45% Anti Hydro accelerating agent, and 1.14% of National Crete fish protein foaming agent, and the foam is injected directly into the hydrated cement slurry in the mixer.

About 16.5 pounds (38 cubic feet) of foam is injected over a period of between 1 and 2 minutes, and the mixer is operative during the entire injection period and for about an additional minute after all of the foam has been added.

The resultant slurry, having a density of about 39 pounds per cubic foot, is then poured into cylindrical molds to produce three castings, each 8 inches long and 4 inches in diameter. The castings are allowed to preset in the molds for about 24 hours, after which they are steam cured in a chamber at a temperature of about 150°F for 7 days.

Upon evaluation, the cured cellular concrete structures that result are found to have an average density of about 33.5 pounds per cubic foot and an average compressive strength under an axially applied load, of about 320 pounds per square foot. The voids are uniform in size and have an average diameter of about 0.025 inch.

Example 2: The procedure of Example 1 is substantially repeated but utilizing 282 pounds of cement, 120 pounds of water and about 6.3 pounds of polyvinyl acetate resin. The admixture is agitated under the conditions specified for a period of about 20 minutes, and 28 pounds (6.3 cubic feet) of foam is injected during a period of 2 to 3 minutes. The density of the resultant slurry is about 44.5 pounds per cubic foot and the evaluated cylindrical castings are found to have an average density of 45.5 pounds per cubic foot, an average compressive strength of about 610 pounds per square inch, and to have uniform voids of about 0.030 inch average diameter.

GYPSUM AND MORTAR

SET RETARDERS

Mercerized Cellulose

B.S.J. Ericson; U.S. Patent 3,936,313; February 3, 1976; assigned to MoDoKemi AB, Sweden describes a method of preparing a calcium sulfate plaster additive composition having water-retaining and retarding properties. The additive composition comprises (1) a water-soluble cellulose ether obtained by reacting cellulose mercerized with alkali hydroxide with suitable reactants, and (2) a retarding constituent. The method includes the step of adding to the resulting cellulose ether containing alkali hydroxide impurity, without prior purification of the cellulose ether to remove residual alkali hydroxide, an acid, the alkali salt of which serves as a calcium sulfate plaster retarder, in an amount approximately stoichiometrically equivalent to the amount of alkali hydroxide remaining from the reaction.

In comparative tests made with a calcium sulfate plaster additive prepared in accordance with the process and a calcium sulfate plaster additive where the cellulose ether had been purified of alkali in the conventional way, and the retarding alkali salt had been prepared separately in a conventional manner and then dry-blended, it was found that the additive according to this process dissolves considerably more rapidly in water which resulted in a more rapid distribution of the additive in the calcium sulfate plaster. The reason is not known, but it is believed that this effect is due to the fact that the additive obtained according to this process is a homogeneous composition of water-retaining and retarding cellulose ether salt components. The process is applicable to any water-soluble cellulose ether commonly used in calcium sulfate plaster compositions.

In a particularly preferred example, the cellulose ether is hydroxyethyl hydroxypropyl cellulose or ethyl hydroxyethyl hydroxypropyl cellulose having a flocculating temperature above about 70°C, preferably above 80°C. The preferred ether usually has an MS (molecular substitution) from about 0.5 to 2 for hydroxyethyl, an MS of from 1 to 2 for hydroxypropyl, and a DS (degree of substitution) for ethyl from about 0.1 to 0.4. These cellulose ethers which may be prepared by adding ethylene oxide and propylene oxide to the alkali mercerized cellulose, and then ethylating with ethyl chloride have been found to exhibit very good water-retaining properties in calcium sulfate plaster compositions, and sodium hydroxide is only consumed during the addition reactions to a small extent.

This means that the sodium hydroxide added in the mercerization step can be utilized to a considerable extent to form retarding sodium salts in situ. The reason why hydroxyethyl hydroxypropyl cellulose and ethyl hydroxyethyl hydroxypropyl cellulose have such excellent water-retaining and working improving action is not known, but presumably it is related to the high tendency of aqueous sodium solution of these cellulose ethers to gel even at high solids contents. Also, the calcium sulfate plaster additives in accordance with this process impart to the calcium sulfate plaster composition a low adhesion to metal surfaces and a reduced tendency to flow. The following examples illustrate the process.

Example 1: Ethyl-hydroxyethyl hydroxypropyl cellulose, EHEHPC (DS_{ethyl} = 0.15; $MS_{hydroxyethyl}$ = 0.05; $MS_{hydroxypropyl}$ = 1.8) was produced by mercerizing 1 part by weight of cellulose with 20% aqueous sodium hydroxide for 30 minutes at room temperature. After the mercerization was complete, the cellulose was squeezed to a press factor of 2.5. The resulting 2.5 parts of alkali cellulose was shredded, after which it was transferred to an autoclave. Air was removed from the autoclave, and a reaction mixture comprising 1.5 parts of ethyl chloride, 1.4 parts of propylene oxide and 0.3 part of ethylene oxide was charged in the autoclave.

After charging the temperature was raised to 70°C in 30 minutes and held at this level for three hours. The reaction was then stopped and the remaining ethyl chloride vented. One half of the resulting cellulose ether which had a flocculating temperature of about 65°C, was neutralized with 0.4 part of citric acid for 1 part of cellulose; then it was ground to form a slightly yellowish white and very loose powder.

The remainder of the cellulose ether was processed in a conventional manner by slurrying and washing the cellulose ether in hot water at about 95°C. The residue of alkali was neutralized with acetic acid. The product was dried and ground in the same way as the product according to the process, resulting in a relatively hard, greyish white and manifestly sintered product. Sodium citrate was mixed with the product in an amount corresponding to the amount of sodium citrate in the product of this process.

The two products were tested and compared with respect to their dissolving characteristics by dissolving 3 g of each product in 147 g of water at 20°C with mild agitation with a glass rod. The calcium sulfate plaster additive according to the process dissolved with rapid increase in viscosity and was considered completely dissolved after about 30 minutes. The control product gave a relatively slow increase in viscosity. It was considered completely dissolved only after about 70 minutes.

The two products were also tested with respect to setting time measured according to Vicat (ASTM C-191-58) in a calcium sulfate plaster. The amount of water was 520 g per 1000 g of calcium sulfate. The results obtained are shown in Table 1 on the following page. It is seen from the setting times that the product prepared according to this process exhibits a substantially longer setting time than the control product, although the amount of retarder is the same in both cases. The reason cannot be fully explained, but it is believed that a contributing factor is that the product according to the process is rapidly dissolved and dispersed in the calcium sulfate plaster without forming gel lumps. For comparison it may be mentioned that calcium sulfate plaster containing only washed cellulose ether (the EHEHPC ether of the control) has a setting time of about 5 minutes.

TABLE 1

Amount of Additive (% by weight based on calcium sulfate plaster)	Setting Time, min Initial	Setting Time, min Final
0.25% EHEHPC (including 0.03% Na citrate) (Example 1)	92	107
0.22% EHEHPC + 0.03% Na citrate (control)	30	45

Examples 2 and 3: In a manner similar to Example 1, hydroxyethyl cellulose and hydroxypropyl cellulose were neutralized, without prior washing, with sufficient citric acid to neutralize sodium hydroxide present from the mercerizing step. The resulting products were added in varying amounts to calcium sulfate plaster the setting time of which was measured according to Vicat. The following results were obtained.

TABLE 2

	Amount of Additive % by wt Based on Calcium Sulfate Plaster	Water, g/1,000 g Calcium Sulfate (to obtain normal consistency)	Added Amount of Acid (g)/kg Cellulose for Complete Neutralization	Setting Time, min Initial	Setting Time, min Final	Consistency
Hydroxyethyl cellulose	0.1	500	563	21	26	Good
Hydroxyethyl cellulose	0.2	490	563	46	51	Good
$MS_{hydroxyethyl}$ = 2.3 (Example 2)	0.3	470	563	63	74	Good
Hydroxypropyl cellulose	0.1	455	262	12	17	Good
Hydroxypropyl cellulose	0.2	450	262	23	26	Good
$MS_{hydroxypropyl}$ = 3.75 (Example 3)	0.3	455	262	27	31	Good

The results show that the presence of cellulose ethers containing sodium citrate considerably increased the setting time of the calcium sulfate plaster. The setting time for a control composition without any additive was five minutes. Storage tests made with the compositions show that they also have excellent storage stability. This indicates that the sodium hydroxide present in the cellulose ether has been well neutralized.

Example 4: Ethyl hydroxyethyl hydroxypropyl cellulose, EHEHPC (DS_{ethyl} = 0.2; $MS_{hydroxyethyl}$ = 0.5; $MS_{hydroxypropyl}$ = 1.0) was prepared according to the same method as in Example 1 by adding to cellulose mercerized with 20% aqueous sodium hydroxide, 1 mol of ethylene oxide and 2 mols of propylene oxide per mol anhydroglucose unit in the presence of ethyl chloride as a reaction medium. After alkoxylation was complete, phosphoric acid was added to the crude cellulose ether until the product was neutralized, after which it was ground. The effect of the resulting product on the setting of calcium sulfate plaster was tested, and the results obtained are shown in Table 3 on the following page.

TABLE 3

Amount of Additive % by wt Based on Calcium Sulfate Plaster	Water, g/1,000 g of Calcium Sulfate	Setting Time, -----min----- Initial	Final	Consistency
0.1	480	27	38	Some stiffening
0.2	480	59	73	Good
0.3	480	75	100	Good

It appears from the data that phosphoric acid can be used advantageously for direct neutralization of crude cellulose ethers and that the resulting composition has a good retarding action on the setting of calcium sulfate plaster.

Diethylenetriaminepentaacetic Acid

R.A. Kuntze; U.S. Patent 3,304,189; assigned to Canadian Patents and Development Ltd., Canada describes retarders which may be defined as water-soluble compounds selected from the group consisting of aliphatic polyamino polycarboxylic acids of the general formula R_2N—(X)—NR_2 where X is an aliphatic chain containing at least 3 but not more than 9 carbon atoms between the terminal amino groups, and at least two R's are selected from the group consisting of lower carboxy acid groups, and their alkali metal, alkaline earth metal, ammonium and triethanolamine salts, while the remaining R's are hydrogen atoms or lower alkyl groups.

The synthetic plaster retarders are used in the approximate concentration of 0.5 to 10 pounds per ton of calcined gypsum (0.025 to 0.5% by weight). An efficient synthetic retarder of the process is diethylenetriaminepentaacetic acid (DTPA) which has a retarding efficiency more than twice that of the usual natural protein retarders and which is available commercially at a price that makes it a preferred choice economically.

Parent Polyamine	Approx No. of Amino Hydrogen Atoms Substituted by Acetic Acid Groups	Pounds Retarder per Ton Calcined Gypsum	Setting Time Hr	Min
Tetramethylenediamine	3	3	18	57
Hexamethylenediamine	4	3	8	15
Diethylenetriamine	2	3	5	45
Diethylenetriamine	5	3	37	17
Diethylenetriamine (as calcium salt)	5	3	41	22
Diethylenetriamine (as magnesium salt)	5	3	28	21
Diethylenetriamine (as ammonium salt)	5	3	49	37
Diethylenetriamine (as triethanolamine salt)	5	3	50	43
Triethylenetetramine	2	3	25	18
Triethylenetetramine	6	3	54	27
Tetraethylenepentamine	2	3	25	7
Tetraethylenepentamine	7	3	8	61
Lysine	4	3	23	32
Cystine	4	3	7	10
1,3-Propylenediamine tetraacetic acid*	4	5	14	26
2-Hydroxy-1,3-propylenediamine tetraacetic acid*	4	5	18	25
Diethylenetriamine pentaacetic acid*	5	2	11	41
2-Ethyl-1,3-propylenediamine (tetraacetic acid)*	4	5	14	35
N,N'-dimethyl-1,3-propylenediamine (diacetic acid)*	2	5	12	10
Di-2-propylethylenetriamine (pentaacetic acid)*	5	3	34	45

*As sodium salts.

Some examples of the retarding efficiency of retarders according to the process are given in the preceding table. The setting times, as determined by the temperature method, e.g., heat of hydration measurements, were obtained by the addition of the specified amount of retarder per ton of calcined gypsum. All retarders were sodium salts unless otherwise stated. The retarder of the process may be employed either alone or in combination with other retarders. Thus, the retarder may be employed in combination with a conventional protein hydrolysate retarder with synergistic results.

Boric Acid and Ammonia

D.F. Smith; U.S. Patent 3,294,087; December 27, 1966 has found that there is a marked synergistic action between ammonium borate and ammonia in delaying the set of plaster in the slurry, as illustrated in the following examples.

Example 1: A solution was made containing the materials shown below:

	Grams
Water	200
H_3BO_3	1.5
29.4 weight percent aqua NH_3	4
K_2SO_4	4.5

The above solution was mixed with 350 g powdered plaster of Paris which did not set until about four hours after it was mixed with the solution. (Within practical limits the amount of plaster relative to solution is immaterial from the standpoint of life of the slurry since a given solution will cause setting of a small amount of plaster in about the same time as with a large amount of plaster, the composition of the solution being determined. The amount of plaster added is controlled by the thickness of slurry required for the coating method used.)

Example 2: A solution was made using the following materials:

	Grams
Water	176.7
H_3BO_3	1.5
29.4 weight percent aqua NH_3	27.3
K_2SO_4	4.5

The above solution was mixed with 350 g of another part of the same plaster of Paris used in Example 1. The plaster did not set until about five hours.

Example 3: A solution was made containing the following materials:

	Grams
Water	200
H_3BO_3	0.35
29.4 weight percent aqua NH_3	4.0
K_2SO_4	4.5

After the above solution was mixed with 350 g of another part of the same plaster of Paris used in Example 1, the plaster set in seven minutes.

Example 4: A solution was made containing the following materials:

	Grams
Water	136
H_3BO_3	0.35
29.4 weight percent aqua NH_3	68
K_2SO_4	4.5

After the above solution was mixed with another part of the same plaster of Paris used in Example 1, the plaster set in 24 minutes.

Example 5: A solution was made containing the following materials:

	Grams
Water	79
29.4 weight percent aqua NH_3	125
K_2SO_4	4.5

After the above solution was mixed with 350 g of another part of the same plaster of Paris used in Example 1, the plaster set in 2¾ hours.

Example 6: A solution was made containing the following materials:

	Grams
Water	79
H_3BO_3	0.35
29.4 weight percent aqua NH_3	125
K_2SO_4	4.5

After mixing the above solution with 350 g of another part of the same plaster of Paris used in Example 1, the plaster set in 19 hours. Example 1 (0.58% NH_3 or 2% aqua ammonia) is illustrative of the Eberl process (U.S. Patent 2,557,083). Example 2 shows a higher ammonia concentration (3.9% NH_3 or 13% aqua ammonia) than used by Eberl. The slurry life of Example 2 is somewhat longer than that of Example 1 but only slightly longer.

Example 3 illustrates the very short slurry life with reduced borate concentration. Example 4 shows a measurable increase in slurry life over that of Example 3 at the same borate concentration, but even when the NH_3 concentration is increased to 9.8% NH_3 (33% aqua ammonia) the slurry life is only 24 minutes. Example 5 shows a moderate slurry life when the NH_3 concentration is increased to 18% (61% aqua ammonia), without borate.

However, in Example 6 where the NH_3 concentration is the same as in Example 5, but the borate (boric acid) concentration is the same in Examples 3 and 4, a very long slurry life is obtained. Thus Example 3 (or even Example 4) shows the very short slurry life due to ammonium borate at a low concentration and Example 5 shows the moderate slurry life due to 18% NH_3 without borate while Example 6 shows a slurry life much greater than can be explained by the sum of the set delay of the borate alone (Example 3) and the NH_3 alone (Example 5).

The examples show that there is a marked synergism in the set delay of borate and NH_3 in the same solution. Thus the use of NH_3 permits the use of much lower borate concentration in the slurry (and thus a lower ratio of borate to plaster) while still obtaining sufficiently long slurry life.

The slurry of Example 1 when spread upon crinoline and dried at 190° to 250°F gives a bandage setting in about 3½ minutes, whereas, similar use of the slurry of Example 6 gives a setting time of 2¾ minutes under similar conditions of test.

1,3,5-Pentanetricarboxylic Acid

P.J. Ferrara and G. Dalby; U.S. Patent 3,598,621; August 10, 1971 describe a class of plaster setting retarders. The preferred retarders according to this process named in their free acid form, include pimelic, azelaic, 1,2,3-propanetricarboxylic and 1,3,5-pentanetricarboxylic acids. The most preferred acid is 1,3,5-pentanetricarboxylic acid.

Example 1: 50 grams of standard plaster were mixed with 0.1 gram (4 lb/ton) of one of each of the following retarders: adipic acid, pimelic acid, suberic acid, azelaic acid, sebacic acid, citric acid, 1,2,3-propanetricarboxylic acid. 38 cc of water were added to the resultant mixture and the batch carefully mixed and observed for setting at 10 to 15-minute intervals. Setting time was determined by the conventional methods as for example by means of a thermocouple inserted in the batch for indicating an increase in temperature or by insertion of a probe into the batch whereby the degree of crystallinity could be evaluated. The results in each instance are set out below in the table.

Example 2: Example 1 was repeated but there was additionally included in the dry mix of acid and plaster 75 grams of sand. The results obtained were substantially the same as those reported in Example 1.

Example 3: Example 1 was repeated but using 0.05 gram of the 1,2,3-propanetricarboxylic and 1,3,5-pentanetricarboxylic acids (equivalent of 2 lb/ton). The results are set out in the table which follows.

Example 4: Example 1 was repeated but with the 1,3,5-pentanetricarboxylic acid present in an amount equivalent to 0.4, 0.2, 0.1 and 0.01 lb/ton of plaster respectively. The results are set out in the table which follows.

Example 5: Example 1 was repeated using 0.1 gram (4 lb/ton) of hoof meal as described in U.S. Patent 2,865,905. The results of this example are also set out in the table below.

Acid	Quantity (lb/ton)	Setting Retardation (hr)
Adipic	4	½
Pimelic	4	4½
Suberic	4	5/6
Azelaic	4	2
Sebacic	4	5/12
Citric	4	3½
1,2,3-propanetricarboxylic	2	1⅓
1,2,3-propanetricarboxylic	4	2½
1,3,5-pentanetricarboxylic	0.2	3¾
1,3,5-pentanetricarboxylic	0.4	8¾
1,3,5-pentanetricarboxylic	2	24
1,3,5-pentanetricarboxylic	0.1	1½
1,3,5-pentanetricarboxylic	0.01	⅔
Hoof meal	4	5¾

From the table, it can be seen that the acids tested in each instance retard the setting of the plaster. As the acids are free of the disadvantages associated with the use of hydrolyzed proteinaceous materials, they constitute a preferred retarder. In the instance of 1,3,5-pentanetricarboxylic acid amounts as low as

0.01 lb/ton produced retardation of the setting time, an amount of 0.2 lb/ton of the aforesaid producing a retardation of setting time equivalent to that of 4 lb/ton of hoof meal or, conversely, 0.20 lb of 1,3,5-tricarboxypentane is the equivalent of 20 times that amount of the conventional retarder.

It has been found that the reaction product of an aliphatic polycarboxylic acid as set forth above with a polyamine, such as a di- or trialiphatic amine, preferably of low molecular weight, is a superior set retarder for plaster, stucco, cement, mortar and the like. Particularly, the reaction product of 1,3,5-pentanetricarboxylic acid and ethylenediamine or diethylenetriamine is an exceptional set retarder. Another aspect of this process lies in the combined use of chelating compounds with the ethylenediamine derivative of the polycarboxylic acids listed in the table of acids shown on the preceding page. These chelates should be used in a proportion of about 1 to 5 pounds per pound of polycarboxylic acid.

Sodium Salt of Styrene-Maleic Anhydride Copolymers

In a process described by *R.J. Pratt and D.W. Young; U.S. Patent 3,544,344; December 1, 1970; assigned to Sinclair Research, Incorporated* a plaster composition is prepared having significantly increased setting time. The composition comprises a settable gypsum plaster and a minor amount, sufficient to retard setting of the plaster, of a water-soluble salt, such as the sodium salt, of the ester formed by reacting styrene-maleic anhydride copolymer with an alkoxy polyalkylene glycol.

The styrene-maleic anhydride copolymer employed has a molecular weight of about 500 to 5,000 and the preferred alkoxy polyalkylene glycol is methoxy polyethylene glycol having a molecular weight of about 300 to 5,000. The preparation of the styrene-maleic anhydride-alkoxy polyalkylene glycol ester is illustrated in Example 1.

Example 1: The preparation of the ester is as follows. A mixture of 350 g (1.0 equivalent) of methoxy polyethylene glycol having a molecular weight of 350 and 225 grams (1.0 anhydride equivalent) of styrene-maleic anhydride copolymer (1:1 monomer ratio) having a molecular weight of about 1,800 was heated at 185° to 195°C for 2.5 hours under a nitrogen atmosphere. The product was analyzed by alkali titration as a 70% half-ester. That is, 35% of all carboxyls were converted to ester. Example 2 illustrates the preparation of the alkali metal salt of the styrene-maleic anhydride-alkoxy polyalkylene glycol ester of this process.

Example 2: Preparation of a solution of sodium salt of 70% half-ester is as follows. A mixture of 12.0 grams of the ester of Example 1 was stirred with 88.0 grams of water at 160°F (steam bath) until the ester was dispersed. Sodium hydroxide as a 10% solution was added dropwise until the pH stabilized at 7.0. About 0.7 to 1.0 gram of alkali was required. Solid alkali may also be used. This solution was approximately 12% solids.

The plaster composition can be prepared conveniently by adding the neutralized ester salt either in dry form or as a solution or paste to the gypsum plaster prior to, with or after the addition of sufficient water to cause the plaster to set. Thus the ester salt can be removed from the water and the dried ester salt mixed with the dry gypsum plaster to form plaster compositions having increased setting time when subsequently mixed with water.

Example 3: Twenty-five grams of water were mixed with 0.75 gram of the sodium salt solution of Example 2. The resulting solution was added to 45 grams of molding gypsum plaster having a setting time of 10 minutes. The resulting slurry was stirred for five minutes and then poured into an aluminum weighing dish. The setting time was the time interval between mixing the ingredients and the hardening of the slurry. The table summarizes the results obtained and compares them with results obtained by varying the proportions and the composition of the set-retarding additive.

(Methoxy polyethylene glycol) mol. wt. ethylene oxide units (Y)	Styrene-maleic anhydride glycol, half-ester level	Concentration of retarder (as sodium salt), percent	Setting time (minutes)
(1) 350, Y=7	70	0.4	300
(2) 350, Y=7	60	0.4	480
(3) 350, Y=7	60	0.2	55
(4) 350, Y=7	60	0.1	20
(5) 750, Y=16-17	32	0.2	240
(6) 750, Y=16-17	32	0.05	25
(7) 5000, Y=113	6	0.2	20
(8) 350, Y=7	0	0.4	10
(9) None	0	0	10
(10) Ethylene glycol capped with n-butyl group.	100	0.2	10
(11) 167, Y=3	70	0.2	15
(12) 350, Y=7	50	0.2	10

As may be seen in the table above, the longest setting times (Composition 1 and Composition 2) were achieved using 0.4 weight percent of the sodium salt of the 60 and 70% half-esters of methoxy polyethylene glycol having 7 ethylene oxide units and molecular weight of 350. Good results were also obtained in Composition 5 using only a 32% half-ester but of a methoxy polyethylene glycol having 16 to 17 ethylene oxide units and a molecular weight of 750. The use of the glycol alone (Composition 8) of 0.2% of the 50% half-ester of a methoxy polyethylene glycol having 7 ethylene oxide units and a molecular weight of 350 (Composition 12) produced no change in setting time.

SET ACCELERATORS

Ground Calcium Disulfate Dihydrate and Sucrose

W. Kinkade and E.E. O'Neill; U.S. Patent 3,573,947; April 6, 1971; assigned to United States Gypsum Company describe a process for producing an accelerator for a calcium sulfate plaster which comprises the steps of grinding together calcium sulfate dihydrate and sucrose at a temperature of up to about 145°F until the surface area of the mixture is substantially increased, and heating the ground mixture under conditions which essentially preclude the formation of a dewpoint condition in the mixture to a final moisture content from about 6% to 14% by weight.

Example 1: A ball mill was preheated to 115° to 125°F and charged with 66.5 pounds of land plaster with a Blaine surface area of 2,400 cm^2/g (100% through 30 mesh), 3.5 pounds sucrose (5% by weight of the gypsum-sugar mix) and 550 pounds of $^{13}/_{16}$ inch by $^{13}/_{16}$ inch Burundum balls. The mill was closed and the charge milled for 1½ hours, at which time its temperature was about 145°F. The product when removed had a Blaine surface area of 12,000 cm^2/g. This milled land plaster-sugar mixture was transferred to a circulating air oven where it was placed in shallow trays to a depth of about ¾ inch and calcined at 250°F for 23 hours. The product had a combined moisture content of 12%. During the calcination step the white milled plaster developed a light tan color and an odor of caramel. When viewed under the microscope the particles appeared to have a uniform and continuous tan colored exterior. The tan color may be used as an indicia of completion of the heating or baking step.

Portions of the baked accelerator were each added to 200 g fresh molding plaster in the amounts shown below, the accelerated plaster mixed with 180 ml water for 7 seconds in a Waring Blendor and its vicat and temperature rise set determined. Separate portions of the baked accelerator were exposed for 40 hours in a cabinet maintained at 80°F and 80% relative humidity. After exposure the accelerative potency was determined as above with the results reported below.

Example 2: The procedure of Example 1 was repeated except that the charge comprised 63 lb land plaster and 7 lb sucrose (10% by weight of the gypsum-sugar mix). After milling for 1½ hours, the gypsum-sugar mix had a surface area of 11,100 cm^2/g. The accelerative potency of samples withdrawn from this material was determined as milled and after exposure in the 80°F 80% RH cabinet. The main body of the milled mix in each was heated at 250°F for 24 hours in shallow trays and its accelerative potency determined in the freshly baked condition and after exposure in the humidity cabinet with the following results.

		Grams Added					
		0.05		0.10		0.20	
Composition	**Treatment**	**Vicat***	**Tr***	**Vicat**	**Tr**	**Vicat**	**Tr**
95% land plaster	Example 1	6:20	17:05	5:20	15:30	4:30	14:05
5% sucrose	Example 1 plus 40 hr at 80°F and 80% RH	8:50	20:50	7:15	18:45	6:10	17:00
90% land plaster	Example 2 milled	6:45	18:05	5:35	16:10	4:30	14:35
10% sucrose	Milled-humidified	10:30	22:40	8:40	20:55	6:50	18:15
	Milled-baked	6:10	16:25	5:00	14:35	4:10	13:35
	Milled-baked-humidified	7:00	18:20	5:45	16:20	4:45	14:15

*Vicat and temperature rise (Tr) set times given in minutes-seconds.

The accelerative potency must be measured at various levels of concentration because the relationship of setting time reduction per unit of accelerator addition is not a straight line function. In making the setting time determinations, the indicated quantity of accelerator, prepared according to Example 1, was added to 200 g aged molding plaster and 180 ml water, and mixed for 7 seconds in a Waring Blendor. Initial slurry temperature was 80°F.

It is clear from the results on the following page that at a low concentration of accelerator, a relatively small increase produces a much greater decrease in setting time than that from an increase of a similar size after the concentration has passed 0.1%. A further increase of accelerator concentration to 1% produced only a minor further reduction in setting time. Accelerator made according to Example 1 was successfully used in wallboard manufacture and also in formulating set stabilized wall plasters. It was one of few materials found capable of shortening the setting time of Keene's cement. The increase in accelerator activity which resulted with a higher proportion of sugar to stucco in the mill charge is illustrated by the example below.

Accelerator added, grams	Vicat, minutes: seconds	Setting time temperature rise, minutes: seconds
0.005	11:30	24:10
0.01	9:15	21:30
0.02	7:15	19:00
0.03	6:30	18:20
0.04	5:55	17:45
0.05	5:40	17:00
0.125	4:20	14:50
0.25	3:30	13:20
0.375	3:10	12:50
0.5	2:45	12:00
1.0	2:20	10:30

Example 3: The method of Example 1 was repeated except that the gypsum-sugar mix in the charge comprised 25% by weight sucrose. The comparative accelerative activity of the products of Examples 1 and 3 was determined by adding equivalent quantities of each accelerator to 200 grams of molding stucco and determining the setting time. The comparative data are shown in the table below.

Accelerator	Grams added to 200 grams molding stucco	Vicat set minutes: seconds	Temperature rise set minutes: seconds
Example 1, 5% sugar	0.05	6:20	17:05
	0.10	5:20	15:30
	0.20	4:30	14:05
Example 3, 25% sugar	0.05	6:00	16:35
	0.10	5:10	15:35
	0.20	4:05	13:40
Control	0.0	32:30	45:05

Zinc Sulfate

D.F. Smith; U.S. Patent 3,316,901; May 2, 1967 has found that zinc sulfate is an excellent set-accelerator for plaster of Paris. Furthermore, when the dry plaster of Paris products contain zinc sulfate and/or the hydrates, the delay and/or variation in the set is much reduced. This, apparently, is due to the fact that when such products are wet in water to make a cast, any ammonium salts therein hydrolyze to produce ammonia and this ammonia combines with Zn^{++} ion to form the relatively stable complex $Zn(NH_3)_4^{++}$ ion. This effect then results in continued hydrolysis of the ammonium salts with substantial reduction in the amount of ammonium salt remaining to delay the set of the plaster. Other compounds having similar effect are $CdSO_4$, $NiSO_4$, $CoSO_4$ and $CuSO_4$.

The zinc or other sulfate may be dissolved in the slurry liquid in amounts from 0.1 to 2.5% (in terms of $ZnSO_4$) of the weight of the plaster, with enough extra ammonia to form the complex ion.

Example 1: 0.7 pound of casein is stirred into 2 pounds of water. Stirring is continued while 3 lb of 28 weight percent aqueous ammonia are added and the stirring continued for about 20 minutes to dissolve the casein. Then 8.5 lb of a 10 weight percent solution of cooked starch are added with thorough stirring. Then 7 lb more of the aqueous ammonia are added and the solution made up with water to a total weight of 96.5 lb. Then 170 lb of finely ground plaster of Paris (steam-calcined gypsum) are stirred into the solution and finally 3.5 lb of finely ground $ZnSO_4 \cdot 6H_2O$ are added and the whole mix thoroughly stirred.

The slurry is applied with a roll coater to a 32 x 28 mesh surgical gauze at a rate to yield a dry plaster of Paris bandage weighing 46 grams per square foot. The coated gauze is then dried at 200° to 260°F in a circulating air oven. The adequacy of drying can be judged visually by observing the absence of wet spots as the bandage emerges from the drier and is confirmed by allowing the dry bandage to stand protected from moisture for 3 to 4 hours and then wetting it in water and smoothing the wet bandage with the hands to assure absence of grittiness in the wet bandage before it sets.

The drying should not be continued substantially beyond the point where the stored and wetted bandage shows no grittiness. The bandage so produced showed a setting time of 3 minutes by the method of Federal Specification GG-B-101d, June 2, 1959, and otherwise met the requirements of this Federal Specification.

Example 2: 1 lb of boric acid (U.S.P.) was dissolved in 11 lb of 28 weight percent aqua ammonia. 50 lb of water are added and then 2.2 lb of polyvinyl acetate homopolymer aqueous dispersion (emulsion) containing 55 weight percent solids (du Pont Elvacet). Three pounds finely ground $ZnSO_4 \cdot 7H_2O$ are then dissolved in the solution and it is made up with water to a total weight of 100 lb. Then 170 lb of finely ground steam calcined plaster of Paris are thoroughly stirred into the solution and the slurry is coated upon gauze as in Example 1. The slurry coated gauze was dried at 240°F as before. The final, dry product showed a setting time of 3½ minutes by the method of Example 1 and otherwise met the Federal Specification cited.

WATER CONTROL AND WORKABILITY

Hydroxyethylcellulose

In a process described by *G. Selden; U.S. Patent 3,243,307; March 29, 1966; assigned to Tile Council of America, Incorporated* a dry mix for producing thinset hydraulic cement mortars having good water retentive properties, good bonding properties and flow characteristics is produced by mixing with the hydraulic cement which may consist of Portland cement with or without sand and with or without other additives such as glycerin, perfume and polyvinyl alcohol, a small amount, up to a maximum of about 4%, of hydroxyethylcellulose. The preferred dry mortar composition, contains finely divided inorganic fibers, such for example as mineral wool, glass fibers, and the like, in amount of from 1 to 2%.

The incorporation of the inorganic fibers improves the thixotropic properties of the mortar, i.e., the use of inorganic fibers prevents sag of the mortar. Using a mortar containing inorganic fibers to bond ceramic tile on a vertical wall, the tile will not tend to slip down the wall during setting of the mortar. In the case of mortar not containing inorganic fibers, it is necessary to first bond the lowermost row of tile to the wall and then bond the successively higher rows so that an underlying row prevents sag or travel of a higher tile on the wall.

Thus the thin-set hydraulic cement dry mixture of this process consists essentially of from 24 to 99.4% hydraulic cement; from 0.6 to 4% hydroxyethylcellulose, preferably from 0.6 to 2%; from 0 to 75% sand; from 0 to 2% inorganic fiber, and small amounts of other conventional additives, for example, perfume, glycerin, polyvinyl alcohol, etc.

Example 1:

	Percent
Hydroxyethylcellulose, viscosity 3,500 to 6,500 cp	1.2
Inorganic fiber	1.7
Hydraulic cement	97.1
Total	100.0

Example 2:

	Percent
Hydroxyethylcellulose, viscosity 15,000 cp	0.6
Hydraulic cement	99.4
Total	100.0

Example 3:

	Percent
Hydroxyethylcellulose, viscosity 15,000 cp	1
Polyvinyl alcohol	1
Hydraulic cement	48
Sand	49
Inorganic fiber	1
Total	100

In the above examples, the constituents are mixed dry and the dry mixture is mixed with water in amount to produce a mortar which will remain on a hawk (a plasterer's tool). Approximately two gallons of water are mixed with 40 pounds of the dry mixture; a high-strength thin-set mortar results.

The mortars of this process have the advantage, among others, over mortars previously known containing methylcellulose, that they require an appreciably smaller amount of hydroxyethylcellulose as compared with the amount of methylcellulose, to produce equivalent high-strength mortars. Moreover as noted, mortars containing hydroxyethylcellulose can be applied over a wide temperature range including temperatures of 100° to 110°F and result in bonds of high shear strength.

Extrusion Shaping Using Bentonite Clay and Added Alcohol

A process described by *T. Yano, H. Onishi and I. Tanaka; U.S. Patent 3,872,204; March 18, 1975; assigned to Onoda Cement Co., Ltd., Japan* relates to a method for continuously manufacturing shaped gypsum articles by means of extrusion. The process involves mixing insoluble anhydrite with an ordinary accelerator for the setting, bentonite, methyl or ethyl alcohol and a proper amount of water; kneading the mixture into a doughy paste; continuously shaping the paste by forcing it through an extruder; and cutting the extruded product in proper size, followed by curing and setting.

For extrusion of paste of insoluble anhydrite containing a solidification accelerator, it is necessary to render the paste water-retentive and plasticizable, and impregnate the paste with such a shaping agent that is least likely to decrease the mechanical strength of the final product. As the result of various experiments, it has been found that clays, particularly bentonite, is most suitable as such a shaping agent.

The reason is that addition of small amounts of bentonite, for example, 3 to 10 parts by weight, preferably 4 to 7 parts by weight based on 100 parts of insoluble anhydrite imparts water retentivity and plasticity to the paste to permit its very smooth extrusion, and moreover does not exert any substantial harmful effect on the mechanical strength of final product.

However, paste prepared by adding bentonite to insoluble anhydrite containing a setting accelerator presents difficulties in obtaining a final product with dimensional precision, because addition of bentonite which tends to present noticeable swelling gives rise to the prominent initial expansion of the paste. Further, the swelling of bentonite leads to the occurrence of fine hair cracks in the paste, rendering the final product considerably brittle.

Studies have been conducted to eliminate the above-mentioned drawbacks and it has been found that addition of both bentonite and methyl or ethyl alcohol to the paste prominently minimizes its initial expansion and saves the paste from fine hair cracks, and further that incorporation of the alcohol decreases the drying shrinkage of wet bentonite. Impregnation of bentonite and methyl or ethyl alcohol provides an excellent final product which is prominently elevated in dimensional precision and free from fine hair cracks and displays great mechanical strength and low hygroscopicity.

Example 1: One hundred parts of insoluble anhydrite powders derived from phosphogypsum, which were calcined 10 minutes at a temperature of 900°C, were mixed 25 minutes with 5 parts of bentonite and 1.5 parts of polyvinyl chloride fibers of 15 deniers and 10 mm long in a Nautamixer so as to fully disperse the fibers in the calcined powders. To this mixture were added 1.5 parts (anhydride base) of potassium alum, 1.4 parts of methyl alcohol and 23 parts of water. The mass was kneaded 2 minutes first in an Eirich-type batch mixer by rotating it at high speed, and then in a clay kneader until there was formed a suitable doughy paste.

The paste was supplied to a screw-type extruder and passed through a degassing vacuum chamber evacuated to -720 mm Hg, and then through a hollow die 60 mm high and 480 mm wide at a pressure of 10 to 12 kg/cm^2, obtaining a continuously extruded band of the dough. The band was moved at a speed of 80 m/hr by a belt conveyer, and, one hour after extrusion, was continuously cut off in proper length. Two hours after extrusion each cut panel solidified to such an extent that it could be carried by a fork lift truck. The panels were left in a product yard in heaps each consisting of ten superposed pieces. Seven days later, the physical properties of the panels were tested, the results being presented in the table below.

Physical Properties of the Product

Compressive strength	3020 t/m²
Bending strength	261 kg-m
Shrinkage	0.035%
Hygroscopicity	7.6%
Dimensional precision	1.1 mm
Sound absorptivity	37.5% (500 Hz)
Permeation loss of sound	–35 dB (500 Hz)
Apparent bulk density	735 kg/m³
Fire resistance (30 min)	Successful
Impact resistance	10 kg-m

Example 2: One hundred parts of insoluble anhydrite powders used in Example 1 were mixed 10 minutes in a Nautamixer with 5 parts of bentonite and 3.0 parts of chopped strand glass fiber of 6 deniers and 10 mm long so as to fully disperse the fibers in the calcined powders. To the mixture were added 1.5 parts (anhydride base) of potassium alum, 1.6 parts of methyl alcohol and 23 parts of water.

The mass was mixed 10 minutes in a special paddle-type mixer by rotating it sufficiently slowly to prevent the glass fibers from being broken, and then in a clay kneader until there was obtained a suitable dough. Later, there were produced panels having the same shape as those of Example 1 through the same operation. Tests were made on the physical properties of the panels which had been allowed to stand for seven days after extrusion, the results being given in the table below.

Physical Properties of the Product

Compressive strength	2490 t/m²
Bending strength	288 kg-m
Shrinkage	0.040%
Hygroscopicity	7.8%
Dimensional precision	0.9 mm
Sound absorptivity	37.5% (500 Hz)
Permeation loss of sound	–35 dB (500 Hz)
Apparent bulk density	735 kg/m³
Fire resistance (30 min)	Successful
Impact resistance	6 kg-m

Latex and Fibers for Improved Extrusion

J.C. Yang; U.S. Patent 3,852,083; December 3, 1974 describes a process for the production of fiber-containing plastic-formed plaster of Paris products which have superior hardness, surface smoothness and flexural strength as well as good extrusion characteristics and physical integrity. In the process a composition comprising specified concentrations of plaster of Paris, natural or synthetic fiber, latex, hydromodifier and water is formed into a plastic tractable mixture which may be extruded, molded, or otherwise plastic-formed under pressure and then set to produce the desired shaped product. Typical compositional ranges for the products are as shown on the following page.

Suitable Concentrations
(wt % dry basis)

Component	Broad	Preferred
Plaster of Paris	30 - 80	45 - 75; 50 - 70
Fiber: natural	2 - 30	3 - 23; 5 - 20
synthetic	0.1 - 2.5	0.3 - 1.5; 0.5 - 1.2
Hydromodifier	0.50 - 10	0.10 - 1.5; 0.15 - 0.70
Latex	0.5 - 10.0	0.9 - 6.0; 1.0 - 5.0
Siliceous auxiliary cementing agent	0 - 60	5 - 50; 5 - 30 or 30 - 50

The concentrations are individually selected to total 100 weight percent, and also include water in a water:solids weight ratio in the range of from about 0.20:1 to 0.80:1, preferably from 0.25:1 to 0.65:1; with the proportionate amounts being such as to cause the mixture to be a plastic tractable mixture having shape-retaining characteristics.

It has been found that optimum results are obtained when natural inorganic fibers are used. Chief among these is asbestos. All types and grades of asbestos fibers may be used either alone or in mixtures, but chrysotile asbestos (serpentine group) is preferred. Other types of asbestos are also quite suitable including the crocidolite and amosite varieties of the amphibole group of fibers.

Another suitable class of fibers are the natural organic fibers. These include such materials as fibers of cotton, paper, wood, wool, sisal, jute, hemp, flax, and the like. It is generally preferable to use a lower concentration of water with these fibers than with the asbestos or glass fibers. Similarly, organic synthetic fibers such as nylon, rayon or other synthetic polymers capable of being formed into strands which can be chopped to size may be used.

Although the natural polymer emulsions are satisfactory for use in this process, it is preferred to use the synthetic polymer emulsions. These include a wide variety of rubber-like polymer materials some of which are closely related to natural rubber. Those related to natural rubber generally include the butadiene polymers and the various polymers of butadiene derivatives. These include butadiene-styrene copolymer, butadiene-acrylonitrile copolymer, polychloroprene (Neoprene), as well as the many variations and derivatives of these materials such as the various GRS-type synthetic rubbers and the Buna N and Buna S rubbers.

Methyl cellulose, hydroxypropylmethyl cellulose, and ethylene oxide polymers are preferred hydromodifiers. Of those commercially available, the methyl cellulose material available as Methocel MC (4,000 cp), the hydroxypropylmethyl cellulose material available as Methocel 65 HG (4,000 cp), and the ethylene oxide polymer available as Polyox WSR 301 have been found to give highest strength results; however, all grades of commercial methyl cellulose and substituted methyl cellulose, such as hydroxypropylmethyl cellulose appear to be usable. Other hydromodifiers which are particularly useful include hydroxyethyl cellulose, e.g., Natrosol 250, partially hydrolyzed acrylamide polymer, e.g., Separan NP 10, and acrylamide polymer, e.g., Superfloc 16.

The mixing and handling of the compositions prior to extrusion or other mechanical plastic forming may be carried out in a variety of ways. Generally, it is advisable to form a dry furnish, i.e., mixture, of all dry ingredients as a preliminary step in the operation. However, if special mixing or conveying equipment is available, it is possible to mix the plaster of Paris, latex, fiber, hydromodifier, water and other ingredients, if any, together all at once and pass this mixture immediately to the extruding or other mechanical forming equipment. This type of arrangement can be used to provide a completely continuous operation in which the individual ingredients for the final products are charged into suitable hoppers or containers in the equipment line, from which they are discharged by suitable weighing or measuring devices into a continuous mixer and thence to the extruder or other plastic forming equipment.

AIR ENTRAINMENT

Cellulosics and Sodium Pentachlorophenolate

K.E. Lindgren, K.K.G. Andersson, and H.G.R. Johansson; U.S. Patent 3,489,582; January 13, 1970; assigned to Aktiebolaget Gullhogens Bruk, Sweden describe a binding agent which can be used for the production of all required qualities of mortar by adding different quantities of sand. The agent is made up of hydraulic cement and from 5 to 50% by weight of a finely divided inorganic filler such as finely divided stone. Surface tension influencing air entraining agents may be added as also may consistency improving agents. The following examples illustrate the process.

Example 1: 80% Portland cement clinkers, 5% of gypsum, 5% of powdered limestone with 50% having a grain size of less than 2 μm, 10% of powdered limestone with 100% having a grain size finer than 500 μm were ground together with 0.03% of tall oil and 0.025% of ethylhydroxyethylcellulose with a viscosity of 8,000 to 12,000 cp and 0.003% of sodium pentachlorophenolate in a ball mill.

Example 2:

	Percent
Portland cement	90
Powdered limestone with 30% having a grain size finer than 1 μm	10
Tall oil	0.03

Example 3:

	Percent
Gypsum-slag cement	80
Powdered limestone with 50% finer than 2 μm	20
Tall oil	0.03

Example 4:

	Percent
Portland cement	50
Powdered quartz with 100% having a grain size finer than 500 μm	45
Powdered quartz with 50% having a grain size finer than 2 μm	5
Ethylhydroxyethylcellulose with viscosity 8,000 to 12,000 cp	0.06
Tall oil	0.03

The binding compositions of Examples 1 through 4 may with advantage be used for the production of ready-to-use plaster and mortar, so-called dry mortar which is mixed with water at the work site. In order to illustrate the effect of the process, a binding agent according to Example 1 is compared with various binding agent mixtures containing Portland cement and powdered lime wherein the mortar is produced by mixing with standard sand having the following distribution of grain sizes.

Percent	Millimeter
100	<2
67	<1
33	<0.5
12	<0.15
2	<0.08

The workability of the mortar is given as the water retaining ability and the strength at an age of 28 days is determined according to standard methods. The results are set out below. In this table the qualities of the mortar are indicated by A through D, where A indicates the highest quality; K stands for lime and C stands for cement, and the numbers in order give the parts by volume of lime, cement, and sand.

Mortar Quality	Mortar Designation	Binding Agent/Sand Ratio by Volume	Water Retaining Ability, %	Strength at 28 Days, kg/cm^2 Compression	Flexing
A	KC 1:4:20	1:4	31	266	50
B	KC 1:1:8	1:4	56	105	31
C	KC 2:1:12	1:4	67	60	22
D	K 1:5	1:5	80	<5	<2
A	Ex. 1 according to the process	1:2.5	76	218	46
B	Ex. 1 according to the process	1:5	63	116	35
C	Ex. 1 according to the process	1:7	56	83	28
D	Ex. 1 according to the process	1:9	51	70	25

The table shows that the mortar according to the process may be used in all classes of mortar quality by mixing in 2.5 to 9 times as much sand without losing the acceptable workability and strength. The water retentivity, which must not fall below 30%, is seen to lie very near the acceptable limit for KC 1:4:20 mortar while all mortars according to the process show very good values. In standard class C one can reduce the amount of binding agent by 35% according to the process. This class is the one normally used in moderate climate zones.

In related work *B.S.J. Ericson; U.S. Patent 3,215,549; November 2, 1965; assigned to Mo och Domsjö Aktiebolag, Sweden* has found that when a water-soluble nonionic cellulose ether is added together with a water-soluble salt of a chlorophenol to a mortar or paste, i.e., a mixture containing in addition to water one or more filler materials and optionally one or more inorganic binders, one obtains a mortar or paste having a considerably improved water retention, adhesion and strength properties.

The chlorinated phenol derivative added according to the process may be any water-soluble salt of ortho-, meta- or parachlorophenol, 2,3-, 2,4-, 2,5-, 2,6-, 3,4- or 3,5-dichlorophenol, 2,3,4-, 2,3,5-, 2,3,6-, 2,4,5-, 2,4,6- or 3,4,5-trichlorophenol, 2,3,4,5-, 2,3,4,6- or 2,3,5,6-tetrachlorophenol or pentachlorophenol. Pentachlorophenol is preferred, because the desirable effects can be obtained with a lower addition of this product and also because the lower chlorinated phenols have a strong odor which makes them undesirable for many uses. Water-soluble salts of these chlorinated phenols include primarily the alkali metal and ammonium salts, in particular sodium, potassium and lithium salts.

Water-Soluble Salts of Styrene-Maleic Anhydride Copolymers

In a process described by *R.J. Pratt, M.A. Stram, and D.W. Young; U.S. Patent 3,563,777; February 16, 1971; assigned to Sinclair Research, Inc.* a plaster composition is prepared having superior air entraining properties and reduced bulk density subsequent to setting. The composition comprises a settable gypsum plaster and a minor amount of a water-soluble salt of the ester formed by reacting styrene-maleic anhydride copolymer with a monohydric alcohol of about 1 to 12 carbon atoms. Amounts of about 0.05 to 5 weight percent of the ester salt, based on the weight of the plaster, have been found to be effective in increasing air entrainment and thereby lowering bulk density in the plaster composition.

Alkyl Benzene Sulfonate and Alcohol Frothing Agent

A process described by *R.B. Doan, R.C. Taylor and M.H. Reynolds; U.S. Patent 3,912,528; October 14, 1975; assigned to Atlantic Richfield Company* relates to gypsum slurry and to a method for producing gypsum board which utilizes as a frothing agent a mixture comprising 5 to 40 weight percent of a compound selected from the group consisting of monohydric alcohols, glycols, glycol ethers, ketones and esters having 9 or less carbon atoms, up to 40 weight percent water and 20 to 70 weight percent of an alkyl benzene sulfonate having a branched or linear secondary alkyl side chain with an average carbon content ranging from 6 to 15 and a maximum carbon number spread of 8 carbon atoms.

Preferably, the average carbon content ranges from 9.5 to 12.5 and the carbon number spread is a maximum of 5 carbon atoms. This frothing agent is generally utilized in amounts ranging from 0.01 to 0.1 part per 100 parts by weight gypsum.

With respect to the examples which follow, water was used having a hardness of 500 parts per million and a temperature of 40°F. These are generally considered extreme conditions, not conducive to foam formation. As a basis for comparison, 2 grams of the frothing agent was placed in solution with distilled water to form 100 ml of a 2% solution. 5 ml of this solution was mixed with 155 ml of the hard water and 8 g of plaster of Paris was added.

Upon mixing, one could note the time in seconds to generate a specific amount of foam such as 500, 600, or 700 cc. At 120 seconds, a final reading of the amount of foam generated was taken. To determine stability of the foam, readings were taken on the amount of water which settled out at one, two, and five minute intervals after the mixer was stopped. The more rapid the rate of accumulation of water would tend to indicate the breakdown of the froth or foam.

Example 1: With respect to a solution using only alkyl benzene sulfonate as the frothing agent under the aforementioned conditions, it took 55 seconds to produce 500 cc of foam; and at the end of 2 minutes the total foam produced was 550 cc. With respect to drainage, over a 5-minute period, it was noted that 40 cc of water drained out in 1 minute, 75 cc in 2 minutes, and 120 cc in 5 minutes, leaving a final foam height of 545 ml.

Examples 1 through 31: These were made using a frothing agent containing 60 weight percent of a linear alkyl benzene sulfonate having an average of 10.3 carbon atoms in the chain with a maximum spread of 3 carbon atoms. The additives used in Examples 2 through 31 and the amount of the additive is indicated by weight percent.

Where 40 weight percent of the additive is indicated, water is generally present as a diluent as indicated in the procedure outlined above. The results from Example 1 are repeated in tabular form for ease of comparison with the runs using the various additives. Examples 30 and 31 utilized as the frothing agent, 100% of two of the more effective additives with no alkyl benzene sulfonate present and no foam generated.

Example	Additive	Wt. %	Sec. to Generate cc. 500	600	700	Total Foam	Drainage at 1 m.	2 m.	5 m.
1	None		55			550	40	75	120
2	Ethanol	5	41			560	30	60	112
3	Ethanol	20	34			585	15	40	100
4	Isopropanol	5	35			590	10	35	100
5	Isopropanol	20	21	47	120	700	8	10	55
6	Isopropanol	40	8	36	75	780	5	10	40
7	n-Butanol	20	21	40	86	760	3	10	45
8	Isobutanol	20	21	40	90	730	5	10	50
9	t-Butyl Alcohol	5	28	82		650	5	15	70
10	t-Butyl Alcohol	10	22	48	112	700	5	10	58
11	t-Butyl Alcohol	40	14	38	57	820	3	10	40
12	Crude t.B.A.	20	19	37	52	770	5	10	48
13	Crude t.B.A.	20	83	56	103	710	5	10	40
14	Amyl Alcohol	20	31	70		680	8	15	70
15	Ethylene Glycol	20	50			550	40	68	118
16	Propylene Glycol	5	43			540	35	68	118
17	Propylene Glycol	40	32			570	20	50	108
18	Dipropylene Glycol	5	35			580	15	40	100
19	Dipropylene Glycol	20	17	47		695	5	10	60
20	Dipropylene Glycol	40	22	40	90	730	5	10	42
21	Tripropylene Glycol	20	28	47	106	700	5	10	52
22	D.P.G. BTMS.[1]	10	37	83		660	5	15	70
23	D.P.G. BTMS/t.B.A.	10/10	17	37	91	740	5	10	45
24	Glycerine	20	53			550	38	68	118
25	Hexylene Glycol	20	22	40	90	730	3	10	45
26	Acetone	20	30	100		600	10	20	80
27	Methylethyl Ketone	20	30	63		690	5	10	50
28	Ethyl Acetate	20	28	55		650	8	12	60
29	Sodium N.T.A.	10	48			540	35	60	115
30	No LAS - t-Butyl Alc	100	—	20cc Foam Disappeared Immediately					
31	No LAS - D.P. Glycol	100	—	No Foam					

[1]Still bottoms containing approximately
58% dipropylene glycol
35% tripropylene glycol
7% heavier glycols and unknown

Examples 32 through 39: These were made using a frothing agent containing 60 weight percent of a linear alkyl benzene sulfonate having an average of 11.3 carbon atoms in the chain with a maximum spread of 4 carbon atoms. The procedures followed were as indicated for Examples 1 through 31. For comparative purposes, Example 32 was run using as the frothing agent a linear alkyl benzene sulfonate without the additives of this process.

Example	Additive	Wt. %	Sec. to Generate cc. 500	600	700	Total Foam	Drainage at 1 m.	2 m.	5 m.
32	None					300	150	150	152
33	t-Butyl Alcohol	20	31	71		660	5	15	70
34	Crude t-B.A.	20	45	92		620	8	15	70
35	Dipropylene Glycol	20	45			540	30	55	110
36	D.P.G. BTMS[2]	20	56			580	10	20	80
37	D.P.G. BTMS[2]	20	41			595	10	25	90
38	Tripropylene Glycol	20	33	83		600	10	15	70
39	Hexylene Glycol	20	31	68		610	10	15	70

[2]See footnote 1 from Examples 1 through 31.

WATER RESISTANCE

Resinous Polymer

J. Williams; U.S. Patent 3,869,415; March 4, 1975; assigned to Temec Limited, England describes a plaster composition having water-resistance properties comprising 30 to 60% by weight of calcium sulfate plaster solids, 0.25 to 15% by weight of a water-proofing resin selected from the group consisting of vinyl toluene-butadiene copolymers, polyvinyl chloride, polyvinyl acetate, petroleum and coal tar hydrocarbon polymer resins and styrene and acrylic copolymers, 4 to 48% of an organic solvent for the resin and 20 to 50% by weight of water. The composition preferably includes a cellulose ether suspending agent for the resin solution. Up to 50% by weight of the plaster solids may be replaced by a hydraulic cement.

The resin is used in the compositions to impart water-resistance to the set compositions. A number of hydrophobic natural, synthetic and modified resins can be used for this purpose, e.g., ester gums, maleic anhydride modified rosin, phenol/formaldehyde modified rosin, rosin Ca/Zn resinate, copolymers of styrene, vinyltoluene/butadiene copolymers, polyvinyl chloride, polyvinyl acetate, acrylic copolymers, and petroleum and coal tar hydrocarbon resins.

The most suitable resins are the hydrocarbon resins such as polyindene, which is produced by the polymerization of unsaturates derived from the deep cracking of petroleum, or coumarone-indene resins derived from the coal tar naphtha fraction boiling between 168° and 175°C. The preferred resins are hard solids at ambient temperatures and have a low molecular weight with melting points between 80° and 140°C (ring and ball); they are also neutral and unsaponifiable.

They are furthermore capable of producing solutions of low viscosity with a high resin solids content when dissolved in an appropriate solvent and are highly resistant to alkali and to any change caused by atmospheric exposure. The following are illustrative compositions of the process.

Example	Percent by Weight
(1) Water	42.50
Methyl hydroxyethyl cellulose (Methofas) — medium viscosity range	0.5
Thiophene-free xylene	8.0
Hydrocarbon resin (Piccopale) (melting point 100 ± 3°C)	4.0
Plaster solids	45.0

(continued)

Example		Percent by Weight
(2)	Water	42.875
	Hydroxypropyl methyl cellulose (Methofas)	0.125
	Thiophene-free xylene	8.0
	Hydrocarbon resin (Necires)	4.0
	Plaster solids	45.0
(3)	Water	42.0
	Ethyl hydroxyethyl cellulose (Modocoll E)	1.0
	Commercially pure xylene	8.0
	Oxidized polyindene resin (Panarez)	4.0
	Plaster solids	45.0
(4)	Water	32.0
	Hydroxyethyl cellulose (Natrosol)	0.375
	Hydrocarbon resin (Epok)	4.0
	Commercially pure xylene	8.0
	Plaster solids	55.625
(5)	Water	45.875
	Substituted cellulose derivative (Natrasol)	0.125
	Coumarone-indene resin (Epok)	2.0
	Thiophene-free xylene	4.0
	Plaster solids	48.0

The plaster compositions can be precolored to a particular shade of color if required. A fungicide or bactericide can also be incorporated into the predispersed liquids if required to protect the liquids from bacterial attacks during storage. Although the organomercury compounds can be used, it is preferable not to use these compounds as fungicide or bactericide additives because of their high toxicity. Satisfactory fungicides and bactericides include sodium pentachlorophenate, p-chloro-m-cresol, sodium ortho-phenylphenate, terpineol, etc. Generally speaking, the preferred compositions of this process comprise the following components in the indicated proportions:

	Percent by Weight*
Water	20 - 50
Substituted cellulose derivative	0 - 2
Organic solvent	4 - 48
Polymer resin	0.25 - 15
Plaster solids	30 - 60
Cement	0 - 50

*Based on the amount of plaster solids

Calcium Sulfate, Boric Acid and Glycerol

A process described by *K.R. Larson; U.S. Patent 3,393,116; July 16, 1968; assigned to United States Gypsum Company* involves a composition comprising calcium sulfate hemihydrate, boric acid and glycerol which is convertible into a formable mixture that is capable of setting to a hard gypsum upon mixture with an amount of water in excess of that required to hydrate the hemihydrate to the dihydrate. The hard gypsum product is characterized by substantial resistance to dehydration at a temperature of 150°F. It has been found that if calcium sulfate hemihydrate is combined with minute quantities comprising approximately 0.05 to 0.65% of boric acid and 0.05 to 1.25% of glycerol based on the weight of the hemihydrate, gypsum products formed from such combination after mixing the same with a quantity of water necessary to effect formation of the dihydrate crystalline structure will tenaciously retain such water of crystallization.

Since substantially all major sources of gypsum rock contain some impurities, such impurities should be deducted from the hemihydrate weight in the course of ascertaining the weight of glycerol and boric acid additives based thereon. Thus, although gypsum water of crystallization is normally driven off at a temperature of about 105° to 110°F, gypsum products formed in combination with the above additives will release substantially no water whatsoever, even after being subjected to a temperature of 150°F for two hours. Such gypsum products thus are most gainfully employed in radiant heating installations in which temperatures of 125°F are imparted to gypsum materials disposed adjacent the radiant heating means.

Organohydrogenpolysiloxane Emulsions

In a process described by *S. Nitzsche, E.H. Pirson, and M. Roth; U.S. Patent 3,455,710; July 15, 1969; assigned to Wacker-Chemie GmbH, Germany* masonry articles formed by standard techniques employing gypsum materials are rendered water-repellent by incorporating into the gypsum material an aqueous emulsion of an organohydrogenpolysiloxane.

Example 1: Emulsion A is prepared as follows. 80 g of a fluid organopolysiloxane consisting of about 80 mol percent monomethylsiloxane and 20 mol percent dimethylsiloxane units with 8% by weight ethoxy groups and 6% by weight of Si-bonded hydroxyl groups as well as 20 g of a methylhydrogenpolysiloxane with a viscosity of 25 cs/25°C obtained by pouring a 30% by weight solution of methyldichlorosilane (CH_3SiHCl_2) in toluene into the five-fold quantity by weight of water and distilling off the toluene from the organic layer, are dissolved in 100 g toluene.

This solution is emulsified in 299 g water, which has been mixed with 1 g of a commercially available emulsifier, nonylphenylpolyoxyethylene ether. An Emulsion B is prepared in the same manner for purposes of comparison, using 100 g of the resins previously described, of monomethylsiloxane and dimethylsiloxane units free of Si-bonded hydrogen instead of the organopolysiloxane mixture of 80 g organopolysiloxane without Si-bonded hydrogen and 20 g organopolysiloxane with Si-bonded hydrogen.

50 g of building gypsum (building gypsum is prepared by heating naturally occurring gypsum, $CaSO_4 \cdot 2H_2O$ to 180° to 700°C) are mixed with 35 g of a mixture of water and various quantities of Emulsions A and B. The mixtures thus obtained are poured into lead cups. After one hour the water repellency of the gypsum slabs thus obtained is measured from the rate of penetration of water droplets of 0.5 ml each. The following results are obtained.

Emulsion	Weight Percent Emulsion Based on Dry Gypsum	Rate of Penetration, sec
Control	0.0	6
A	0.5	5,400
A	1.0	7,380
A	2.0	11,880

(continued)

Emulsion	Weight Percent Emulsion Based on Dry Gypsum	Rate of Penetration, sec
A	4.0	16,980
B	1.0	13
B	2.0	15
B	4.0	78

Example 2: An emulsion is prepared in the manner described in Example 1 for Emulsion A, with the exception that instead of 80 g of Si-bonded hydrogen-free organopolysiloxane, 60 g of this resin are used and instead of 20 g of methylhydrogenpolysiloxane, 40 g of methylhydrogenpolysiloxane are used. The preparation of gypsum slabs and testing of the water repellency of the slabs is carried out as described in Example 1. The following results are obtained:

Weight Percent Emulsion Based on Dry Gypsum	Rate of Penetration, sec
0 (control)	7
0.2	168
2.0	19,500

Example 3: An emulsion is prepared in the manner described in Example 1 for Emulsion A, with the exception that instead of using 80 g organopolysiloxane free of Si-bonded hydrogen, 90 g of this resin are used and instead of 20 g of methylhydrogenpolysiloxane, 10 g of methylhydrogenpolysiloxane are used. The preparation of slabs of gypsum and the testing of the water repellency of the slabs is carried out as described in Example 1. The results obtained are shown below.

Weight Percent Emulsion Based on Dry Gypsum	Rate of Penetration, sec
0 (control)	7
0.2	28
2.0	3,420

Polyvinyl Alcohol and Thermosetting Resins

According to a process described by *A. Nishioka, M. Sakakibara, A. Itabashi and K. Komatsu; U.S. Patent 3,915,919; October 28, 1975; assigned to Japan Synthetic Rubber Co., Ltd., Japan* a gypsum composition which is obtained by molding and drying an aqueous slurry comprising calcined gypsum, at least one polyvinyl alcohol, at least one thermosetting resin and at least one metal compound is excellent in wet strength and wet hardness when immersed in water.

The gypsum composition of this process may be used as a building material and a decorative material. When used in the form of a so-called gypsum board as a ceiling board or a wall board for the purposes of sound absorption and fire-proofing, the shaped articles obtained from the composition have an advantage in that they need not be overlaid with paper on either side, as with conventional gypsum boards, and can be used as produced.

In the examples, the testing of physical properties of the gypsum composition was carried out in the following way: A test specimen was prepared by drying a solidified shaped article in an air stream at 60°C for 48 hours, and then keeping the dried article at 20°C and 45 to 55% relative humidity for 2 days or more. The testing for flexural strength and compressive strength was carried out according to JIS R 5201. Izod impact strength was tested on a test piece of 1.27 by 1.27 by 6.35 cm (unnotched). Following the procedure of JIS-K 5401, pencil hardness was expressed as the minimum hardness of a pencil which can scrape off the surface of the gypsum composition. The testing was conducted with pencils of 9H to 6B by use of a pencil scratch tester (9H shows the maximum hardness and 6B the minimum hardness).

Example 1: In 500 ml of water were dissolved 10 g of PVA having a degree of polymerization of about 1,500 and a degree of saponification of about 86% and 5 g (solids) of a melamine-formaldehyde precondensation product (synthesized by suspending melamine in formalin in a ratio of 3 mols of formaldehyde per mol of melamine and heating at 80°C for 20 minutes).

To the resulting solution was added 0.454 g of cupric acetate (corresponding to 1 mol per 100 hydroxy groups in PVA which is referred to as Cu/OH = 1/100) and uniformly dissolved. The resulting solution was mixed with 500 g of calcined gypsum, stirred for two minutes by means of a mixer, poured into a mold, and allowed to harden. In the table are shown properties of the hardened article which had been dried at 60°C for 48 hours.

Example No.	Specific gravity	Compressive strength (kg/cm^2)	Flexural strength (kg/cm^2)	Surface hardness (pencil)	Water absorption (%)	Surface hardness after immersion	Change in appearance after immersion in water at 20°C. for 1 hour
1	0.90	102	85	8H	4	2H	None
2	0.90	99	87	8H	6	2H	None
3	0.90	100	84	8H	7	2H	None
4	0.90	110	90	9H	4	3H	None
5	0.90	100	85	8H	3	2H	None
6	0.90	98	85	8H	5	2H	None
Ref. 1	0.90	80	70	6H	50	**6B**	**Swell-up**
Ref. 2	0.90	45	50	3H	50	**6B**	**Bleeding** of PVA

Example 2: In 500 ml of water were dissolved 10 g of the PVA used in Example 1 and 5 g (solids) of a melamine-formaldehyde resin emulsion (prepared by suspending melamine in formalin in a ratio of 3 mols of formaldehyde per mol of melamine and heating at 80°C for 1 hour). To the resulting solution were added 11 ml of a 5% aqueous solution of titanium sulfate (corresponding to Ti/OH = 1/100) and 500 g of calcined gypsum. A molded article was obtained from the mixture according to the procedure mentioned in Example 1. In the table are shown properties of the molded article.

Example 3: In 500 ml of water were dissolved 10 g of the same PVA as used in Example 1 and 5 g (solids) of a water-soluble urea-formaldehyde resin (synthesized from a mixture of 1 mol of urea and 1.6 mols of formaldehyde, which had been adjusted to a pH of 7 with sodium hydroxide, by heating at 100°C for 45 minutes). To the resulting solution were added 11 ml of a 5% aqueous solution of titanium sulfate and 500 g of calcined gypsum. A molded article was obtained from the mixture in a manner similar to that in Example 1. In the table are shown properties of the molded article.

Example 4: The same procedure as in Example 1 was repeated, with the exception that calcium acetate was substituted for the cupric acetate to obtain a gypsum composition. In the table are shown properties of the molded article.

Example 5: The same procedure as in Example 1 was repeated, except that silicon oxide was substituted for the cupric acetate to obtain a gypsum composition. In the table are shown properties of the molded article.

Example 6: The same procedure as in Example 1 was repeated, except that magnesium iodide was substituted for the cupric acetate to obtain a gypsum composition. In the table are shown properties of the molded article.

Referential Example 1: In a manner similar to that in Example 1, a gypsum molded article was prepared from 15 g of PVA, 0.454 g of cupric acetate, 500 g of calcined gypsum, and 500 ml of water. The test results of properties of the article were as shown in the table.

Referential Example 2: In a manner similar to that in Example 1, a gypsum molded article was prepared from 10 g of PVA, 5 g of melamine-formaldehyde resin (Sumirez Resin 613), 500 g of calcined gypsum, and 500 ml of water. The test results of properties of the article were as shown in the table.

Vinyl Monomer and Sulfite Ion

T. Yamaguchi, T. Ono and H. Hoshi; U.S. Patent 3,950,295; April 13, 1976; assigned to Mitsui Toatsu Chemicals, Incorporated, Japan describe a hardenable molding composition comprising hydratable gypsum, a vinyl monomer, a source of hydrogen sulfite ions and water. With this molding composition a hardened product can be obtained having high strength due to the combined effect of the hydraulic reaction of the gypsum and the polymerization reaction of the vinyl monomer.

The term vinyl monomer means a compound capable of polymerization into a vinyl polymer in the presence of a radical polymerization initiator, for example, acrylic acid, methacrylic acid, acrylate, methacrylate, acrylic ester, methacrylic ester, acrylamide, N-methylolacrylamide, acrylonitrile, styrene, vinyl chloride, vinyl acetate, divinylbenzene, etc. The quantity of this monomer is not particularly limited, but in case high strength and fire-resistance are aimed at, it is desirable to select an amount within a range of 1 to 15 parts by weight for 100 parts by weight of gypsum.

The composition can be prepared very simply compared to the prior art impregnation method. It can be prepared by merely kneading the requisite components together without the need for any special procedures such as heat treatment to obtain the hardened product or any special equipment. Also, since a hardened product of high strength is obtained with a small polymer content, it is possible to greatly improve the fire resistance.

The composition can be used exactly in the same manner as the prior art resin impregnated gypsum. For example, it may be used as construction material, for industrial art work and for daily commodities, sundry goods and furniture. Also, it can find useful application as a base material and as plaster.

Example 1: 2.0 g of methyl methacrylate were suspended in 55 ml of water at room temperature. 2.0 g of sulfur dioxide were then blown into the monomer suspension, and 100 g of half dehydrated gypsum were further added. The resultant mixture was kneaded for 2 minutes into a paste and then quickly charged into a mold and left there. Subsequently, unreacted monomer was removed under reduced pressure and the bending strength and compression strength of the resultant product were measured. For comparison, gypsum alone was added to 55 ml of water at room temperature and the paste prepared by kneading the resultant mixture for 2 minutes was quickly charged into the same mold and left there. Subsequently, the bending strength and compression strength of the resultant product were measured.

	Curing Time (days)	Bending Strength (kg/cm^2)	Compression Strength (kg/cm^2)
Composition of process	1	81	167
	2	80	170
	7	87	179
Comparison composition	1	38	81
	2	36	79

The measured values in the table are averages of 5 to 8 measurements.

Example 2: 2.0 g of methyl methacrylate were suspended in 50 ml of water at room temperature, and 6.0 g of sulfur dioxide were blown into and dissolved in the monomer suspension.

Then, 80 g of half dehydrated gypsum and 20 g of calcium hydroxide were added to the resultant reaction liquid and the resultant mixture was kneaded for 2 minutes into a paste and quickly charged into a mold and left there. Subsequently, unreacted monomer was removed and the bending strength and compression strength of the resultant product were measured.

For comparison, 80 g of gypsum and 20 g of calcium hydroxide were added to 50 ml of water at room temperature and the paste prepared by kneading the resultant mixture for 2 minutes was quickly charged into the same mold and left there. Subsequently, the bending strength and compression strength of the resultant product were measured.

	Curing Time (days)	Bending Strength (kg/cm^2)	Compression Strength (kg/cm^2)
Composition of process	1	71	149
	4	74	154
	7	77	161
Comparison composition	1	32	80
	7	36	77

The measured values in the table are averages of 5 to 8 measurements.

Example 3: 0.6 g of polyvinylidene chloride and 2 g of polyvinylidene chloride-polyvinyl chloride latex were added to 55 ml of water solution containing 0.2 g of a nonionic surface active agent.

Then, 2.0 g of methyl methacrylate monomer were added to form an emulsion and 2.0 g of sulfur dioxide were blown into and dissolved in the emulsion, to which were further added 100 g of half-dehydrated gypsum. The resultant mixture was kneaded for 2 minutes and then quickly charged into a mold and left there. Subsequently, unreacted monomer was removed and the bending strength and compression strength of the resultant product were measured.

Curing Time (days)	Bending Strength (kg/cm^2)	Compression Strength (kg/cm^2)
1	79	160
4	81	163
7	85	165

The measured values in the table are averages of 5 to 8 measurements.

Example 4: 2 g of the polymer latex used in Example 3, 0.2 g of nonionic surface active agent and several drops of silicone were added to 55 ml of water. Then, 2.0 g of methyl methacrylate monomer were added to form an emulsion and 1.8 g of sulfur dioxide were blown into and dissolved in the emulsion, to which were further added 100 g of half-dehydrated gypsum.

The resultant mixture was then kneaded for 2 minutes and then quickly charged into a mold and left there. Subsequently, unreacted monomer was removed and the bending strength and compression strength of the resultant product were measured.

Curing Time (days)	Bending Strength (kg/cm^2)	Compression Strength (kg/cm^2)
1	101	185
4	116	229

The measured values in the table are averages of 5 to 8 measurements.

Example 5: 2 g of the polymer latex used in Example 3, 0.2 g of nonionic surface active agent and several drops of silicone were added to 55 ml of water and then 2.0 g of methyl methacrylate monomer were added to form an emulsion. 0.6 g of sulfur dioxide were added to the emulsion, to which were further added several drops of hydrogen peroxide and 100 g of half-dehydrated gypsum. The resultant mixture was kneaded for 2 minutes and then quickly charged into a mold and left there. Subsequently, unreacted monomer was removed and the bending strength and compression strength of the resultant product were measured. The measured values in the table are averages of 5 to 8 measurements.

Curing Time (days)	Bending Strength (kg/cm^2)	Compression Strength (kg/cm^2)
1/12	45	89
1	71	150
3	109	193
7	125	246

OTHER ADDITIVES AND PROCESSES

Metal Carbonate-Fatty Acid Product and Lignin Sulfonate-Polyvinyl Acetate

A.S. Minicozzi and A.S. Minicozzi, Jr.; U.S. Patent 3,826,663; July 30, 1974 describe a plaster additive which is a necessary component of the improved lime plaster and cement plaster which comprises component (A) and component (B).

Component (A) consists essentially of the saponified reaction product of at least one alkali metal carbonate with at least one fatty acid having about 16 to 18 carbon atoms, about 5.5 to about 8.5 parts of aluminum potassium sulfate, and about 25 to about 70 parts silica, the total of the reaction product of the alkali metal carbonate and the fatty acid being about equal to the total two other noted components.

Component (A) preferably also contains between about 5 and about 25.5 parts of aluminum silicate (kaolin). When kaolin is included in the additive composition, the total of the reaction product of the carbonate and the fatty acid is preferably about equal to the total of the silica, aluminum potassium sulfate, and aluminum silicate.

Component (B) comprises at least two materials selected from the group consisting of (1) polyvinyl alcohol, (2) polyvinyl acetate, and (3) an alkali metal or ammonium salt of lignin sulfonate. Each of the components (1), (2) and/or (3) when included in component (B) are in an amount of at least 3% of the component (B). The ratio of component (A) to component (B) in the plaster additive is from about 16:1 to about 16:16 and preferably between about 16:2.5 and 16:5. The preparation of component (A) is described in U.S. Patent 3,366,502 issued January 30, 1968.

Component (B) preferably contains at least 30% of an alkali metal or ammonium salt of lignin sulfonate. Lignin sulfonate is generally prepared by the treatment of wood chips with sulfurous acid and is often obtained in admixture with sugars. Compositions containing ammonium lignin sulfonate plus the residual wood sugars are useful as the lignin sulfonate source. Other lignin sulfonates, e.g., the sodium salt, or the acid per se, may be used in the preparation of component (B).

The polyvinyl alcohol is available as a powder and preferably has a molecular weight of between about 3,000 to 15,000, e.g., about 10,000. The polyvinyl acetate is preferably utilized as a solid powder, in the form of an emulsion. It preferably has a molecular weight between about 20,000 and 40,000.

The following are two preferred component (B) compositions: lignin sulfonate, 40 to 60%, and polyvinyl acetate, 60 to 40%; lignin sulfonate, 90 to 97%, and polyvinyl alcohol, 10 to 3%. The preferred component (B) compositions may also contain such additional additives as the alkaline earth and alkali metal salts of long chain fatty acids, such as calcium stearate, and gummy materials such as gum arabic.

A preferred component (B) composition contains between one-half and 3 parts of the lignin sulfonate, between 1 and 5 parts of polyvinyl alcohol, between 1 and 10 parts of polyvinyl acetate, between 2 and 8 parts of calcium stearate, and between 1¼ and 4 parts of gum arabic.

A lime plaster according to the process containing for example 50 parts of hydrated lime, 20 parts of limestone, and 1.5 parts of the additive are blended and packaged. The package is delivered to the building site and stored until needed.

The usual practice is to slake the mixture by soaking in water about one day before it is to be applied. Gauging plaster (plaster of Paris) may be admixed by the plasterer and the plaster applied. The amount of gauging plaster added is usually between 10 and 80 parts per 50 to 80 parts of hydrated lime; the amount added depends upon the setting time of the composition, upon the amount of plaster to be applied in a given time which is partially dependent upon the area of wall surface covered and thickness applied, and in part upon the number of workmen applying plaster from the same source.

In practice it has been found that a single coat may be applied in thicknesses up to three-fourth inch. There is excellent adhesion of the plaster to the wall. There is excellent joining and adhesion of later applied plaster to earlier applied plaster. The plaster is readily worked by the plasterer. It dries readily without checking.

Example 1: Five pounds of component (A) were prepared using the components and following the procedure set forth in Example 1 of U.S. Patent 3,366,502, except that a fatty acid which was a mixture of fatty acids (largely oleic acid and including some glycerides) was used instead of the specified "white olefin."

Example 2: One pound of a component (B) mixture was prepared by dissolving 0.6 lb of polyvinyl acetate and water and then adding 0.4 lb of a commercial ammonium lignin sulfonate containing residual wood sugars to the water and boiling until a homogeneous mixture was prepared. Portions of this solution were then added to portions of the paste of Example 1 and mixed vigorously while the entire mixture was boiling to insure a homogeneous composition. The resulting paste was then dried and pulverized. The ammonium lignin sulfonate used was the product sold under the name "Orzan A." Similar products were prepared using a commercial sodium lignin sulfonate.

Example 3: The procedure of Example 2 was followed except that 0.95 lb of the ammonium lignin sulfonate was admixed with 0.5 lb of polyvinyl alcohol.

Example 4: The procedure of Example 2 was followed by mixing in boiling water one part of ammonium lignin sulfonate, two parts of polyvinyl alcohol, two parts of gum arabic, and four parts of calcium stearate until a homogeneous solution was formed. This was then added to the paste mixture of component (A) described in Example 1.

Example 5: Eight parts of polyvinyl acetate was mixed with two parts of polyvinyl alcohol by adding both to boiling water and mixing until homogeneous. The homogeneous mixture was then added in a desired amount to the paste composition of component (A) described in Example 1.

Example 6: A lime plaster was prepared by mixing 50 parts of hydrated lime; 20 parts of limestone having a density of 80 to 90 lb/ft^3 with 99% retained on a No. 40 sieve screen, at least 70% retained on a No. 60 sieve screen and with at least 90% retained on a No. 100 sieve screen; and 1 part of an additive prepared by mixing 5 parts of component (A) from Example 1 and 1 part of component (B) from Example 4. A similar lime plaster was prepared except that component (A) differed from that of Example 1 only that it did not contain the potassium carbonate and the kaolin.

Example 7: A cement plaster was prepared by mixing 30 parts of Portland cement; 10 parts of limestone; ½ part of an additive obtained by mixing 16 parts of component (A) from Example 1 and 1 part of component (B) from Example 5, and ½ part of a second additive prepared by mixing 1 part of component (A) from Example 1 and 1 part of component (B) from Example 4; and ¼ part of a solid commercial cement-setting retarder which is a carboxylic acid derivative.

Calcium Cyanamide

F. DeMaria; U.S. Patent 3,503,766; March 31, 1970; assigned to American Cyanamid Company has found that Portland cement compositions exhibit increased strength when they contain from 0.5 to 15%, based on total weight, of an additive selected from the group: cyanamide, cyanamide dimer, alkali and alkaline earth metal salts of cyanamide dimer.

Example 1: In order to demonstrate the effect of black calcium cyanamide ($CaCN_2$) on cement paste, for characteristics such as the time period required for the initial set, the time period required for the final set, quantity of water per unit weight of cement, and the normal consistency, tests were run using varying amounts of additive. The results are tabulated below.

Sample	Hours, Minutes		Water added to 500 Grams Cement, cc.	Normal Consistency (mm.)
	Initial Set	Final Set		
Blank	2, 25	4, 6	119	10
1% additive	2, 43	3, 51	130	10
	3, 7	3, 57		
3% additive	3, 11	4, 3	133	9.5
	3, 6	4, 2		
5% additive	3, 25	4, 33	143	11
	3, 42	4, 42		
10% additive	3, 10	3, 56	152	9
	3, 19	4, 5	(*)	

*Specification 9 to 11 mm.

When preparing the test samples for the time of the studies of this example, all of the paste examples were adjusted to within the normal consistency specification of 9 to 11 mm (depth of penetration of a standard-sized needle in a standard time). The adjustment is accomplished by increasing or decreasing the water/cement ratio in the paste mix.

The data demonstrate a series of rate of set tests of cement paste, comparing a standard control blank to which no additive was added, to other samples, to which 1, 3, 5, and 10% of additive were added, respectively. The initial set time for mixtures containing the varying percentages of additive are greater than the corresponding initial set time for the control blank.

Also, the cubic centimeters of water per 500 grams of cement is greater for the varying percentages of additive. The water retention characteristics of the mixture are better at percentages less than 5% of additive. For the mortar tests of Examples 2 and 3, standard graded silica sand, type C-109, from Ottawa, Ill., was used. The grading of this sand is given in the table below.

Sieve Size	Percent Passing
No. 30	99.6
No. 50	24.6
No. 100	1.4

These data illustrate that (1) if the silica sand is graded through a screen of No. 30 mesh sieve size, 99.6% of the sand passes through the screen, (2) if the silica sand is graded through a screen of No. 50 mesh sieve size, 24.6% of the sand passes through the screen, and (3) if the silica sand is graded through a screen of No. 100 mesh sieve, 1.4% of the sand passes through the screen.

Example 2: To demonstrate the individual and comparative effects of two different additives, a series of tests were run, using calcium cyanamide ($CaCN_2$) for certain tests and hydrogen cyanamide (H_2CN_2) for other tests. These tests were compared with a control blank of mortar which contained no additive.

Using calcium cyanamide, separate tests were run using 2.5, 5 and 7.5% of additive, respectively, based on the total weight of all ingredients after incorporation. For each of the percentages illustrated, test runs were performed, and for each, the tensile strength and compressive strength was determined, in pounds per square inch. The triplicate castings, performed for each percentage, were made and cured in a water-saturated atmosphere for 11 days at 50°C.

		Average of 3 Castings	
Additive	Parts/Weight of Cement, percent	Tensile Strength, psi	Compressive Strength, psi
--	--	183	2,450
$CaCN_2$	2.5	183	2,450
$CaCN_2$	5	198	2,600
$CaCN_2$	7.5	214	2,850
H_2CN_2	1.25	195	2,550
H_2CN_2	2.50	218	2,650

In Example 2, tensile strength and compressive strength obtained from various percentages of additive incorporated in mortar are compared to each other, and compared to a control blank. The additives used were calcium cyanamide for three demonstrations, and hydrogen cyanamide for two demonstrations.

The maximum improvement in strength for about 7.5% of calcium cyanamide is about 16% improvement for both the tensile strength and the compressive strength. The free cyanamide (hydrogen cyanamide) has a greater effect on the tensile strength, as contrasted to calcium cyanamide, and the free cyanamide has a comparable effect on the compressive strength, as contrasted to calcium cyanamide.

Dextrin Adhesive Grout

J.V. Fitzgerald and M.J. Kakos; U.S. Patent 3,486,960; December 30, 1969; assigned to Tile Council of America, Incorporated describe an adhesive grout which is prepared by mixing 11 to 40% by weight of water with a dry composition comprising hydraulic cement and nontoxic dextrin, the composition having a water retentivity between 15 and 35.

It has been found that the dextrin-containing mortars of this process require surprisingly small amounts of water to attain workable viscosities. Since water in excess of that required for hydration of Portland cement is usually lost through evaporation and contributes to undesirable shrinkage, the low water requirement is an advantage.

Water retentivity values were obtained on Portland cement containing various amounts of the modified starches. This property was measured by placing a ⅛" layer of the mix previously slurried with the specified amount of water on the porous side of a quartered 4¼" x 4¼" Commercial Standard 181 glazed wall tile. A thin glass slide was placed over the mortar layer and the assembly positioned under a microscope lens.

As the water left the mortar travelling into the porous bisque of the tile, the mortar layer contracted thereby causing the slide to be displaced downward. This displacement could be accurately measured with a microscope and plotted against the square root of time.

The slope of the straight line divided into 1,000 yielded the retentivity values listed in the table. Starch is expressed as percent of total mix. Water is expressed as percent of weight of dry mix. Most proprietary dry-set mortars measured in this manner have retentivity values in the range between 35 and 50, whereas dry wall grouts generally measured between 15 and 35.

Retentivity Values for Various Starch-Grey Portland Cement Combinations

Starch Identification	Amount, percent	Starch	Water requirement, percent	Retentivity
Modified tapioca Dextrin	5.0	Crystal gum [1]	25	45.0
Yellow corn Dextrin	5.0	C.P. 8051 [2]	20	91.0
50% soluble white Dextrin	5.0	600 Dextrin [3]	25	33.0
85% soluble white Dextrin	5.0	653 Dextrin	22	71.0
95% soluble yellow Dextrin	5.0	700 Dextrin	25	118.0
White corn Dextrin	5.0	7071 Globe Dextrin [2]	21	52.0

[1] National Starch and Chemical Corp.
[2] Corn Products Co.
[3] Clinton Corn Processing Co.

Example 1: The dry mixture, shown on the following page, was slurried with 28% by weight of dry mix, of water. The mix became prematurely stiff, presumably because of the absence of a wetting agent. Consequently, another 3% of water was necessary to obtain the desired consistency again. After 30 minutes the mix was remixed.

	Percent
Grey portland cement	93.0
White dextrin, 85% soluble (Clinton Dextrine 653)	5.0
Dimethylol urea	1.0
Calcium chloride	1.0

It was then trowelled onto a vertical, rigidly supported gypsum wallboard surface using a ⅛" square notched trowel with ⅛" flats, so as to obtain an average mortar thickness of ¹⁄₁₆". At 5-minute intervals a standard grade 4¼" x 4¼" glazed wall tile (water absorption of about 13%) was pressed onto this mortar surface and twisted through a 90° angle. Open time was recorded as the longest time after application of the mortar that a tile was retained on the surface when so applied. When carried out at 70°F and 50% RH, the open time for this mix was 50 minutes, which is quite acceptable.

Example 2: Using the same composition presented in Example 1 but allowing the mortar to mix to slake for an additional hour, the following test was performed. The mortar was trowelled onto the surface as described in Example 1. Immediately thereafter, 10 tiles described in Example 1, were pressed on the mortar with a 3" space between each tile.

At 5-minute intervals successive tiles were twisted through an angle of 90° and back to the original position. Adjustability of the mortar was then designated as the longest time that tile remained affixed to the mortar when so tested. For the test at 70°F and 50% RH, the adjustability for the mortar was recorded as 40 minutes which is quite acceptable.

Example 3: The dry mixture:

	Percent
Grey portland cement	91.3
(Corn Products' 8051 Dextrine) yellow dextrin 95% soluble	4.0
Calcium chloride	1.0
Asbestos	2.0
Alkylaryl polyether alcohol (Triton X-120)	0.5
Dimethylol urea	1.2

and 2 parts mason's sand and 1 part of this dry mix were slurried with enough water to give a consistency which, when trowelled on a dry substrate, with a notched trowel, gave a rigid, nonflowing rib formation.

Shear test specimens were prepared from pieces of glazed ceramic tile of dimensions 4½" by 2⅛", which were halved sections of standard 4¼" tile of about 13% water absorption. A ⅛" unsanded mortar bonding layer was used. In preparing the specimens the long side (factory-finished edge with lugs ground off) was offset about ¼" so that 8 square inches of each tile were covered with mortar.

The specimens were allowed to cure and were then shear tested after 7 days, 28 days, and 7 days dry plus 7 days water soak. The shear test was performed by compression loading (2,400 pounds per minute) on the offset edge of the vertically placed specimen.

The sanded mortar was used to prepare vitreous tile shear bond specimens, 2" x 2" natural clay ceramics with water absorption of about 1.5% being used. The sample preparation and testing method were similar to those described above. Results of the shear bond tests for both sets of samples are summarized in the table below.

	7 Days	7 Days Dry plus 7 Days Water	28 Days
Average 4 wall tile (non-sanded mortar), p.s.i.	94	48	133
Average 4 vitreous tile (sanded mortar), p.s.i.	57	36	37

Wheat Paste

C. Polis; U.S. Patent 3,519,450; July 7, 1970 describes a binding composition which comprises a gypsum plaster, a lightweight granular aggregate and wheat paste. The wheat paste is present in an amount of 3 to 20% based on the dry weight of the total composition. The following is a typical formula for a ready-mixed mortar made following this process.

Formula	Pounds
Calcined gypsum plaster and perlite	80
Wheat paste	5

In order to obtain the most desirable mortar for application to the surface to be decorated, the above composition is dry mixed in a rotating drum for about 15 minutes. Approximately 6 pounds of the mixture is then mixed with about 3 quarts of water to form a workable mixture for the final application on the wall surface.

Aluminous Cement Molding Composition

A process described by *S. Ingrassia and J. Orsini; U.S. Patent 3,472,669; October 14, 1969; assigned to Société Civile d'Études et de Brevets "Inor," Monaco* involves an aluminous mortar which may be molded without shrinkage and comprising an aluminous cement, a detergent and a silicone.

Example: 20 parts of Silicone V.M., a solution in deionized water of a potassium salt of siliconic acid having the pH of about 14, and an amount of water to give 1,000 parts of an aqueous mixture, i.e., 780 parts of water, are added to 200 parts of Teepol, salts of secondary sulfuric acid alkyl esters.

The aqueous mixture is mixed at a temperature of 15°C with cement and aggregates in the proportions of 50 parts of aggregates to 16 parts of aqueous mixture. The resulting mortar is poured into a water-tight open mold of a material which is electrically nonconductive. The mold is immersed in water for 8 to 12 hours until the mortar has set. The mold and molding it contains are withdrawn and left until the surface of the molding is dry and the molding is removed.

Calcined Gypsum and Quicklime

A process described by *J. Dean; U.S. Patent 3,232,778; February 1, 1966* relates to a plaster coating composition characterized by an ability to bond as a hard, shrink-resistant, moisture-proof coating to a wide variety of materials and surfaces.

The plaster coating composition consists essentially of a base mixture comprising in the major proportion a mixture of Portland cement and calcined gypsum, and in minor proportion calcined lime or quicklime. Preferably, the plaster coating composition also contains a small amount of diatomaceous silica to enhance workability. It may also contain casein or other agent to assist in bonding, and a retarder to inhibit the setting of the plaster. In preparing the plaster composition, the ingredients are advantageously mixed dry (e.g., in an agitated mixer) to provide a dry mix or blend capable of use upon addition of water in appropriate amounts.

For example, a trowelable mix can then be prepared on the job by admixing a small amount of water with the dry mix or, if desired, a wet mix suitable for brushing, rolling, or spraying can be prepared by adding slightly larger amounts of water. Alternatively, the ingredients can be premixed with water, and the composition suitably packaged for sale.

The plaster coating composition can be applied to virtually any surface, whether vertical or horizontal, including sheetrock, plywood, painted surfaces, fiber glass, Styrofoam, or similar foamed materials, and with substantially a permanent bond. The shrink-resistant characteristics of the plaster composition make it particularly suited to application on sheetrock, since the joints can be filled or the entire wall surface covered, in a single operation. The composition also dries to a hard, scuff-resistant, moisture-proof surface (due primarily to the combination cement and gypsum base), which is suited to a wide variety of constructional purposes.

Example 1: One very satisfactory plaster coating composition has been formulated as follows:

	Lb	Oz
White Portland cement	8	0
Calcined gypsum	6	0
Calcined finish lime	3	6
Diatomaceous silica (Celite, Johns-Manville)	1	6
Casein	0	6
Powdered expanded vermiculite (Zonolite, The Zonolite Co.)	0	2
Retarder (Amadex, Corn Products, Inc.)	-	1

In formulating the ingredients shown in the above table, about equal proportions by volume of the dry cement, gypsum, lime, silica and vermiculite were mixed together in an agitating container, the casein and retarder then added, and the mixture agitated further until suitably mixed for placing in cans, sacks, or other containers.

Example 2: A wet coating composition suitable for troweling on a wall surface was prepared by mixing the dry ingredients of Example 1 with an approximately equal amount of water, as follows: 40 lb of dry mix and 5 gal of water. When applied in conventional manner upon an interior wall surface constructed of sheetrock panels, the resulting plaster coating composition filled the open joints between the panels and provided a smooth wall surface resembling conventional plaster in every way. The compositiion set to a hard surface in about 2 hours without any evidence of shrinkage.

During a succeeding period of about 1 year, the plaster coating composition was observed to retain a hard, scuff-resistant surface which was free of cracks and which was waterproof. The coating composition was similarly applied to an existing exterior wall surface, where it was exposed to normal weathering. At the end of a year, the surface was substantially in its original form, and showed no visible sign of weathering.

Disintegratable Calcined Gypsum

M.K. Lane and J.S. Sheahan; U.S. Patent 3,359,146; December 19, 1967; assigned to United States Gypsum Company describe a process which involves making gypsum casts having an increased strength to density ratio by the steps of providing a calcined gypsum which disintegrates upon mixing with water, forming a slurry and mixing it until the calcined gypsum particles disintegrate, adding accelerator to the slurry and casting it while the calcined gypsum has a combined moisture content of less than 9% by weight.

The process is based in part upon the finding that the strength of set gypsum plaster increases as the particle size of the hemihydrate prior to setting decreases, and in part on the provision of a process to enable the advantage of the discovery to be employed in finished gypsum products.

The ability to develop very fine particles, particularly in the presence of water as in a slurry is a property which will be called dispersibility and is affected by a number of operations which go into the preparation of the calcined gypsum before its contact with the gauging water to form a slurry.

The disintegration of the particles can be observed with the naked eye by dropping particles of calcined gypsum into cold water, and its extent may be determined by mixing the calcined gypsum briefly with water and then interrupting the hydration reaction by dilution of the slurry with ethanol or isopropanol and quickly filtering, washing with alcohol and drying at about 110°F. The particle size before and after slurrying with the water may be measured by the Blaine air pêrmeability apparatus (ASTM-C-204).

The source and purity of the gypsum, and the fineness to which it is reduced before and after calcination, affect this property, and a major influence is exerted by the type of calcination utilized. The most dispersible calcined gypsums are those known as the beta variety which are produced by atmospheric-pressure calcination in a kettle or kiln, the particles of which contain many cracks and incipient fractures which cause the particles to rupture upon contact with water.

The calcined gypsum considered most desirable for the process is that produced from a rock having a high percentage of calcium sulfate dihydrate, that is more than about 85% and preferably above 88%, and having a low percentage of salt so that it is not aridized or chemically preaged in the calcination step. In some cases, it has been found desirable to wash the rock to lower the salt content before calcination. Precipitated gypsum of adequate purity may also be used.

Example 1: The following will illustrate the procedure of the standard run. The mixer employed was a Hobart Model A-200 equipped with a wire whip agitator. About 1,500 cc of gauging water was placed in the 12-quart mixing bowl in preparation for a run, the exact amount being determined empirically to give the proper consistency to the slurry.

Just prior to starting the agitator, 20 grams of ground gypsum block accelerator, 1,600 grams of calcined rock, 21.6 grams of paper fiber reinforcing agent, 10.8 grams of a cereal grain binding agent, and 2.3 grams of a dispersing agent or consistency reducer, such as Orzan A were added. Orzan A is an ammonium lignosulfonate, containing wood sugars.

The mixer was turned on at 365 rpm, and after 5 seconds, about 650 cc of foam with a density of 13 lb/ft^3 were added and the mixing continued for a total of 20 seconds. Five test cubes were cast and as soon as they had set, they were taken from the molds and placed in a kiln at 350°F where they were dried to 70% of their wet weight. Drying was then completed at 110°F and the compressive strength and density measured in the usual manner. The average of the 5 cubes was reported as the result of the run.

At the time the cubes were cast, about one part of slurry was added to two parts of reagent grade isopropyl alcohol, with rapid stirring, so that the isopropyl alcohol interrupted the hydration. The particles were largely the hemihydrate and were filtered off, washed with isopropanol and dried at 110°F. The cellulose fiber in the dry product was removed by brushing through a No. 50 sieve and the dried and sieved powder was analyzed by a standard procedure for combined moisture (ASTM Method C-471) and the surface area and therefore particle size were determined by the Blaine air permeability apparatus which is described in ASTM method C-204.

Referring to the table, the results shown for Run No. 1 were obtained by following the procedure of Example 1 using a good quality calcined gypsum. For Run No. 10, the procedure of Example 1 was repeated using good calcined gypsum rock from another source. The strengths and densities approximated those obtained in preparing the standard samples.

Example 2: The increased strength obtainable with greater fineness, in this case as a result of careful dry grinding, is demonstrated by Run 2, in which the procedure of Example 1 was repeated using the same calcined gypsum with the exception that it had been ground to a much finer particle size so that dry fineness was 12,500 cm^2/g. This resulted in increased slurry fineness, a shorter set time and produced a surprising increase in the strength of the cast product. Run 11 was made with the same calcined gypsum of Run 10 after it had been ground to about 12,000 cm^2/g, further illustrating the increase in strength realized by increasing the slurry fineness before casting the gypsum product.

Effect of Grind on Cast Strength

	Run 1	Run 2	Run 10	Run 11
Fineness, dry	4,000	12,500	4,000	12,040
Fineness, slurry	13,000	16,700	10,700	16,600
Water:hemihydrate ratio	0.93	1.03	0.91	1.03
Percent hydration as cast	7.5	9.2	7.4	8.5
Set, minutes	15	12.5	15	13
Strength, percent of standard	103	125	97	115

Example 3: Using calcined gypsums from different sources, Runs 30 and 40 were made according to the standard procedure of Example 1. Duplicate runs, No. 31 and 41, were made except that the ground gypsum block accelerator was not added until several seconds, 8 and 5 respectively, after mixing began. As reported below, only a very slight increase in slurry fineness was measured, but the increase in strength was substantial and was accompanied by an increase in accelerator efficiency, as indicated by the reduction in set time.

Effect of Delayed Accelerator Addition

	Run 30	Run 31	Run 40	Run 41
Fineness, dry	4,500	4,500	5,000	5,000
Fineness, slurry	14,500	14,500	14,000	14,700
Water:hemihydrate ratio	0.90	0.90	0.90	0.90
Accelerator delay, sec.	0	8	0	5
Percent hydration as cast	7.5	7.8	6.8	6.75
Set, minutes	13.0	11.0	13	10.5
Strength, percent	95	110	91	107

The advantages of the process were also obtained with a continuous type mixer customarily used for preparing slurry for the manufacture of plasterboard in commercial operations.

In related work *M.K. Lane; U.S. Patent 3,652,309; March 28, 1972; assigned to United States Gypsum Company* describes a process of making a gypsum plaster possessing a high degree of stability over extended periods of time and having a set time of at least 4 hours in aggregated mortar applications which involves providing a specially sized calcined gypsum having a particle size that is essentially all finer than 32 microns, and mixing the specially sized gypsum with retarder, accelerator and other additives to give the desired properties when mixed with water and aggregate.

Ceramic Frit and Sand

J.E. Waters, Jr.; U.S. Patent 3,917,489; November 4, 1975 describes products which are comprised of calcined gypsum in one of the forms of plaster of Paris combined with frit. That combination, when water is added, produces a castable material which may be placed in a kiln and fired immediately upon being removed from a mold without any need for drying.

It need not be, indeed it may not be, cast in a water-absorbent mold. Like ordinary plaster of Paris, it can be cast in a flexible mold. However, unlike ordinary plaster of Paris, the material can be fired into a vitreous body. It becomes a ceramic material.

The bare combination of calcium sulfate and frit presents the problem that the ceramic body produced on firing is subject to a considerable degree of shrinking. Depending upon the quantity of water in the original slip the degree of shrinkage may be 15, or even as high as 20%. That difficulty is overcome in part by adding a filler material such as sand, talc and other common materials such as sand, talc and other common materials such as grog and asbestos. Sand is preferred because of its lower cost except when a special texture is desired in the finished piece. When fillers other than siliceous materials are added, the requirement for frit is increased.

One of the advantages of the process is that materials calculated to alter the color of the finished ceramic body may be added to the composition without deleterious effects. For example, the addition of chrome oxide, Cr_2O_3, will color the body green. That material gives rise to the formation of microscopic bubbles throughout the body which tend to increase its volume and reduce its density. However, the result is a closed cell structure which is nonporous. The addition of red iron oxide results in a red coloring. The addition of black iron oxide will color the body black. However, the process frees oxygen so that silica should not be used in more than small amounts as a filler when black iron oxide is used as a coloring agent.

Some examples of the process are set out below. These are formulas of the dry ingredients. They are used by mixing them to achieve uniformity and by adding water and stirring until a consistency is reached which the plaster worker characterizes as a "cream." At that point, the material is poured into a mold and allowed to set in the manner of plaster.

When the body has set, the mold is removed and the body is placed in a kiln and fired. These formulations result in the production of a vitreous ceramic body at temperatures in the range of Cone 07 to Cone 04. If the amount of flux is reduced, vitrification will occur at a somewhat higher temperature in the manner with which the ceramist is familiar. All of these formulations, except the last one, specify plaster either in the form of casting plaster or Hydrostone (proprietary plaster mixture).

A lower density product results from the use of casting plaster and the highest density product results from the use of Hydrostone. A product of intermediate density results from the use of a plaster product having a density intermediate casting plaster and Hydrostone.

Examples 1 through 5: The following are compositions which have been used in the process. The proportions are given by volume.

Example	Ingredients
1	48 parts plaster as casting plaster or Hydrostone 12 parts frit
2	48 parts plaster as casting plaster or Hydrostone 12 parts frit

(continued)

Example	Ingredients
	12 parts feldspar 12 parts talc
3	48 parts plaster as casting plaster or Hydrostone 12 parts frit 24 parts feldspar or nepheline syenite 32 parts silica sand
4	48 parts plaster as casting plaster or Hydrostone 21 parts frit 24 parts nepheline syenite 30 parts silica sand 2 parts chrome oxide
5	48 parts plaster as casting plaster or Hydrostone 18 parts frit 48 parts nepheline syenite 1 part black iron oxide

Porous Aluminum Granules

S.J. Rehmar; U.S. Patent 3,579,366; May 18, 1971; assigned to Intrusion-Prepakt describes an additive for cementitious grout and concrete. The additive is a porous granule characterized by a skeletal structure and a network of interstitial voids formed by in situ gas liberation, the skeletal structure consisting essentially of a dispersion of finely divided, oxide-free aluminum particles in a matrix formed of a powdered inorganic extender and a water-soluble binder which is inert to the aluminum particles.

Example: The following ingredients were mixed with gentle blending until the mixture presented a homogeneous appearance:

	Pounds
Aluminum flakes (standard varnish leafing 100 mesh powder)	45
Limestone dust, 100 mesh	300
Hydrated aluminum oxide	90
Methylcellulose	20
Ammonium carbonate	3.5
Water	175

When the mixing was stopped, the mixture had a pasty consistency, which was then molded into pellets. The pellets were baked at a temperature within the range of about 300° to about 350°F, whereby the water was driven off, and the ammonium carbonate decomposed with generation of carbon dioxide and ammonia gases.

The resulting pellets were sufficiently strong to resist damage by abrasion when incorporated with the dry ingredients to a cementitious grout. Upon addition of water to the mixture of pellets and cementitious grout, the water-soluble binder dissolved readily and the high alkalinity of the environment rapidly attacked the oil film on the aluminum particles, permitting the aluminum to oxidize with the evolution of hydrogen gas.

The concentration of pellets in a cementitious hydratable grout or concrete, can be varied over wide limits, and will depend upon the amount of desired expansion or gas entrainment of the grout or concrete. In general, however, the granules or pellets should be present in an amount providing the quantity of metallic aluminum required to react with alkali in the grout or concrete, and produce hydrogen gas in a volume sufficient to control expansion and bleeding of the grout or concrete.

Granules or pellets produced in accordance with the process may be stored in a relatively dry place almost indefinitely without degradation. Similarly, mixtures of the granules or pellets and a cementitious hydratable grout or concrete can be stored for extended periods of time, if kept essentially dry.

Fluid Coke

H.N. Babcock; U.S. Patent 3,519,449; July 7, 1970; assigned to U.S. Grout Corporation has found that the shrinkage of aqueous hydraulic cement mixtures is eliminated by incorporating a fluid coke with a controlled amount of moisture less than about 3% by weight. The amount of fluid coke that may be used to eliminate the shrinkage is less than 10% based on the weight of the cement in the mixture.

In these examples, the performance of the fluid coke mixture was judged by the expansion and contraction of the cementitious system as soon as it was mixed with water and cast in a cylindrical mold with about 10% of exposed surface. The expansion and contraction of the cast was determined by the vertical movement of the top surface.

For the purpose of higher accuracy, a light test was used to measure the movement of the top surface. The test consists of using a focused light beam to project a shadow of the top surface onto a screen equipped with a vertical graduation. The magnification is 72 times. The movement of the top surface on the screen is recorded in every 10 to 20 minutes for each cast until final set which usually takes about 3 to 4 hours.

A thin layer of water was added to the mold for cast setting under no evaporation condition. To facilitate the detection of the movement of the top surface, a marble was placed on the top of the surface and the expansion or contraction of the cast was determined by the movement of the apex of the shadow projected on the screen

Example 1: In this example, various cement-sand mixtures with different amounts of fluid coke were cast in cylindrical molds measuring 2 inches in diameter by 3⅞ inches in height. The casts were allowed to set under normal evaporation condition. In the first set of casts, the following aqueous hydraulic cement mixture was used of which only the moisture content of the fluid coke was changed in different casts.

	Grams
Type 1 cement	146
Sand	293
Fluid coke (15% of cement)	21.9
Water (3.8 gallons per 94 pounds sack)	49

The fluid coke used was dried in an oven at 250°F overnight and cooled in dry air until it reached an ambient temperature of about 75°F. The difference in weight between right after drying and subsequent to cooling was 0.89% which represents the amount of air or water regained by the fluid coke. For the purpose of convenience, this fluid coke is identified as Standard.

Samples of fluid coke with different moisture content were prepared from the Standard by adding thereto 1, 2, 3 and 4% by weight of water with care to avoid water evaporation. When the water was added to the Standard, slight lumpiness was observed. This condition, however, disappeared overnight and the fluid coke was free-flowing.

Six casts were made using mixtures prepared according to the proportions set forth in the table above using in each cast samples of fluid coke with a different moisture content and in one cast no fluid coke for the purpose of comparison. The results of light test for determining the expansion and contraction of the casts are tabulated below.

Cast	Amount of moisture added to "Standard," percent	Light test growth after 4 hours (inches)
A	0	+3¼
B	1	+2¾
C	2	+1½
D	3	−1
E	4	−1¼
F	No fluid coke	−2⅜

The results of the light test show that growth or expansion can be realized with fluid coke containing 2% added moisture or computed on an absolute scale, less than 2.89% moisture. It is interesting to note that the difference of 1% of moisture between casts C and D causes a change of 2½ inches in the test scale whereas the difference of 1% between casts B and C or D and E leads to changes of 1¼ and ¼ inch, respectively. In the second set of casts, mortar with proportions set forth in Example 1 was prepared in a manner similar to the first set and with the exception that the amount of fluid coke used was 10% by weight of the cement. Light tests of the casts show that shrinkage was eliminated when the moisture content in the fluid coke was below 3%. In the third set of casts, mortar with the following proportions was prepared in a manner similar to the first set:

	Grams
Type 1 cement	230
Sand	470
Fluid coke (8% cement)	18.4
Water (3.8 gallons per 94 pounds sack)	77.5

Light test shows shrinkage was prevented with fluid coke containing less than 2% of moisture. In still another set of casts using 6% by weight of cement in a manner similar to the first set, light tests show that shrinkage was prevented with fluid coke containing less than 1% of moisture.

Example 2: In this example, the no evaporation condition was used for setting the mortar. The mortar used has the following composition:

	Grams
Type 1 cement	146
Sand	293
Water	50

The fluid coke has less than 0.89% by weight of moisture. The results of the light test on various casts using different amounts of fluid coke is tabulated below:

Percent Coke	Volume Change After 3 Hours, inches
0	1 3/8
1	-3/4
2	No shrinkage
3	+3/4

In related work *R.W. Gaines and H.N. Babcock; U.S. Patent 3,503,767; March 31, 1970; assigned to U.S. Grout Corporation* have found that the shrinkage of aqueous hydraulic cement mixtures is eliminated by incorporating a fluid coke with a particle size predominately finer than 100 mesh. The amount of fluid coke that may be used to eliminate the shrinkage is less than 10% based on the weight of the cement in the mixture.

Catalyzed Gypsum Wallboard

M.E. Gerton; U.S. Patent Reissue 26,362; March 12, 1968; assigned to Fibreboard Corporation describes a process for interior wall construction where gypsum wallboard or similar material is utilized as the wall base having a gypsum setting catalyst applied to its surface and a special thin coat of gypsum plaster applied to the wall base to form a finished wall surface.

The construction of the process comprises a conventional wallboard or lath nailed or affixed by suitable means to the wall studding. The wallboard or lath will generally be of the paper-surfaced, gypsum-core type or of a compressed fiberboard material suitable for use as plaster backing. The wallboard or backing differs from that conventionally found in the art in that a suitable catalytic agent has been impregnated into the outer surface.

The composition of this catalytic agent is critical to the construction. Thus, it has been found that a catalytic agent comprising either potassium sulfate or mixtures of potassium sulfate and aluminum sulfate are particularly suitable for the catalytic agent. While potassium sulfate alone may be used, best results are obtained with mixtures of potassium and aluminum sulfate. Thus, on a dry basis, the catalyst may comprise 100 to 60 parts by weight of potassium sulfate and 0 to 40 parts by weight of aluminum sulfate.

Generally, for convenience, the catalyst mixture is applied to the wallboard during the manufacturing process. However, under certain circumstances, the catalyst may be applied to the wallboard at any time subsequent to the manufacture and prior to the plastering operation.

Slaked Lime, Hardwall Plaster and Gypsum Board Joint Filler

A process described by *J. Adolf; U.S. Patent 3,895,018; July 15, 1975* provides a flexible plaster composition which when applied to most prepared and painted surfaces, adheres readily. Such surfaces include rough and porous surfaces such as plaster, gypsum wallboard, concrete, finished plywood sheet, wallpaper, metal and plastic. Plaster according to the process, when spread to a thickness of from one-sixteenth to one-eighth of an inch does not readily crack, spall or check.

The plaster composition of the process includes, as a minimum, a mixture of sand and latex sealer and a primary filler ingredient which can be hardwall plaster, or slaked lime, or a gypsum wallboard joint filler. Filler ingredients affect drying time which is dependent on hydraulic setting characteristics of the filler ingredient when combined in the mix.

Working properties of the plaster composition are improved by mixing all five ingredients above, in suitable proportions, with water, producing also a low-cost composition. Secondary fillers of a particle size similar to the primary filler and having suitable drying characteristics may also be included.

Example:

	Parts
Hardwall plaster	2-5
Slaked lime	1-2
Gypsum wallboard joint filler	1-2
Latex sealer	1-2
Sand	5
Water	3-4

A hardwall plaster is Paristone (Domtar Construction Materials Ltd.). A slaked lime is Hydrated Lime Type S (Western Lime and Cement Co.) complying with ASTM Specifications C206-49 and C207-49 Type S and Federal Specification SS-L-351 Type M (Masonry).

A gypsum wallboard joint filler is Gyproc Joint Filler General Purpose (Domtar Construction Materials Ltd.). An alternative gypsum wallboard joint filler is Synko Topping Cement (Synkoloid Co. of Canada Ltd.). Other well-known makes of generally similar wallboard joint filler are available and in this specification, such fillers are termed joint fillers. The hardwall plaster slaked lime and joint filler act as filler ingredients and affect drying characteristics of the mix. Other filler ingredients can be used as later explained.

A latex sealer is Latex Emulsion Sealer 3660 (Brandrum Henderson Co.). This is a polyvinyl acetate copolymer internally plasticized latex resin emulsion, pigmented with titanium dioxide and extenders. A further latex sealer is Interlux (International Paints, Canada). Specific ingredients above give satisfactory results when used in the above composition.

Polyvinylpyrrolidone and Melamine-Formaldehyde Resin for Bandages

D.F. Smith; U.S. Patent 3,671,280; June 20, 1972 has found that minor additions of polyvinylpyrrolidone and melamine-formaldehyde resins to plaster of Paris bandages permit making "casts" of improved strength and water-resistance with good physiological properties.

Example 1: 81 parts of high-density, low-consistency (see U.S. Gypsum Co. Bulletin IGL No. 19 for methods of testing plaster) plaster of Paris made by steam-calcining gypsum (see U.S. Patent 1,901,051) is slurried in 36 parts of a liquid comprising 80% toluene and 20% monomethyl ether of ethylene glycol in which is mixed 10 parts uncured melamine-formaldehyde resin prepared with a mol ratio of melamine to formaldehyde of 1:2, and 10 parts of dry polyvinylpyrrolidone of average molecular wieght 40,000.

To the slurry was then added 0.4 part powdered potassium sulfate, 1.06 parts powdered potassium chloride, 1 part powdered ammonium sulfate and 1 part water-insoluble polyvinyl acetate (as used in U.S. Patent 2,655,148). The mix was well stirred until all soluble materials dissolved and the resin softened. The resulting slurry was then coated on 32 x 28 mesh surgical gauze to form a discontinuous coating and dried at 200° to 230°F for 10 minutes to yield a dry product weighing 210 to 240 grams per 5 square feet.

The product was slit into 4 inch by 5 yard strips, rolled into a bandage and packaged in aluminum foil or polyethylidene (Saran) sheet in order to protect it from atmospheric moisture. When tested by the methods of Federal Specification GG-B-101d, the bandage showed a setting-time of 5 minutes and low loss of plaster both dry and when wet and squeezed out to make a cast.

A cast made from a 4 x 5 bandage gave a strength about 200 pounds greater than an ordinary cast after 1 day's drying and broke as a unit exhibiting a structure in which plaster is bonded to itself and to the backing, in contrast to an ordinary cast that shows little or no bonding to the backing. Repeated use of these bandages did not result in skin reactions. When a cast was dried for 3 days and then soaked in water for 24 hours it retained 95% of the strength of a cast 4 days old and not soaked in water. An ordinary cast in this test would retain only 50 to 60% of its strength.

Example 2: 276 grams of the dry, uncured resin and 75 grams of the polyvinylpyrrolidone of Example 1 were placed into a wide-mouthed glass jar with a tight, screwed-on cover (for example, rubber-gasketed). 15 grams of ammonium chloride (polymerization or condensation catalyst for the resin) were placed in a tightly sealed or tightly wrapped water-soluble film packet (for example water-soluble polyvinyl alcohol film) so as to keep it out of contact with the resin and prevent polymerization of the resin in storage.

The packaging was done in a dry atmosphere in order to prevent absorption of moisture by the resin and the polyvinylpyrrolidone, and the jar was kept tightly sealed until its contents were used. When ready to use, the contents of the jar were dumped into a container of a size to permit complete immersion of the bandage to be used and a jarful (1 pint) of tepid water was added and the mix stirred until dissolved. A 4 inch by 5 yard ordinary bandage (a fast-setting bandage made by the methods, for example, of U.S. Patents 3,191,597; 3,282,265;

3,236,232; 3,294,087; or 2,557,083) was moistened, the excess solution squeezed out (for example, leaving about 35% of the weight of the dry bandage of solution) and a cast made according to the above-noted Federal Specification. The set was about 5 minutes; the cast was similar in properties to that of Example 1. If the cast is available in a polyethylene envelope, the solution may be poured into the envelope to wet it thoroughly and the excess returned to its container.

If used economically this amount of solution is sufficient to wet about nine 4 inch x 5 yard or twenty 3 inch x 3 yard bandages. The above procedure makes over 1½ pints of solution which can be used for up to about 2 hours. Repeated use of this solution failed to elicit any skin reactions.

D.F. Smith; U.S. Patent 3,791,837; February 12, 1974 has also found that dextran is a very efficient bonding agent for plaster of Paris bandages.

Surface Treatment Using Radiation Polymerized Polysulfone

According to a process described by *M. Takehisa, H. Kurihara, T. Yagi, H. Watanabe and S. Machi; U.S. Patent 3,873,492; March 25, 1975; assigned to Japan Atomic Energy Research Institute, Japan* a gypsum composition having affinity for a thermoplastic resin is obtained by adhering a polysulfone resin to the surface of a gypsum powder. The composition is suitable for preparing a gypsum-thermoplastic resin composite.

The polysulfone resins used are those having groups shown by (—SO_2—) in the main chain of the polymer. While many kinds of polysulfone resins and their preparation processes are well known, alternate copolymers comprising olefin and sulfur dioxide can preferably be used for the purpose in the process from the viewpoint of their properties and cost.

Copolymers comprising sulfur dioxide and a vinyl compound or an allyl compound are also used. It has been known that the formation reaction of polyolefinsulfone is initiated by any means such as irradiation with an ionizing radiation, irradiation with light, the application of an initiator which releases free radicals upon its decomposition such as azoisobutyronitrile, and the like, to the above mixture of the monomer.

In the polymerization reaction in the presence of gypsum, irradiation with light is not so reasonable since the effective initiation of the reaction is restricted only in the surface layer of the mixture and the reaction cannot be effectively initiated in the inside of the mixture. The initiation of the reaction by means of an ionizing radiation is a preferably method, since an ionizing radiation, especially gamma rays or the like, has a strong penetration power, and results in homogeneous initiation of the reaction in any part of the system of the reaction mixture.

Particularly, this method can effectively be applied to the formation reaction of polysulfone in which the limiting temperature of the polymerization reaction is low, since the irradiation of an ionizing radiation induces initiation of the reaction, independent of the temperature. In the process with application of a radical initiator, the decomposition rate of the initiator is generally increased at higher temperature. On the other hand, the rate of the formation reaction of polysulfone is decreased when the temperature is elevated to temperatures in the neighborhood of room temperature.

Therefore, the effect of the initiator is reduced when once the temperature control is missed. However, this method can be industrially applied with sufficient favor when the reaction apparatus and the reaction system are reasonably selected.

In the polymerization reaction in the presence of gypsum, which is a fundamental concept of this process, the reaction takes place on the surface of the gypsum, or the polysulfone which is formed in the neighborhood of gypsum is adhered or precipitated on the surface of the gypsum during the polymerization reaction, and a very strong bond is formed between gypsum and polysulfone.

Although the study of the reasons for such strong bonding is still in progress, the following is considered to be one of the reasons. That is, it is estimated that, from the general formula shown by

$$-SO_2-[CH_2-\underset{}{\overset{R}{\overset{|}{C}}}H-SO_2]_n-CH_2-\overset{R}{\overset{|}{C}}H-SO_2-CH_2-$$

where R means hydrogen atom, another atom or alkyl group, this material may be physically bonded to gypsum through the $-SO_2-$ group and may be physically bonded to a polymer through the $-CH_2-CHR-$ group.

Example 1: The commercially available gypsum with 0.5 H_2O for food additive use was dehydrated by heating at 150°C for 2 hours. The resulting gypsum with 0.5 H_2O was charged in an autoclave with the inner volume of 200 cc, and the air in the autoclave was sufficiently replaced with gaseous ethylene. Then, 20 grams of commercially available liquid sulfur dioxide was loaded.

Gaseous ethylene was loaded in the autoclave with a pressure of 50 kg/cm^2 (gauge pressure), and the content was irradiated by ^{60}Co γ-rays with a dose rate of 5 x 10^4 rads/hr for 2 hours, at room temperature (23° to 25°C). After the irradiation, unreacted gas was released and the unreacted gas was entirely removed by keeping the system under reduced pressure. The reaction product was in the form of white powder whose appearance was not different from the raw gypsum used. The increase in weight of the gypsum after the above treatment was about 1%.

The gypsum, whose surface properties were improved by the above treatment, was mixed with radiation polymerized polyethylene (density: 0.941, MI: 0.05) in the weight ratio of 50:50, and was blended (= kneaded) for 20 minutes at 160°C using a type 50 Plasticorder roller mixer at a roller rotating speed of 30 rpm.

The blended material was molded into a sheet 1 mm thick, by hot pressing at 160°C. Test pieces shaped in a JIS (Japan Industrial Standards) No. 3 dumbbell were stamped out from the above sheet which had a slight tinge of yellow and had an appearance similar to that of opaque polyethylene sheet, and a tensile strength test was carried out. The results are shown in the table on the following page. The tensile speed rate is 20 mm/min in all examples.

The characteristics of the composite comprising gypsum with 0.5 H_2O, whose surface was treated with poly(ethylenesulfone), and radiation polymerized polyethylene show an improvement.

	Stress at Yielding Point (tensile strength) (kg/cm^2)	Elongation at Breaking Point (%)
Radiation polymerized polyethylene surface treated gypsum with 0.5 H_2O	240	18
Radiation polymerized polyethylene gypsum with 0.5 H_2O	220	12

Example 2: The same gypsum as that used in Example 1 was charged in an autoclave with the inner volume of 200 cc and the air in the autoclave was sufficiently replaced with butene-1, then, butene-1 was loaded with the pressure of 1 kg/cm^2 (gauge pressure) at 20°C. Then, 10 grams of commercially available sulfur dioxide was loaded in the autoclave.

The mixture comprising gypsum with 0.5 H_2O, butene-1 and sulfur dioxide was irradiated by ^{60}Co γ-rays with a dose rate of 5 x 10^4 rads/hr for 0.5 hour at room temperature (20° to 22°C). Unreacted gas was entirely removed by a method similar to that in Example 1 after the irradiation was completed, and the reaction product comprising white powder was obtained, whose apperance was not different from the raw gypsum.

The increase in weight of the gypsum after the above treatment was scarcely observed. The gypsum, whose surface properties were improved by the above treatment, was mixed with commercially available low density polyethylene pellets in the weight ratio of 50:50, and was blended under the same conditions as those in Example 1, and the blended material was molded into a sheet. The test pieces shaped in a JIS No. 3 dumbbell were stamped out from the above sheet and a tensile strength test was carried out. The results are shown below.

The characteristics of the composite comprising gypsum with 0.5 H_2O, whose surface was treated with poly(butenesulfone), and low density polyethylene are:

	Stress At Yielding Point (kg/cm^2)	Elongation at Breaking Point (%)
Low density polyethylene surface treated gypsum with 0.5 H_2O	141	44
Low density polyethylene gypsum with 0.5 H_2O	113	15
Low density polyethylene (100%)	99	830

From the above data, it is obviously shown that the characteristics of the composite comprising low density polyethylene-gypsum with 0.5 H_2O was improved by treating the surface of gypsum according to the process.

Refractory Composition Using Fluxing Component

According to a process described by *H.W. Burr; U.S. Patent 3,841,886; October 15, 1974; assigned to Motus Chemical, Incorporated* calcium sulfate in any of its various forms is transformed so that its degradation characteristics are controlled by combining it with at least 2 weight percent, based on the calcium sulfate content, of a primary fluxing substance, i.e., a material which lowers the fusion point of the calcium sulfate (A.E. Dodd, *Dictionary of Ceramics,* p. 118; *Ceramic Industry Magazine,* January 1970, p. 93). Chemical compounds having a fluxing effect on ceramic materials and which are wholly or partly soluble are not normally used in ceramic production because of adverse results that impair the fire ware.

Fluxing substances are rarely used in refractory products because there is little or no need for them at refractory temperatures. However, it has been found that soluble types of fluxing compounds are very efficient in a calcium sulfate matrix.

Generally, primary fluxes useful herein include inorganic metal compounds having a fusion point below 1450°C when fired with calcium sulfate. In this regard, a number of compounds which have melting points substantially above 1450°C form a eutectic below 1450°C with calcium sulfate when fired therewith and are therefore suitable as primary fluxes. The following examples, in which all parts are by weight, illustrate the process.

Example 1: A test sample was made by intimately mixing and pressing together 100 parts of calcium sulfate dihydrate, 5 parts of sodium hydroxide, and 15 parts of water into a cylindrical form 1⅓ inch in diameter and 2⅔ inches in length. The cylinder was fired for 3 hours to 1000°C and then observed for any apparent deformation, densification, improved surface hardness, moisture degradation and weight change.

It was found that the sample lost about 9.5 parts in weight and had a good structural change in that there was a glassy surface and excellent densification with no cracking observable, nor was there any moisture degradation and the sample displayed improved surface hardness.

Examples 2 through 16: The procedure of Example 1 was repeated except that test samples were prepared using the following materials as flux in place of the sodium hydroxide: (2) lithium chloride, (3) sodium chloride, (4) sodium sulfate, (5) sodium carbonate, (6) sodium metaphosphate, (7) potassium chloride, (8) potassium sulfate, (9) potassium carbonate, (10) potassium hydroxide, (11) potassium tripolyphosphate, (12) calcium chloride, (13) calcium phosphate, (14) strontium chloride, (15) lead oxide, and (16) lead sulfate, In each case, after firing at 1000°C the structural changes were determined as good as defined in Example 1.

Examples 17 through 23: The procedure of Example 1 was repeated except that the test samples were formed with the following flux in place of the sodium hydroxide: (17) copper carbonate, (18) manganese chloride, (19) manganese sulfate, (20) manganese carbonate, (21) ferrous chloride, (22) ferrous sulfate, and (23) titanium sulfate.

In each case, the samples did not exhibit any significant alteration other than weight loss and were therefore refired to 1150°C and reexamined for structural transformation. In each case, after the refiring, the observed structural change was good as defined in Example 1.

Examples 24 through 28: The procedure of Examples 17 through 23 was repeated (i.e., the samples were refired at 1150°C after showing no structural change), but in place of the fluxes therein, the test samples were formed of the following fluxes: (24) zinc sulfate, (25) zinc oxide, (26) barium sulfate, (27) ferric oxide and (28) nickel chloride.

In each case, the observed structural change was fair, signifying that deformation and densification was small, only minor moisture degradation was observed and the sample had somewhat improved surface hardness.

Examples 29 and 30: The procedure of Example 1 was followed except, for comparison purposes, in place of the sodium hydroxide, a test sample was formed using calcium carbonate and another test sample was formed using calcium oxide as the flux. In both cases, upon firing at 1000°C, the test samples disintegrated. The following table sets forth a comparison of the properties of the materials tested in accordance with Examples 1 through 30.

Example	Flux	--- Structural Change --- at 1000°C	at 1150°C	Weight Loss, Parts
1	NaOH	Good	-	9.5
2	LiCl	Good	-	5.5
3	NaCl	Good	-	9.3
4	Na_2SO_4	Good	-	7.4
5	Na_2CO_3	Good	-	7.6
6	$(NaPO_3)_m$	Good	-	16.3
7	KCl	Good	-	13.8
8	K_2SO_4	Good	-	14.0
9	K_2CO_3	Good	-	12.6
10	KOH	Good	-	13.9
11	$K_5P_3O_{10}$	Good	-	16.6
12	$CaCl_2$	Good	-	7.3
13	$CaH_4(PO_4)_2$	Good	-	14.2
14	$SrCl_2$	Good	-	6.9
15	Pb_3O_4	Good	-	9.5
16	$PbSO_4$	-	Good	6.2
17	$Cu_2(OH)_2CO_3$	-	Good	11.5
18	$MnCl_2$	-	Good	11.3
19	$MnSO_4$	-	Good	19.7
20	$MnCO_3$	-	Good	10.8
21	$FeCl_2$	-	Good	7.2
22	$FeSO_4$	-	Good	12.0
23	$Ti(SO_4)_2$	-	Good	4.1
24	$ZnSO_4$	-	Fair	12.8
25	ZnO	-	Fair	10.3
26	$BaSO_4$	-	Fair	10.2
27	Fe_2O_3	-	Fair	10.6
28	$NiCl_2$	-	Fair	7.5
29	$CaCO_3$	Disintegrated	-	-
30	CaO	Disintegrated	-	-

Refractory Mortar

A process described by *R.E. Fisher; U.S. Patent 3,649,313; March 14, 1972;*

assigned to Combustion Engineering, Incorporated provides an air-setting refractory mortar particularly adapted for setting refractory brick in metallurgical process ladles. The mortar is particularly useful with high-alumina brick or other high quality brick. This mortar had a low angle of contact with the brick and thus adheres readily. In contrast, it has a high angle of contact with molten metal and slag and thus is resistant to penetration. The mortar dries readily and adheres tightly, forming a crack-free, noncurling joint. The refractory mortar employs a base of a calcined refactory material with a high-alumina content, such as bauxite or alumina, together with graphite, a phosphate binder, methylcellulose and water. The formulation of the mortar may vary such as within the ranges shown below.

	Weight Percent
Calcined bauxite 200 mesh	0.0 - 70.0
Calcined alumina 200 mesh	0.0 - 60.0
Norwegian graphite	2.0 - 20.0
Aluminum phosphate	1.0 - 20.0
Water	5.0 - 15.0
Methylcellulose	0.10 - 1.00

The percentage of bauxite plus alumina (the alumina-containing refractory particles) is between 50.0 and 80.0%. A typical and preferred composition of the mortar is as follows:

	Weight Percent
Calcined bauxite 200 mesh	65.8
Calcined alumina 200 mesh	5.5
Norwegian graphite	4.5
Aluminum phosphate	14.0
Water	10.0
Methylcellulose	0.2

The calcined bauxite is produced by calcining natural bauxite, which is an alumina hydrate, at a temperature between 2800° and 3100°F, which is sufficiently high to eliminate the shrinkage. The calcined alumina is obtained by calcining relatively pure aluminum oxide to produce a dense, low-porosity, stable product.

Both the calcined bauxite and calcined alumina are crushed and pulverized so as to pass through 200 mesh U.S. Standard sieve. It is possible to use either all calcined bauxite or all calcined alumina in place of the mixture of the two materials. The calcined bauxite is lower in cost but it is also less refractory than calcined alumina.

However, calcined bauxite particles have a rougher surface and tend to react with the matrix, fine particles and glassy phase, to increase the refractoriness and corrosion resistance at liquid metal temperatures. The calcined alumina gradually reacts with the matrix material and forms a more refractory matrix which provides high load-bearing properties at elevated temperatures and greater resistance to slag penetration. The preferred graphite is Norwegian graphite which is a natural material found in Norway. The graphite-bearing ore is crushed and the graphite separated from the gangue by flotation, dried and sized as follows.

U.S. Standard Sieve	Percent Retained
20	0.0
30	0.1
40	0.6
50	5.0
70	13.6
100	21.2
140	21.6
200	17.4
270	5.0
Through 270	15.2

The carbon content of the resulting material is about 90%. The Norwegian graphite is preferred because of the large flake structure and because it is relatively soft and resistant to oxidation in contrast with graphites from other locations. The phosphate binder may be phosphoric acid but aluminum phosphate is preferred. The methylcellulose in the composition is for the purpose of improving the water retention properties of the mortar. Typical chemical analyses of the mortar of the process after drying and after calcining are as follows:

	Dried Mortar, %	Calcined Mortar, %
Silica	4.01	6.45
Alumina	69.34	89.32
Ferric oxide	0.97	1.25
Titania	2.31	2.98
Calcium oxide	Trace	Trace
Magnesium oxide	Trace	Trace
Alkalies	Trace	Trace
Carbon	7.90	--
Loss on ignition	14.47	--

The bonding strengths after heating the brick and the mortar to various temperatures are as follows:

Temperature of Test	Modulus of Rupture
230° F. (110°C.)	800 psi
1500° F. (816°C.)	300 psi
2000° F. (1093°C.)	400 psi
2550° F. (1400°C.)	300 psi
3000° F. (1650°C.)	800 psi

The mortar has excellent resistance to carbon monoxide disintegration. It is unaffected after 1,250 hours exposure to carbon monoxide at 900°F after being prefired at 1000°F according to the ASTM C288-62 test method. No material fusion or shrinkage is evidenced after ASTM refractoriness test C199-45 at 2912°F. Also the slag resistance is high as there is zero penetration of a mortared joint after 5 hours exposure to blast furnace slag at 2912°F. This lack of penetration by the slag as well as by the molten metal is due to the very high angle of contact and the resulting nonwettability of the brick which is coated with the mortar.

SPECIALTY PRODUCTS AND PROCESSES

WATER PERMEABLE POROUS CONCRETE

In underground concrete linings of tunnels, galleries or in aboveground supporting walls on slopes, etc., the problem is to lead off water issuing from the moisture in the rock or soil of wide areas in such a way that the lining proper is not damaged by the hydrostatic pressure of the water and particularly thus becoming likewise permeable to water.

It is known to drain such areas according to the tubing method where the water is led off by means of drain pipes arranged in the concrete lining. In drifts and tunnels, which are later walled with steel reinforced concrete or with sprayed concrete or an anchorage, the areas in question were protected against rock moisture by foil insulation and the water between the lining and the insulation was led off by ribbed foils, tube drainages, etc. In rock sections under pressure the back water can seep more or less uncontrolled through the compressed wood layers of the forms.

It has already been tried to eliminate this inconvenience with special filtering stones which were arranged as a porous wall without an intermediate layer on the rock. On this porous wall was then applied the sprayed concrete lining. The area pressure relief and elimination of the water through the porous wall prevented the uncontrollable issuance of water in larger quantities, but the porous wall of hollow bricks can only be used up to the side wall because the filtering stones cannot absorb any roof pressure.

J. Compernass, E. Grünberger and F.E. Schmidt; U.S. Patent 3,847,630; November 12, 1974; assigned to Henkel & Cie GmbH, Germany describe a method for the production of a water-permeable porous concrete comprising the steps of mixing a concrete mixture containing:

(a) granular mineral aggregates of a uniform grain fraction in the range of 3 to 30 mm with a tolerance range of 4 to 15 mm,

(b) hydraulic cement in the prescribed amount according to the job specifications,

(c) a macromolecular water-soluble compound in an amount of from 0.05 to 2% by weight based on the cement, the macromolecular water-soluble compound being selected from the group consisting of plant gums, starch products, cellulose ethers and synthetic polymers, and

(d) water in such an amount that the water-cement ratio is between 0.32 and 0.48, whereby the concrete mixture is sprayable, and placing the concrete mixture by spraying.

Examples 1 through 11: Various concrete mixtures were produced according to the method. Mixture I consisted of 250 kg of blast furnace cement, 1,500 kg of gravel (grain size 3 to 7 mm) and so much water that the water-cement ratio was 0.38. Mixture II consisted of 300 kg of blast furnace cement, 1,650 kg of gravel (grain size 15 to 30 mm) and so much water that the water-cement ratio was 0.40. Mixture III consisted of 280 kg of Portland cement, 1,600 kg of gra- gravel (grain size 7 to 15 mm) and so much water that the water-cement ratio was 0.39. Other types of hydraulic cements may be employed as is known in the art.

To these mixtures were added various amounts of water-soluble macromolecular substances and wetting agents. The following table contains in the first column the example number of the formulation. This is followed by the concrete mixture used, the amount and type of the water-soluble macromolecular substance, and the wetting agent.

Example	Mixture	Water-Soluble Macromolecular Substance	Wetting Agent
1	I	0.5 kg of hydroxypropylmethylcellulose, having a viscosity in 2% solution of 80,000 cp according to Höppler at 20°C	0
2	I	2.0 kg of methylcellulose, having a viscosity in 2% solution of 300 cp according to Höppler at 20°C	0
3	I	2.0 kg of methylcellulose, having a viscosity in 2% solution of 300 cp according to Höppler at 20°C	0.2 kg of sodium lauryl sulfate
4	I	1.0 kg of methylcellulose, having a viscosity in 2% solution of 300 cp according to Höppler at 20°C	0.1 kg ethylene oxide adducted on nonyl phenol (molar ratio 9:1)
5	II	1.2 kg of methylcellulose, having a viscosity in 2% solution of 4,500 cp according to Höppler at 20°C	0.1 kg of sodium lauryl sulfate
6	II	0.8 kg of hydroxyethylcarboxylmethylcellulose, having a viscosity in 2% solution of 4,000 cp according to Höppler at 20°C	0.12 kg of sodium lauryl sulfate
7	II	1.0 kg of polyacrylamide having a viscosity in 10% solution of 100,000 cp according to Höppler at 20°C	0
8	I	2.0 kg of guar flour, having a viscosity in 2% solution of 10,000 cp according to Höppler at 20°C	0.1 kg of ethylene oxide adducted on nonyl phenol (molar ratio 9:1)
9	III	0.8 kg of hydroxypropylmethylcellulose (according to Example 1)	0
10	III	2.0 kg of methylcellulose (according to Example 2)	0
11*	III	1.0 kg methylcellulose	0.1 kg ethylene oxide adducted on nonyl phenol (molar ratio 9:1)

* To this formula were added 1.2 kg of an aqueous solution containing a 40% solid, which consisted of 80% bitumen and 20% polyvinyl propionate.

All components were vigorously mixed in a pressure mixer and then sprayed pneumatically. In all cases a porous concrete was obtained after setting which had a strength between about 120 and 280 kg/cm^2. The water permeability was excellent.

The substances listed in the foregoing table were added either to the finished mixture or to the gravel mixtures obtained by sizing with no difference in result.

Comparative Tests A to D: Test A — Into a vertical wall of rock material, testing rods with centimeter graduations were struck in at a right angle to the wall. Then mixtures, according to Examples 1 and 3 above, were prepared and sprayed onto the wall to a height of 2 meters and width of 3 meters. A uniform coat was obtained on the wall which was between 20 and 25 cm thick and adhered well.

Test B — The same mixtures were separately filled to a height of 10 cm in an iron pipe of 10 cm diameter and 50 cm length. The pipe was closed at the bottom by a board. After 7 days the board was removed and within 10 seconds, 1 liter of water was poured into the pipe. After 15 more seconds had elapsed, no water stood above the solid layer of the test body. The largest part of the water (more than about three-fourths of a liter) had issued from the bottom of the test body.

Test C — Mixtures, according to I and II above, were prepared without, however, any addition of the water-soluble colloids. It was attempted to spray these mixtures onto a wall, as described in Test A. When this was tried, starting from the bottom, and spraying upwards, the coat thickness decreased gradually, so that on consideration of the total surface, a coat thickness of only 1 to 3 cm was attained. About 80% of the sprayed concrete rebounded, as was determined by measuring the rebound volume.

Test D — Mixtures, according to Examples 1 and 3 above, were prepared. However, instead of the applied grain fraction of between 3 and 7 mm, a sand fraction was applied, which possessed no higher grain diameter than 3 mm. These mixtures were filled into pipes, 10 cm high, according to Test B. After 7 days, after removal of the bottom board, 1 liter of water was poured into the pipes each time. After 1 hour had elapsed, the water level was practically the same as the initial water level.

It can be concluded from the above tests, that both the macromolecular water-soluble compound and the uniform grain-sized aggregate are essential to the production of a sprayable concrete which remains porous on setting. Elimination of the macromolecular water-soluble compound from the recipe effects the sprayability of the concrete, and elimination of the uniform grain-sized aggregate effects the porosity of the set concrete.

CEMENT ASPHALT BALLAST GROUT FOR TRACKS

In the conventional track, correction of irregularity of track and renewal of ballast are often necessary, and a great deal of labor is required in the maintenance of the track. In order to obviate such drawbacks, various maintenance-free track structures have been developed. Among these track structures, the slab track, which is produced by injecting cement asphalt mortar (hereinafter referred to as mortar) under the concrete slab, is a most preferable track because, in the slab track the irregularity of track hardly occurs, and the sinking of track can be restored by injecting mortar, and therefore the maintenance of the slab track is very easy.

O. Torii, T. Mizunuma, I. Mino and T. Ando; U.S. Patent 3,867,161; February 18, 1975; assigned to Japanese National Railways and Denki Kagaku Kogyo KK, Japan describe a cement asphalt ballast grout composition for directly joining-type track, which comprises 100 parts by weight of cement, 20 to 400 parts by weight of an asphalt emulsion, 5 to 20 parts by weight of a calcium sulfoaluminate hydrate-forming mineral, 0.1 to 5.0 parts by weight of an electrolyte, 0.01 to 5.0 parts by weight of a thickener and 0.01 to 0.03 part by weight of a foaming agent. The cement to be used in the process includes Portland series cement, mixed cement and the like.

As the calcium sulfoaluminate hydrate-forming mineral to be used in the process, mention may be made of, for example, type K, type M and type S minerals described in *American Concrete Institute Journal,* 8, 584-589 (1970). Among these minerals, ones having a fineness of a Blaine value of 4,000 to 10,000 cm^2/g are preferably used.

The calcium sulfoaluminate hydrate-forming mineral is used in an amount of 5 to 20% by weight based on the weight of cement. Among the above-described calcium sulfoaluminate-forming minerals, ones having such a property that a cement mortar prepared from 5 to 20% by weight of the mineral and the remainder of cement has a free expansion coefficient of about 0.05 to 0.5% are preferably used.

Furthermore, when a mixture of 85 to 95% by weight of a powdery calcium sulfoaluminate series mineral and 15 to 5% by weight of a powdery mineral consisting mainly of $3CaO \cdot 3Al_2O_3 \cdot CaF_2$ is used in an amount of 5 to 20% by weight based on the weight of cement, a cement asphalt hardened mortar having a higher strength can be obtained.

The calcium sulfoaluminate hydrate-forming mineral forms calcium sulfoaluminate hydrate, that is, ettringite, in the resulting mortar and the mortar expands for about 7 days. Therefore, the sinking of the mortar, just after the placing, can be prevented by the actions of the foaming agent and the ettringite, and further the shrinkage of the hardened mortar due to drying does not occur and cracks do not appear in the hardened mortar.

The mortar shows the thixotropy phenomenon by the actions of the powdery calcium sulfoaluminate hydrate-forming mineral and the thickener. That is, the mortar has a good fluidity during the mixing and injection operations, and in the static state after the mortar is injected, the viscosity of the mortar increases quickly, and the separation of sand can be prevented.

The thickener to be used includes polyvinyl alcohol, carboxymethylcellulose, starch, gelatin, etc., and mixtures thereof. The thickener is used in an amount of 0.01 to 5.0% by weight based on the weight of the cement.

The electrolyte to be used includes sodium chloride, lithium chloride, potassium chloride, calcium chloride, magnesium chloride, barium chloride, etc., and mixtures thereof. The electrolyte is used in an amount of 0.1 to 5.0% by weight based on the weight of cement. The electrolyte prevents the decomposition of asphalt emulsion, and can prolong the time of gelation of mortar.

The foaming agent to be used includes aluminum, aluminum nitride, zinc, tin, calcium silicon alloy, etc., and mixtures thereof. The foaming agent is used in an amount of 0.01 to 0.03% by weight based on the weight of cement.

The asphalt emulsion is used in an amount of 20 to 400%, preferably 100 to 300% by weight, based on the weight of cement. When the addition amount is less than 20% by weight, the elasticity of the hardened mortar is too high, and the characteristic property of the cement asphalt is lost. When the addition amount exceeds 400% by weight, a very long time is required in the hardening of mortar.

The mortar produced from the grout composition has a small variation in consistency, has a low sinking percentage even when the temperature is varied, and further does not cause separation of raw materials during the hardening, and the hardened mortar does not cause dry shrinkage. Accordingly, the grout composition is an excellent cement asphalt ballast grout composition for directly joining-type track.

LOW WATER BLEED TENDON GROUTING COMPOSITION

M. Schupack; U.S. Patent 3,762,937; October 2, 1973 describes a method for preventing water bleed in grout for securing post-tensioned tendons, particularly strand tendons employed in the production of prestressed concrete. The process comprises the use of a gelling agent and a dispersing agent along with cement and water in proportions which provide a pourable, injectable grout composition, the relative proportions of ingredients being selected such that water bleed is controlled or substantially eliminated at pressures less than about 10 psig; at 80 psig, less than 10% water loss occurs. The composition preferably contains an expansion agent resulting in positive expansion as required by the specific circumstances of usage. The composition is capable of reducing water bleed to 0% at 80 psig differential pressure.

The gelling agent may be selected from any of those commercially available. Typical gelling agents are Methocel methylated cellulose and Natrosol hydroxyethylcellulose, all of these being available in many viscosity types. Natrosol 250H has been found to be especially desirable. Natrosol 250H has a Brookfield viscosity of 1,500 to 2,500 cp, measured as a 1% aqueous solution at 25°C.

Commercially available dispersants which are compatible with aqueous systems can be employed in this process. Typical of these are Nopcosant and Lomar D dispersants. Nopcosant is comprised of a sulfonated naphthalene; Lomar D is a highly polymerized naphthalene sulfonate which is supplied commercially as the sodium salt. Lomar D has been found to be most preferable for use in the process.

Example: The filtering action causing bleed results from pressures forcing the water of a grout mixture into the interstices of strand tendons. The size of the interstices are such that they prevent entry of solids, but permit entry of water. The effective driving force of the filtering action is the hydrostatic force created along the vertical rise of the strand tendon and the difference between the density of water (62.4 lb/ft^3) and the density of the solids of grout mixture (approximately 118 lb/ft^3). Thus, for a 100-foot high tendon the bleed filtration pressure would be about 40 psig, and for a 200-foot high tendon the bleed filtration pressure would be about 80 psig. Considering a seven-strand tendon having a diameter of one-half inch, the spaces between each of the six outer wires is on the order of 0.001 to 0.002 inch.

With the noted bleed filtration action, a laboratory-scale pressure filtration funnel in which the grout is forced against a properly-sized filter simulates actual conditions. Accordingly, a test procedure was designed to simulate the bleed filtration phenomenon employing a commercially available pressure filter. The particular device selected was a pressure filtration funnel utilizing a filter member having a Type A fiber glass filter which retains 97% of all particles over 0.3 micron at pressures up to 200 psig. Tests were run under conditions selected to approximate as nearly as possible the actual grouting conditions for a long-strand tendon.

The components of the grout were thoroughly blended in the dry state and then mixed with the desired amount of water by a mixer for from 3 to 5 minutes. After mixing, the grout composition was allowed to set for 10 minutes without agitation to approximate the most severe conditions which would be experienced in actual grouting. For a 200-foot vertical strand tendon which would have a total differential filtering pressure of 80 psig, total pumping time of the grout into the duct would be on the order of 24 minutes. Accordingly, as the grout composition fills the duct in an actual grouting procedure, the grout comes under increasingly greater pressures. This was simulated herein by starting with a pressure of 0 psig and increasing the pressure by 10 psig every 3 minutes until a total of 24 minutes had elapsed. The pressure was applied to the pressure filter funnel in the tests by means of pressurized oxygen.

The results tabulated below summarize tests run according to the foregoing procedure utilizing a Marquette Type II cement. Two different gelling agents, namely, Natrosol 250M and Natrosol 250H, were tested and are identified in the table below as NM and NH, respectively. The dispersing agent in all cases was Lomar D. The expansion agent was Alcoa 606 powdered aluminum. The percentages of the gelling, dispersing and expanding agents are based on the weight of cement added to the mixture. Bleed is reported as the percent separated water, based on the water added to the mixture. The initial bleed pressure indicated is the pressure at which bleed was first noted.

Test	H_2O/Cement, w/w	Gelling Agent, %		Dispersing Agent, wt %	Expanding Agent, wt %	Initial Bleed Pressure, psig	Bleed, % original H_2O at 80 psig
1	0.45	NM	0.5	0	0	20	8.0
2	0.45	NH	0.5	0	0	20	4.0
3	0.45	-	0	0.5	0.01	0	70.0
4	0.45	NH	0.5	0.5	0.015	80	0

(continued)

Test	H_2O/Cement, w/w	Gelling Agent, %		Dispersing Agent, wt %	Expanding Agent, wt %	Initial Bleed Pressure, psig	Bleed, % original H_2O at 80 psig
5	0.45	NH	0	0	0	5	45
6	0.45	NH	0.3	0.5	0.01	50	5.4
7	0.5	NH	0.4	0.5	0.01	70	1.3
8	0.5	NH	0.3	0.5	0.01	30	7.5
9	0.5	NH	0.3	0.3	0.01	30	10
10	0.45	NH	0.3	0.3	0.01	40	4.7
11	0.5	NH	0.4	0	0.01	50	5.0
12	0.5	NH	0.4	0.4	0.01	50	3.5
13	0.5	NH	0.4	0.4	0.01	50	2.9
14	0.5	NH	0.5	0.4	0.01	60	1.5
15	0.45	NH	0.4	0.6	0.01	60	1.3
16	0.45	NH	0.5	0.5	0.01	80	0
17	0.45	NH	0.4	0.5	0.01	50	2.1
18	0.45	NH	0.5	0.5	0.01	50	1.3
19	0.5	NH	0.5	0.5	0.01	80	0.6
20	0.45	NH	0.5	0.5	0.01	60	1.5

From the foregoing it is apparent that gelling agents by themselves do not prevent bleed at 80 psig, but they are capable of preventing bleed at lower pressures. Dispersing agents are ineffective by themselves and in fact increase the amount of bleed over that which is observed with neither dispersing agent nor gelling agent. The combination of sufficient gelling and dispersing agents gives desired bleed rates of less than 10% at 80 psig; the combination of 0.5% gelling and 0.5% dispersing agents being capable of giving a grouting composition with 0% bleed at 80 psig under test conditions. By using grouting compositions containing the combination of dispersing and gelling agents required by the process, desired levels of water retention can be imparted without the disadvantages attendant to the use of relatively high proportions of gelling agents exclusively.

Thus, the grouting compositions with smaller amounts of gelling agents are less viscous and more easily injectable than bleed-resisting tendon grouting compositions which gain their resistance to bleed from gelling agents alone. Additionally, gelling agents increase the amount of air entrapped by a grouting composition. Grouting compositions of this process, in requiring lower concentrations of gelling agents for a given level of bleed resistance, are less prone to entrap air, thus producing cured grouting compositions of greater density and consequently greater strength than known grouting compositions with similar bleed resistance levels.

Still another advantage of the process is that it provides bleed resistant tendon grouting compositions which remain pourable at water-to-cement ratios which result in cured grout compositions having greater strength than grouting compositions with only gelling agents which require higher water-to-cement ratios for the same degree of pourability.

SPRAY COATING

F. Mikoteit and H.-J. Tennstedt; U.S. Patent 3,645,762; February 29, 1972; assigned to Elsa Zement- und Kalkwerke AG, Germany describe a hydraulic mortar or cement composition which is suitable for application by spray coating and useful in sealing the walls of galleries in coal mines or other excavations.

The composition comprises limestone marl containing a small concentration of clay, slightly burnt and completely deacidified cement clinker, and a small amount of an anhydrous calcium salt such as calcium chloride to accelerate initial hardening.

INSULATING CEMENT

C.E. Wallis; U.S. Patent 3,486,917; December 30, 1969; assigned to Continental Capital-Control Centre Establishment, Liechtenstein describes a composition especially adapted for forming a waterproof insulation comprising cement, sand, an additive comprising a montmorillonite mineral, a filler having hydraulic properties, a setting regulator and a surface-active agent. The following examples illustrate the process.

Example 1: A first mixture is made up of 0.2 kg of a trade surface-active agent, 13.0 kg calcinated soda and 3.8 kg tartaric acid. This is then mixed with 13 kg alkali-activated bentonite with a moisture expansion value of approximately 1000% and 30 kg flue ash. The flue ash should, if possible, have an SO_3 content of under 3%. After mixing, a product is obtained which, mixed with 460 kg cement and 480 kg sand, gives an excellent insulating material. In order to achieve optimum results the following grain distribution of the sand which is preferably an arenaceous quartz is necessary: 0.0 to 0.1 mm, 20%; 0.1 to 0.3 mm, 40%; 0.3 to 0.6 mm, 30%; and 0.6 to 0.8 mm, 10%.

The resulting cement mixture is then mixed with water (approximately 25% of the weight of the mixture) and produces a thick slurry which is especially suitable for filling joints or as a surface coating agent against water pressure.

Example 2: One proceeds as in Example 1; however, a first mixture of 0.5 kg surface-active agent, 20.0 kg soda and 6.5 kg tartaric acid is prepared and mixed with 20 kg of the activated bentonite and 100 kg flue ash. The resulting product, mixed with 420 kg cement and 433 kg sand as described in Example 1, and water equalling one-fourth of the total weight, is an insulation material against moisture whether on concrete, natural stone, brickwork or mixed masonry as well as a masonry adhesive.

PROTECTIVE TREATMENT FOR ALKALI-AGGREGATE REACTIONS

In a process described by *R.E. Pickering; U.S. Patent 3,433,657; March 18, 1969* concrete is protected from alkali-aggregate reactions by soaking the concrete with water, and applying to the surface of the concrete structure an aqueous slurry containing an inorganic cement, an acid, for example, an organic hydroxy carboxylic acid such as glyceric acid, which forms complexes with calcium ions or aluminum ions, or a salt of such an acid, and a base such as sodium aluminate, potassium orthosilicate, or borax, which forms an insoluble salt with calcium ions or aluminum ions.

The composition according to the process solves both the problem of alkali reactions in concrete and the problem of corrosion of reinforcement irons. A preferred composition consists of 80 parts by weight of cement, 2 to 10 parts by weight of complex-forming acid or a salt and 10 to 50 parts by weight of base.

In use the composition is suspended in water (about 1 part by weight of powder to 0.8 to 1 part by weight of water) and is then applied to the surface of the object to be protected, for example, brushed on, sprayed on, applied by rough-casting, or smoothed on. Preferably the composition is applied to the surface of the object after the object has been soaked with water.

RADIATION SHIELD

R.E. Vogel; U.S. Patent 3,434,978; March 25, 1969; assigned one-half to Friedrich Marxen, Liechtenstein provides a building material which will provide effective protection, not only against alpha, beta and gamma rays, but also against neutron radiation and which, when in the form of concrete, will possess substantially the same strength characteristics as it would possess in the absence of the radiation protective additive.

The building material having a shielding effect against radioactive radiation includes a cementitious material having an effective amount of at least one compound of a metal selected from the group consisting of lead, bismuth, tungsten, zirconium, iron, tin, cadmium, lithium and barium, with a saturated fatty acid which is solid at room temperature and has at least 9 carbon atoms. The following examples illustrate the process.

Example 1: In a rotary cement mixer, an intimate mixture is formed of the following materials.

	Kg
Moist sand, 5 to 8 mm	6.6
Moist sand, 3 to 5 mm	13.2
Moist sand, 0 to 3 mm	6.4
Hard quartz sand, 0 to 2 mm	5.4
Hard fine sand, 0 to 2 mm	1.2
Finely ground baryte	2.0
Finely ground galena	1.0
Blast furnace cement	7.0
Water	3.6

To the mixture, 5.7 kg lead stearate are then slowly added and, after a homogeneous mixture has been formed, the same is introduced into iron molds or frames and the latter placed for about 10 minutes on a vibrating table. Thereafter, the mixture is pressed in a concrete plate press for a period of between about 2 and 5 minutes at a pressure of between 10 and 100 kg/cm^2.

Example 2: In the manner described above, a mixture is formed of:

	Kg
Sand, 5 to 8 mm	5.96
Sand, 3 to 5 mm	11.88
Sand, 0 to 3 mm	5.78

Separately, 125 grams ammonium stearate are stirred into 4 liters of water (as emulsifier) and thereafter 3.23 kg of lead-montanic acid compounds are stirred into the thus formed mixture. A stiff paste is formed in this manner. The two mixtures formed as described above are combined and further mixed in a concrete mixer. During this further mixing, the following is added and the mixture is then further worked up as described in Example 1.

	Kg
Hard quartz sand, 0 to 2 mm	3.8
Finely ground baryte	2.7
Colloidal lead	1.0
Portland cement	6.5

COMPOSITION FOR FILLING CRATERS

In a process described by *T.S. Ames; U.S. Patent 3,582,376; June 1, 1971; assigned to The Western Company of North America* craters and earth voids are quickly filled with a high-strength, quick-setting cement composition comprising a mixture of cement material consisting essentially of from about 9.5 to about 4 parts by weight of calcium sulfate hemihydrate to about 0.5 to about 6 parts by weight of Portland cement and water which is present in a quantity of from about 22 to about 70% by weight. According to the desired uses, accelerators, retardants or dispersants are added to the mixture. The preferred quick-setting cement composition of this process consists of from 9.5 to 7 parts by weight of alpha gypsum to 0.5 to 30 parts by weight of Portland cement along with from about 22 to about 40% water by weight, based upon the cement material. The following examples illustrate the process.

Example 1: A crater in a runway is filled to about one foot from its top with chunks of concrete, rocks and dirt. This material is pushed into the crater with a bulldozer. After partial filling of the crater with debris, a relatively thin slurry of the quick-setting cement of the process is introduced into the crater pumped from a container through a jet-mixing nozzle.

The quick-setting cement slurry so introduced has the following composition of cement material: 80 parts of plaster of Paris (as prepared by dehydration of gypsum) to 20 parts of conventional Portland cement, both taken by weight. In addition, 1 part by weight, based on the total weights of the cement materials, of Lomar D (the condensed sodium salt of sulfonated naphthalene formaldehyde) is provided as dispersant. Water in the amount of 60% by weight, based on the total amount of cement materials (the plaster of Paris plus the Portland cement) is also provided. The resulting slurry of water and cement has a viscosity of about 30 cp. The flow of water is so adjusted that the proportion of the solids to water is maintained at the approximate quantity ratio enumerated above.

The crater is filled to a point flush with the upper runway surface. In this example, the same cement mixture or slurry is utilized for consolidating debris as is utilized for the upper layer overlying the debris.

The filling of the crater is accomplished in about 1 hour. After the lapse of an additional ½ hour, it is found that the upper surface of the filled crater representing the top surface of the repaired runway portion has set to provide a firm and smooth load-bearing surface. This surface is observed to have a compressive load-bearing capacity well in excess of 500 lb/in^2 (ASTM C-39-61).

Example 2: Example 1 is repeated, except the quick-setting cement slurry utilized is varied after a sufficient quantity has been introduced to consolidate the debris in the crater. On the completion of such consolidation, the flow is varied through the nozzle by adjustment of conventional valve, measuring or other metering means to cause the slurry discharged into a tank to be 50% water, based on the total cement material (plaster of Paris plus Portland cement).

Note, that the ratio of plaster of Paris to Portland cement is maintained the same as was the case in Example 1, i.e., 8:2 parts by weight. Dispersant, in the same quantity as utilized in Example 1, based on the total cement material, is provided for the upper layer. The viscosity of the cement slurry is about 95 cp.

Disposition of the upper layer of quick-setting cement slurry is completed when it becomes flush with the surface of the damaged runway. The surface is smoothed over and the quick-setting cement is allowed to dry for approximately one-half hour. At the end of this time, it is observed that the upper surface of the repaired portion of the runway has a compressive load-bearing capacity in excess of 1,000 lb/in^2 (ASTM C-39-61).

Example 3: A set of tests is run on varied mixtures of plaster of Paris, Portland cement and water. No dispersant, accelerator or retardant is used. The set time (as determined by the Vicat method) is determined and compressive strength tests are run on specimens 30 minutes after pouring. The following table presents the results:

Plaster of paris [1]	Portland cement [1]	Water [1]	Set time (min.)	Compressive strength [2]
100	0	70	8	850
90	10	68	3	1,300
80	20	65	3	1,250
70	30	63	3	1,000
60	40	60	3	800
50	50	58	3	525

[1] Percent by weight based on total cement materials, i.e., plaster of paris plus portland cement.
[2] Pounds per square inch, after thirty minutes from pouring (ASTM C-39-61).

Example 4: In some instances, it is desired that the setting time be accelerated. This can be of particular value if a great emergency exists. A water solution in quantity of 65 parts of a 5% by weight solution of potassium sulfate (based on the total quantity of cement solids) is introduced into and mixed with 100 parts of cement material. The cement material consists of a mixture of 80 parts by weight of plaster of Paris and 20 parts by weight of Portland cement. It is observed that the resulting slurry sets within 1 minute (Vicat test method). After 30 minutes, a compressive strength of 550 lb/in^2 (ASTM C-39-61) is observed.

IRON PLATED CONCRETE FLOORS

A process described by *H.J. Horvitz; U.S. Patent 3,668,150; June 6, 1972* provides flooring with a high resistance to wear. Concrete floors are the traditional flooring for most industrial plants. When a particularly hard wear surface is desired, the builder often resorts to topping floors to provide a hard aggregate at the surface to resist wear; it is also well known to apply dust coats of metallic aggregate to a concrete floor to harden the wear surface. The latter are often referred to as iron plated concrete floors.

One other reason for incorporating iron fillings in a concrete floor is to provide an electrically conductive flooring. Such a process is discussed in U.S. Patent 3,166,518. The electrical conductivity of such a floor deteriorates over a period of time and for reasons which are unrelated to wear per se.

The deterioration problem with floors using ground metal as an aggregate stems from the rather quick corrosion of the metal. The fact of corrosion creates a problem because the wearability of corroded metal is much lower than that of the pure metal in its uncorroded state.

Another aspect of this problem is with white concrete floors which are becoming more important in various aspects of industry. Clearly, when corrosion starts in a white concrete floor the inherent color change will quickly dissipate the light reflectivity of the white flooring and since this is one of the prime reasons for having a white floor it quickly loses its effectiveness.

Thus, the isolated problem is to minimize the corrosion of the metallic aggregate mixture for the purpose of maintaining a longer lasting floor and in some instances a resulting floor which is whiter for a longer period of time. Customarily the initial pH in a concrete substrate is approximately 12. However, over a period of time carbon dioxide from the air permeates the matrix and reduces the pH below about 9. With a pH of about 9, chlorine within the matrix accelerates corrosion and oxidation of the ferrous metallic particles embedded therein. Additionally, at a pH of about 11 there is a tendency toward corrosion of the metal, but the reason for this is not known with certainty.

It was found that the addition of sodium nitrite to concrete retards the corrosion of embedded iron. Thus, to minimize the corrosion of the ground metal aggregate, sodium nitrite is added to dry mixtures of cement and ferrous metal aggregate. Preferably, this dry mixture is applied as a topping to a freshly laid concrete base by the well-known dusting or shaking method.

The basic dry coating mix includes approximately 28 to 38% cement, 62 to 72% powdered ferrous metal and 0.01 to 0.80% sodium nitrite. Cement in this case is the binder for holding the aggregate in the hardened matrix. The cement used in this process is conventional and is usually Type I cement although it is not limited to Type I cement.

The corrosion inhibiting additive, sodium nitrite, is particularly important when powdered ferrous metal aggregate is used in the manufacture of a white floor. Clearly, rusting and corrosion of any kind will quickly destroy the light reflective characteristics of the floor.

The use of sodium nitrite in the dry mix applied to the top of the concrete flooring is particularly unique in this aspect and has contributed greatly to the success of long standing white flooring.

LOW FLUORINE CONTENT CEMENT CLINKER

H. Stich, K. Ruckensteiner, W. Binder and J. Hutter; U.S. Patent 3,652,308; March 28, 1972; assigned to Osterreichische Stickstoffwerke AG, Austria describe a process for the manufacture of cement clinker of low fluorine content from phosphoric acid by-product gypsum according to the gypsum-sulfuric acid process, by heating to a temperature above 500°C before reduction in the calcination furnace, which comprises heating the by-product gypsum, by itself, to a temperature within the range of 500° to 900°C, preferably 600° to 800°C, in the furnace waste gases containing sulfur dioxide and subsequently mixing known additives with the hot material without loss of heat immediately prior to entering the calcination furnace.

The defluorination of the gypsum according to the process may be conducted in various types of apparatus. Thus, for example, the gypsum can, in the moist or predried state, be subjected to the heat treatment, under the influence of the furnace waste gases containing sulfur dioxide in a fluidized bed. However, the multistage cyclone preheaters known for heating the furnace powder can also be employed very advantageously for the process if care is taken that the gypsum is in a free-flowing state as a result of appropriate predrying. In this case, the additives are mixed in one of the lower stages of the preheater, or at the end, depending upon the treatment temperature. The following examples illustrate the process.

Example 1: By-product gypsum from a phosphoric acid installation, having a fluorine content of 0.8%, was heated in the anhydrous state to 800°C, in a furnace atmosphere having a sulfur dioxide content of 8% by volume of sulfur dioxide. At a weight loss of 0.5% the fluorine content after the end of the treatment was 0.09%.

Example 2: Waste gypsum with a fluorine content of 0.33% was heated to 500°C under the same conditions as in Example 1. After the end of the treatment the fluorine content was 0.14%. After mixing 5% by weight of each of the additives which are customary for the gypsum-sulfuric acid process, clay, coke and sand, with the heated by-product gypsum, that is to say the material which is at about 800°C in the former case and about 500°C in the latter case, a cement clinker which meets standard specifications was obtained after carrying out the usual calcination process.

SEALANT FILLER

M.J. Basile; U.S. Patent 3,917,771; November 4, 1975 describes a two-package sealant filler which includes as a first paste component a dispersion of sand in a water-based acrylic latex and as a second highly viscous component, a dispersion of bentonite and Portland cement in glycerin. In use, the first component and a minor amount of the second component are intimately mixed and applied to the work area where it cures throughout into an integral elastomeric mass having the appearance of concrete.

The first component contains between 1 and 3 parts by volume of a water-based acrylic latex to sand, with the latex containing about 14% water and about 86% non-volatile solids, that is about 0.14 to 0.42 parts water and 0.86 to 2.58 parts acrylic resin in a latex state, for example, 3 parts of the acrylic latex and 1 to 3 parts by volume of sand, for example 1 part, the sand and latex being uniformly mixed. The second component includes 1 to 5 parts by volume of bentonite, for example, 1 part bentonite and 1 to 5 parts by volume of Portland cement, for example, 2 parts, the bentonite and Portland cement being suspended in a polyhydric alcohol such as glycerin, glycol, polyglycols, for example, glycerin, to a paste or thick liquid.

The two components are separately packaged in an airtight condition and the filler mass is produced by intimately mixing the desired amount of the first component with a minor amount of the second component, for example, 40 to 60 parts by volume of the first component with 1 part by volume of the second component. The mass is then applied to the opening to be filled or patched.

ELASTOMERIC SEALANT

In a process described by *L.E. Toy and R.G. Weisz; U.S. Patent 3,487,038; December 30, 1969; assigned to Standard Oil Company* aqueous mixtures of concentrated elastomer emulsions and liquid alpha-methylstyrene polymer are combined with hydraulic cements to form solidifiable cement modified elastomeric compositions especially for use as calks, sealants or expansion joint materials.

Illustrative of the thixotropic mixtures of the process are those produced by using a thickened aqueous butyl rubber emulsion having a viscosity of about 20,000 cp and containing 66% solids mixed with from 14 to 53 parts by weight of liquid alpha-methylstyrene polymer per 100 parts by weight of the emulsion. This mixture was very high in viscosity, similar to mayonnaise, and was thixotropic.

The addition of from about 61 to about 77 parts by weight of Portland cement to the thixotropic elastomeric compositions produced sealant compositions that had a working life of from 1 to 5 hours, were sag-free, did not form a tacky surface when exposed to ultraviolet light, and had a Shore A hardness ranging from 20 to 50. The cement to water ratio in the above sealant compositions was from 1.8 to 2.3.

COMPANY INDEX

The company names listed below are given exactly as they appear in the patents, despite name changes, mergers and acquisitions which have, at times, resulted in the revision of a company name.

INVENTOR INDEX

U. S. PATENT NUMBER INDEX

NOTICE